Textbook of
FLUID MECHANICS

THE AUTHOR

Dr. S K Gupta, Born (1949), is INAE Distinguished Visiting Professor, ICAR-NDRI, Karnal and an independent Research Engineer with specialization in Soil and Water Resources Engineering and Civil Engineering. He obtained B. Tech (Agril. Engg.) from the Punjab Agricultural University (PAU), Ludhiana, in 1970 and Master of Agricultural Engineering from the same University in 1976. He obtained Ph.D. in Civil Engineering from the Jawahar Lal Nehru Technological University, Hyderabad, Telangana, India in 1984. Dr. Gupta is Fellow of the National Academy of Agricultural Sciences, Indian National Academy of Engineering and Indian Society of Agricultural Engineering. Since his joining CSSRI in 1971, besides holding scientific positions, he has been Head, Division of Drainage and Water Management, Head, Indo-Dutch Network Project, Head, Division of Irrigation and Drainage Engineering and Project Coordinator (AICRP on Management of Salt Affected Soils and Use of Saline Water in Agriculture). After superannuation, Dr. Gupta worked as Emeritus Scientist (ICAR) and INAE Distinguished Professor at the same institute. Dr. Gupta has been engaged in conducting researches in surface and subsurface drainage for land reclamation, irrigation water management, hydrology of salt affected soils, leaching and use of saline water in agriculture. He has published more than 150 research papers in high impact factor journals and published 15 books. Dr. Gupta has been consultant to WAPCOS-Louis Berger, MoWR, MoRD, UP-Dutch Tube Well Project, ActionAid International, Synergics Hydro India-Oromia Water Works Enterprise, Ethiopia and others. He has been bestowed with Rafi Ahmad Kidwai Award by ICAR. Besides he is recipient of several awards from NAAS, ISAE, IE (India), CBIP and MoWR among other organizations. He is the Chief Editor of Journal of Water Management of the Indian Society of Water Management.

Textbook of
FLUID MECHANICS

– Author –

S.K. Gupta

INAE Distinguished Visiting Professor
ICAR-National Dairy Research Institute, Karnal

Former Head, Division of Irrigation and Drainage Engineering
Project Coordinator, Use of Saline Water in Agriculture
ICAR-Central Soil Salinity Research Institute, Karnal

2025

Daya Publishing House®
A Division of
Astral International Pvt. Ltd.
New Delhi – 110 002

ISBN: 978-93-5919-584-1

Published by : **Daya Publishing House®**
A Division of
Astral International Pvt. Ltd.
– ISO 9001:2015 Certified Company –
4736/23, Ansari Road, Darya Ganj,
New Delhi-110 002
Ph. 011-43549197, 23278134
E-mail: info@astralint.com
Website: www.astralint.com

Preface

Fluid mechanics concerns with the mechanics of fluids, be it liquids or gases and the forces acting on them either at rest or in motion. Fluid mechanics has unquestioned scientific and practical importance because the vast majority of the observable mass exists in a fluid state. The atmosphere and the oceans covering this planet are fluids. Its allure crosses disciplinary boundaries having applications in hydrology, geophysics, oceanography, meteorology, astrophysics, micro-fluidics, aerodynamics, aerospace, navel architecture, geophysics, biophysics, cardiovascular studies, magneto-hydrodynamics, super and hypersonic flows, ecological studies, and quantum mechanics. As an illustration, the blood flow in human cardiovascular system, arteries, vein, and capillaries, can only be studied using the principles of fluid mechanics. Such and other numerous applications make fluid mechanics a subject of interest across engineering disciplines including civil, mechanical, electrical, agriculture, dairy, structural, industrial and biomedical.

This book addresses the need of a textbook on fluid mechanics for students of various conventional engineering disciplines. Initiative to write this book emerged from the lack of a good textbook that could cater to the needs of agricultural engineering, dairy engineering and civil engineering students. The purpose of the book is to make the students aware of the fundamentals of engineering fluid mechanics. After completing the course, students end up with fundamental knowledge of fluids, knowledge of the fundamental laws of mechanics as applied to fluids and are capable of applying the relevant theory to solve practical problems.

The book, divided into three sections with well-defined 13 chapters, begins with the Introduction to fluid mechanics taking the readers through brief history of its development. Other chapters include Pressure and pressure measurement, Archimedes law and stability of floating bodies, Fluid kinematics and visualization, Fluid dynamics: mass and energy conservation, Flow measurement: weirs notches and orifices, Open channel flow, Pipe flow, Hydraulic losses in pipes, Dimensional analysis and similitude, Introduction to fluid machinery, Reciprocating pumps and Hydrodynamic pumps. Contents of the chapters dovetail with the requirement of the agricultural/Dairy/civil engineering students who compulsorily take this subject. For agricultural/Dairy engineering, the contents are as per the Dean's committee recommendations made to ICAR from time to time. The salient features of the book are:

- ✰ The book is highly illustrated with relevant diagrams and photographs for easy to understand reading
- ✰ Theoretical questions appear at the end of each chapter while objective type questions for all the chapters are included in one section at the end
- ✰ Special emphasis is placed on numerical examples with solved problems included at relevant places throughout the book
- ✰ A list of related major books and papers are included in the references
- ✰ A subject index is included.

While writing this book, it emerged that there is no better way to learn a subject than to teach it. The current work has richly benefitted from the hundreds of students whom I have had the good fortune to teach through this material, both at the ICAR- National Dairy Research Institute and Karnal Institute of Technology and Management, Karnal. These students have patiently sat through and have asked insightful questions that made me to improve over the write-up. This book exists and bears the stamp of these students.

The book heavily draws from direct and indirect contributions of many national and international organizations and individuals. During the teaching of the subject, I referred e-courses of fluid mechanics brought out by ICAR and several textbooks recommended by Kurukshetra University. I take the first opportunity to whole-heartedly thank Dr. R.R.B. Singh, Director, ICAR-NDRI, and Dr. Sachin Wadhwa, Principal Director KIT and M, for according permission to deliver lectures and Dr. A. K. Singh, Head Dairy Engineering, ICAR-NDRI and Er. Janmejaya, Head Civil Engineering, KIT and M, Karnal for coordinating my visits to their respective Divisions/Departments. I am personally thankful to Er. Sunil and Er. Muhal for their support in the teaching activities at ICAR-NDRI, Karnal. The research and teaching material drawn from various individuals and organizations is thankfully acknowledged. Although, it is difficult to name each and everyone in this small brief, I attempted to include several important contributions by referring the work in the references. I might have inadvertently missed few of them. I place on record my appreciation for all the contributors who have painstakingly researched and developed the relevant processes, technologies and teaching material. Finally, I like to thank my family and friends for their support. My wife and children have seen me spent days together on this book. There have been occasions when I sacrificed their time to pursue this work. They never complained and remained supportive and encouraging to the end.

I believe that this book will prove to be a good resource to the teachers and students to have a better teaching/learning experience. I am sure that the book will be a cherished possession of the faculty and students of agricultural/Dairy/Civil engineering students and faculty. Besides, it should be a valuable academic asset for libraries of colleges and universities worldwide.

S.K. Gupta

Contents

FUNDAMENTALS OF FLUID FLOW

TURBO-MACHINES

Chapter 1

Introduction to Fluid Mechanics

INTRODUCTION

Fluid mechanics comprises two words namely fluid and mechanics. Word fluid means a substance having particles, which readily change their relative position. Thus, fluid is defined as a substance, which deforms continuously under the action of shear stress, regardless of its magnitude. Mechanics on the other hand is the science that deals with both stationery and moving boundaries under the influence of forces. Fluid mechanics is thus, concerned with the behaviour of fluid (liquids, gases, and plasmas) at rest (fluid statics) and in motion (fluid dynamics/kinematics) and the forces on them. It also deals with the interaction of fluids with solids or other fluids at the boundaries. Fluid mechanics comprises three different branches (Figure 1.1).

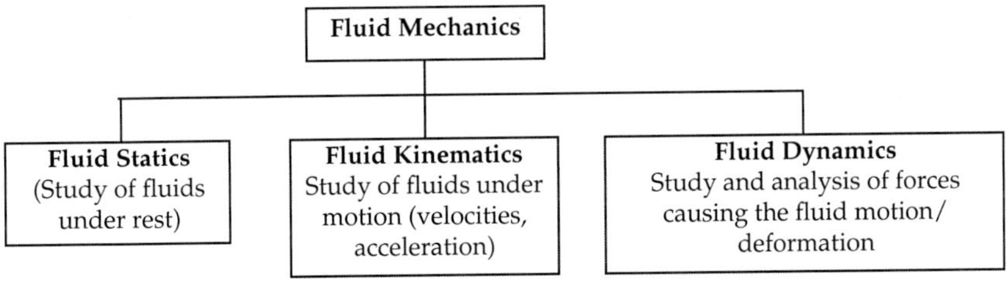

Figure 1.1. Branches of Fluid Mechanics.

Fluid Statics

Many fluid problems do not involve motion. Rather these are concerned with the pressure distribution in a static fluid (fluid statics). When the fluid velocity is zero, known as hydrostatic condition, the pressure variation is due to weight of the fluid. The important areas of fluid statics include:

- ✰ Pressure distribution in atmospheres and oceans
- ✰ Design of manometers, the pressure instruments
- ✰ Forces on submerged flat and curved surfaces
- ✰ Buoyancy on a submerged body
- ✰ Behaviour of floating bodies

Fluid Kinematics/Dynamics

Fluid dynamics/kinematics is the study of the behaviour of fluids under the action of applied forces. Herein, one is interested in finding the power necessary to move a fluid through a device, or the force required moving a solid body through a fluid. Of great importance then are the speed of the resulting motion, and the pressure, density and temperature variations in the fluid. The principles of dynamics and thermodynamics applied to the motion of fluids help to develop equations that describe the conservation of mass, momentum, and energy, which further help to find these quantities of interest. Like any mathematical model of the real world problem, fluid mechanics also makes some basic assumptions about the materials being studied. These assumptions turned into equations must be satisfied, if the assumptions are to hold true.

Fluid Continuum

Strong inter molecular cohesive forces compel the fluid to behave as a continuous mass. The fluid behaviour in which fluid properties are continuous field variables, either scalar or vector, throughout the control volume is known as continuum. Since the fluid flows continuously, any method and technique developed to analyze flow problems should take into consideration the continuity of the fluid. Normally, two approaches are used (Also see Chapter 5).

Lagrangian approach

In this approach, a material of the fluid is identified (or labelled); it is tracked (or followed) as it moves, and changes in its properties monitored. The properties may be velocity, temperature, density, mass, or concentration, *etc.* in the flow field. For example, one uses tracers into moving water bodies to determine currents. Think of a temperature sensor attached to a balloon, both having negligible mass and floating in the atmosphere. It is then possible to record the atmosphere-temperature or the temperature of the flow-field. This approach is more suitable and widely used for particle and solid mechanics.

Eulerian approach

In this approach, we monitor changes in the properties as different materials pass through a certain labelled fixed location in the flow field. As an illustration, recording of temperatures associated with the fixed location in the flow-fluid, as different fluid elements pass over time. Therefore, fluid properties and parameters are computed as field functions, *e.g.* $T(x,y,z,t)$. This approach is more appropriate for use in fluid flows.

BRIEF HISTORY

Archimedes (285-212 BC) postulated the parallelogram law of vector addition in the third century BC. Archimedes also invented the laws of buoyancy and applied them to floating and submerged bodies. He formulated the famous law now known as the Archimedes' principle, the first major work on fluid mechanics. Leonardo Da Vinci (1452-1519) derived the equation of conservation of mass in one dimensional steady state flow field. He experimented with waves, jets, hydraulic jumps, eddy formation and built the first chambered canal lock near Milan. Evangelista Torricelli (1608-1647) invented barometer. Blaise Pascal (1623-1662) formulated the Pascal's law. Isaac Newton (1642-1727) made significant contribution to fluid viscosity and came out with the Newton Law of Viscosity. Daniel Bernoulli (1700 – 1782) started mathematical fluid dynamics and published "Hydrodynamica". He introduced the word "hydrodynamics" to encompass various topics of fluid statics and dynamics. Leonhard Euler (1707 – 1783) first formulated modern fluid mechanics for ideal fluids, in a well-posed mathematical form. He developed both the differential equations of motion and their integral form, now called Bernoulli equation. Jean Louis Marie Poiseuille (1799-1869) explored the viscous flow concepts. Claude-Louis Navier (1785-1836) and George Gabriel Stokes (1819-1903) derived well known Navier–Stokes equations. Osborne Reynolds (1842-1912) and others advanced the understanding of fluid viscosity and turbulence. Ludwig Prandtl (1875-1953) pointed out that fluid flows with small viscosity, such as water flows and airflows can be divided into a thin viscous layer (or boundary layer) near solid surfaces and interfaces, patched onto a nearly inviscid outer layer, where the Euler and Bernoulli equations apply. In spite of the fact that most theoretical equations were available by the end of the 18th century, it was not possible to obtain their solutions because of the lack of solution methodologies. All methods to solve the above set of equations for engineering flow problems appeared only during the second half of the 20th century. It is owing to the tremendous advancements in numerical methods and also to huge developments in computer power. Now the fluid mechanics is well developed and one of the areas of modern fluid mechanics is Computational Fluid Mechanics, which deals with numerical solutions using computers. The other developments include flow visualization, laser Doppler anemometry, phase Doppler anemometry, particle image velocimetry, *etc.* It has made it easier to carry out detailed studies of laminar and turbulent flows.

WHY STUDY FLUID MECHANICS?

Fluid mechanics is one of the basic courses in Engineering. It is a bridge course between what the students already studied in physics and core B. Tech. courses, which will be taught after 1st year of the degree programme. One of the goals of this book is to let the students understand how the fluid moves or objects swim or fly? The second goal is to do with applications of fluid mechanics: to be able to make correct fluid mechanics estimates to build instruments and conduct research. Introduction to such issues is essential to an engineer, who has to deal with flows, which are also vital for the world to exist. Without flows, our natural and technical world would be different, and might not even exist at all. Besides what is stated, mechanics is the basis of the whole of physics in terms of intuition and mathematical methods. Concepts introduced in the mechanics of particles find applications in optics, electromagnetism, quantum mechanics *etc.*

The study of fluid mechanics is quite important to engineers, whose main interest is in the applications of fluid mechanics to solve industrial problems. Civil engineers may be interested in designing irrigation canals, dams, and water supply systems (Gupta and Gupta, 2020). Pollution control engineers may be interested in saving our planet from the constant dumping of industrial sewage into the atmosphere and the ocean. Mechanical engineers may be interested in designing turbines, heat exchangers, and fluid cooling studies. Chemical engineers may be interested in designing efficient devices to mix industrial chemicals. Aerospace engineers may be interested in designing planes that have low resistance and, at the same time, high "lift" force to support the weight of the plane. Dairy engineers will come across the principles of fluid mechanics while handling various types of fluids such as milk, water, air, refrigerants, steam *etc.* It is very important to learn the behaviour of fluid under various initial and boundary conditions to design the system for handling these fluids in dairy plants. To an agricultural engineer, fluid mechanics finds applications in studies related to moisture flow and heat transfer in the soils, roots, plant stem and canopy, food processing, storage rooms, dairy farms, plant factories, greenhouses, screen houses, broilers, fish ponds, silos, grain storage, solar collectors, heat exchangers, and other environments related to the agriculture sector. Design of irrigation channels, drainage canals and soil and water conservation structures requires the application of fluid mechanics.

Some of the common areas of interest where fluid mechanics knowledge finds applications are:

- Water supply and treatment system
- Pumps used for handling of different fluids
- Ships, submarines, aero planes, automobiles
- Storage tanks (milk silos, tankers, feed tanks, balancing tanks *etc.*)
- Piping systems for various utilities, pipe fitting and valves, flow meters *etc.*
- Flow measuring instruments
- Cleaning-in-Place (CIP) systems for optimum performance
- Heat transfer behaviour in processing equipments (such as HTST pasteurizers, spray dryers *etc.*)
- Hydro-power generation

FLUIDS

As already defined, fluid is a substance, which has no definite shape. It cannot preserve its shape unless it is restricted into a particular form depending upon the shape of its surroundings. It deforms or flows upon application of an external force. There are two classes of fluids namely liquids *e.g.* water and milk and gases *e.g.*

steam, air *etc.* Former are composed of relatively close packed molecules with strong cohesive forces. Gases, on the other hand are composed of molecules widely spaced with negligible cohesive forces. Liquids and gases can coexist in a two phase system such as steam water mixtures. Although liquids and gases share some common characteristics, yet have many distinctive characteristics of their own (Table 1.1).

Table 1.1. Distinctive Characteristics of Liquids and Gases

Liquids	*Gases*
Liquids are hard to compress. In general, liquids are incompressible fluids not affected appreciably by temperature and pressure change	Gases are easy to compress and as such categorized as compressible fluids, greatly affected by changes in temperature and pressure
A given mass of the liquid occupies a fixed volume, irrespective of the size and shape of the container (Figure 1.2)	A gas has no fixed volume and expands continuously unless restrained by the containing vessel
It changes its shape according to the shape of its container with an upper free surface (Figure 1.2). Liquids will always have a free surface, if the volume of the container is greater than that of the liquid	A gas does not have a free surface as it completely fills any vessel in which it is placed (Figure 1.2)

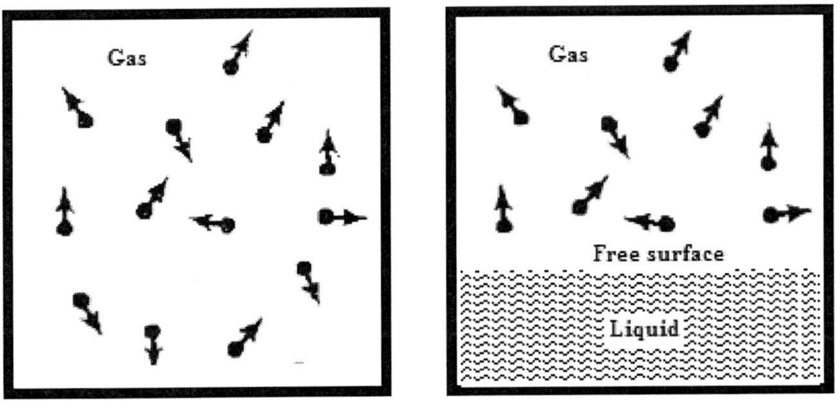

Figure 1.2. Illustration of Free Surface in Liquids.

It may be useful at this stage to look at the characteristic differences between solids (another form in which substances exist) and liquids (Table 1.2).

FLUID CHARACTERISTICS

Characteristics of a system either intensive (mass independent) or extensive (depends on size of system) are its properties. These properties of the fluid describe the state of the system. State postulates govern the number of properties required to fix the state of the system. Some of the fluid properties discussed in the forthcoming sections are:

- ☆ Density and specific gravity

Table 1.2. Distinctive Characteristics of Solids and Liquids

Solid	*Fluid*
More compact structure	Less compact structure
Attractive forces between the molecules are larger therefore, more closely packed. Solids can resist tangential stresses in static condition	Attractive forces between the molecules are smaller therefore, more loosely packed
A solid subjected to shear stress undergoes a definite deformation or breaks. Deformation is proportional to shear stress up to some limiting condition	A fluid subjected to shear stress shows no fixed deformation. It continues to deform until the shear stress is applied
Solid may regain its original shape partly or fully upon removal of the tangential stress	Once distorted by the shear stress, fluid never regains its original shape

- ☆ Specific weight and specific volume
- ☆ Viscosity
- ☆ Surface tension
- ☆ Vapour pressure
- ☆ Compressibility
- ☆ Temperature

Density (ρ)

Density, ρ of a fluid is its mass per unit volume expressed in units of kg/m³.

$$\rho = \text{mass}/\text{volume} = m/V \tag{1.1}$$

Here m is the mass, kg and V is the volume, m^3. Typical values of water and air are 1000 kg/m^3 and 1.23 kg/m^3. Changes in temperature and pressure can slightly change these values. For the same pressure, air density varies from a low of 0.946 kg/m^3 at 100°C to 1.293 kg/m^3 at 0°C. Table 1.3 reports the changes in density along with other properties of water as affected by temperature at the atmospheric pressure. Most properties reported in Table 1.3 either decrease or increase with increasing temperature except the isothermal compressibility. It initially decreases up to 45° C and later increases with increasing temperature beyond 45° C.

Specific Gravity or Relative Density (S_g)

Specific gravity is the ratio of the density of a substance to the density of water at a specified temperature. Specific gravity tells us how heavy something is with respect to the standard fluid (water). It is dimensionless and expressed as:

$$S_g = \rho/\rho_w \tag{1.2}$$

Here S_g is the specific gravity of the substance, ρ is density of the fluid or substance (kg/m^3) and ρ_w is the density of water - normally at a temperature of 4°C (kg/m^3). Specific gravity is also the ratio of the specific weight of a liquid to the specific weight of water at standard temperature and pressure (STP) (20°C, 1 atm). When calculating the specific gravity of solids, water with its density at 4°C (39°F) is used as a reference since water at this point has its highest density of 1000

Table 1.3. Select Properties of Water as Influenced by Temperature

Temperature	Density	Specific Weight	Isothermal Compres-sibility	Dynamic Viscosity	Vapour Pressure	Surface Tension
(°C)	(kg/m³)	(kN/m³)	((1/pa)x10⁹)	(mPa-s)	(kPa)	(N/m)
0	999.84	9.806	0.501	1.7921	0.6113	0.07564
10	999.70	9.804	0.478	1.3077	1.2281	0.07422
20	998.21	9.789	0.458	1.0050	2.3388	0.07275
30	995.65	9.765	0.446	0.8007	4.2455	0.07118
40	992.22	9.731	0.441	0.6560	7.3814	0.06956
50	988.03	9.690	0.440	0.5494	12.344	0.06791
60	983.20	9.642	0.443	0.4688	19.932	0.06619
70	977.78	9.589	0.449	0.4061	31.176	0.06442
80	971.82	9.530	0.457	0.3635	47.373	0.06261
90	965.35	9.467	0.468	0.3165	70.117	0.06075
100	958.40	9.399	0.480	0.2838	101.325	0.05885

kg/m³. S_g of water has the same numerical value as its density expressed in kg/m³ or Mg/m³. Air is the reference for calculating specific gravity of gases - being the ratio of the density of the gas to the density of the air - at a specified temperature and pressure.

$$S_g = \rho_{gas}/\rho_{air} \tag{1.3}$$

Such that S_g is the specific gravity of gas, ρ_{gas} is the density of gas (kg/m³) and ρ_{air} is the density of air (at NTP - 1.204 kg/m³). NTP stands for normal temperature and pressure given as 20°C (293.15 K) and 1 atm (101.325 kN/m² or 101.325 kPa).

Specific Weight (γ)

Weight is a force arising from the existence of a gravitational force. Weight per unit volume is specific weight. The SI unit for specific weight is (N/m³). Specific weight (or force per unit volume) is

$$\gamma = \rho\, g \tag{1.4}$$

Such that γ is the specific weight (N/m³) and g is the acceleration of gravity under normal conditions (9.807 m/s²). Typical values of water and air are 9814 N/m³ and 12.07 N/m³. Unlike density, specific weight is not a fixed property of a material. It depends on the value of the gravitational acceleration, which varies with location.

Specific Volume (V)

Specific volume of a substance is the volume per unit of its mass. One can write it as:

$$V = 1/\rho = \text{Volume/mass} \tag{1.5}$$

The unit of specific volume is m³/kg and its dimension is L^3/M.

Example 1.1

Calculate the density of a ball of 42 mm diameter having a mass of 50 g.

$V = (4/3)\,\pi\, r^3 = (4/3)\,\pi\,(42 \times 0.001/2)^3 = 3.8 \times 10^{-5}\ m^3$

The density of the ball is

$\rho = (50 \times 0.001)/(3.8 \times 10^{-5}) = 1315.5\ kg/m^3$

Viscosity (μ)

The ease with which a fluid pours is an indication of its viscosity. Fluid with a high viscosity such as syrup pours and deforms more slowly than fluid with a low viscosity such as water. Similarly, flow ability of honey is poor as compared to milk, honey being highly viscous. Viscosity, μ, is a measure of resistance to fluid flow because of intermolecular cohesion. In other words, viscosity is the internal friction to fluid motion, which can lead to energy loss. It expresses the magnitude of internal friction in a fluid, as measured by the force per unit area resisting uniform flow. Effect of viscosity is apparent when the fluid is in motion. Viscosity represents the internal resistance of a fluid to motion (*i.e. fluidity*). In a fluid flow, when the fluid elements move with different velocities, each element experiences some resistance due to fluid friction within the elements. Therefore, shear stresses develop between the fluid elements moving with different velocities. The relationship between the shear stress and the velocity field given by Sir Isaac Newton, in the form of law of viscosity is known as Newton's law of viscosity.

Kinematic viscosity (η)

Kinematic viscosity (η) is the ratio of the dynamic viscosity to the fluid density

$$\eta = \mu/\rho \qquad (1.6)$$

This property is important when significant viscous and gravitational forces exist. Units of kinematic viscosity are m^2/s. CGS units is Stoke such that 1 stoke = 0.0001 m^2/s. Typical values for water and air are $1.14\times10^{-6}\ m^2/s$ and $1.46\times10^{-5}\ m^2/s$.

Newton Law of Viscosity

Consider a fluid contained between two parallel plates such that distance between the plates is y units (Figure 1.3 top). Let us also say that the area of the plates is A. The plate AD is stationary while plate BC is moving, although initially it is also at rest. If a shear force, F (force applied tangentially and parallel to a surface) is applied to top plate at point B, then the upper plate starts moving and attains a velocity say u m/s. Now the position changes from ABCD to AB′C′D (Figure 1.3 bottom).

The upper layer, which is moving faster, tries to draw the lower slowly moving layer along with it by means of a force F along the direction of flow. Similarly, the lower layer tries to retard the upper one, according to Newton's third law, with an equal and opposite force F on it. The distribution of fluid velocity from the top plate to the bottom is known as velocity gradient or velocity profile (Figure 1.4). Such a fluid flow where x-direction velocities change with y-coordinate is called shear

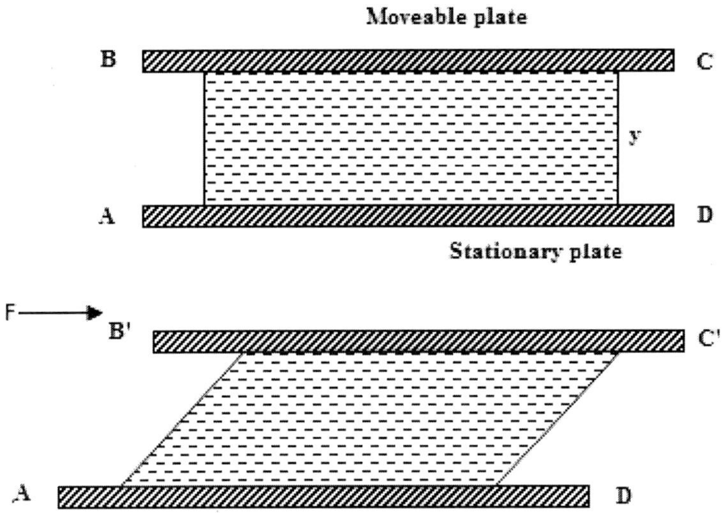

Figure 1.3. Initial State (Top) and State After Application of Shear Force to Plate BC (Bottom).

flow of the fluid. Thus, a tangential force F on the respective layers experiences the dragging effect of one layer on the other. If F acts over an area of contact A, then the shear stress τ is defined as $\tau = F/A$. Here τ, pronounced as Tau, is the symbol of shear stress. Newton postulated that τ is proportional to the quantity $\Delta u/\Delta y$ where Δy is the distance of separation of the two layers and Δu is the difference in their velocities. In the limiting case of $\Delta y \rightarrow 0$, $\Delta u/\Delta y$ equals du/dy, the velocity gradient at a point in a direction perpendicular to the direction of the motion of the layer (rate of shear deformation). Thus

$$\tau \ \alpha \ du/dy \tag{1.7}$$

Or

$$\tau = \mu(du/dy) \tag{1.8}$$

Here, μ is the constant of proportionality and is known as the coefficient of viscosity or simply dynamic viscosity, which is a property of the fluid and depends on its state. The magnitude of the velocity gradient (du/dy) has no effect on the magnitude of viscosity. Sign of τ depends upon the sign of du/dy. For the profile shown in Figure 1.4, du/dy is positive everywhere and hence, τ is positive. Both the velocity and stress are positive in the positive direction of the coordinate parallel to them. The viscosity of a fluid defined in this manner is the Newton's law of viscosity (eq. 1.8). It states that, the shear stress between adjacent fluid layers is proportional to the velocity gradients between the two layers.

For a given temperature and pressure, the ratio of shear stress to shear rate is constant, and is defined as the viscosity or coefficient of viscosity. The units of viscosity are Ns/m^2 or kg/m-s. CGS units of dynamic viscosity is poise such that

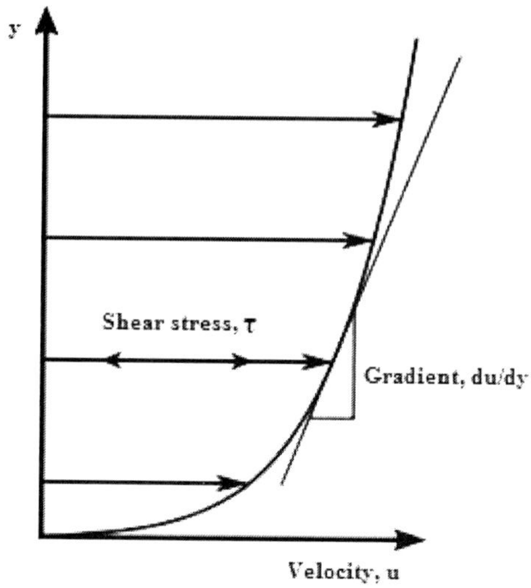

Figure 1.4. Definition Sketch of Shear Stress.

1 poise = 0.1 Ns/m^2 or centipoise such that 1 centipoise (cP) = 0.01 poise. Typical values for water and air are 1.14×10^{-3} kg/m-s and 1.78×10^{-5} kg/m-s. The viscosity is a function of the condition of the fluid, particularly its temperature. In general, viscosity of liquids decreases with increase in temperature (Table 1.3), whereas viscosity of gases increases with increase in temperature.

Newtonian and non-Newtonian fluids

Newtonian fluids are named after Sir Issac Newton (1642 - 1726) who described the flow behaviour of fluids with a simple linear relation between shear stress [mPa] and shear rate [1/s]. This relationship is now known as Newton's Law of Viscosity and is given as:

$$\tau = \mu\,(du/dy)$$

Here u is the flow velocity along the boundary and y the height above the boundary. Fluids, *viz.* water, air, oil, gasoline, alcohol, kerosene, benzene, glycerine and mercury all obey Newton's law of viscosity and are known as Newtonian fluids. Viscosity of a Newtonian fluid remains constant, no matter the amount of shear applied for a constant temperature. It means a linear relationship exists between shear stress and the velocity gradient (rate of shear) (Figure 1.5). Other classes of fluids, *viz.* paints, different polymer solutions, blood that do not obey the typical linear relationship, of τ and du/dy are known as non-Newtonian fluids. For non-Newtonian fluids, slope of the curves varies (Figure 1.5). In non-Newtonian fluids viscosity itself may be a function of the velocity gradient as well as the condition of the fluid.

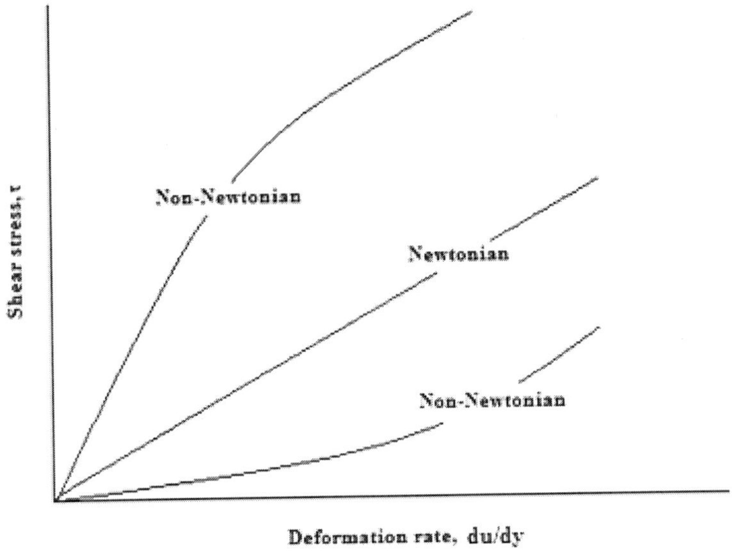

Figure 1.5. Definition Sketch of Newtonian and Non-Newtonian Fluids.

Ideal and real fluids

A non-existent, assumed fluid without either viscosity or compressibility is called an ideal fluid or perfect fluid. In an ideal fluid, there is no friction and the fluid can 'slide' tangentially along the solid boundary. An ideal fluid is incompressible (so that its density does not change), flows at a steady rate, non-viscous (no friction between the fluid and the container through which it is flowing), and flows irrotationally (no swirls or eddies). A fluid with compressibility but without viscosity is occasionally discriminated and called a perfect fluid, too. Furthermore, a gas subject to Boyle's-Charles' law is called a perfect or ideal gas.

A real fluid, on the other hand, possesses friction (or viscosity) such that it cannot 'slide' along boundary – no slip boundary condition. In a real fluid, viscosity produces resistance to motion by causing shear or friction forces between fluid particles and between these and boundary walls. Thus, velocity component perpendicular to the wall must be the same as that of the wall. It also means that tangential velocity will be zero if the wall is at rest. As such, fluid tends to 'stick' to solid surfaces and have stresses within their body. The inclusion of viscosity allows the existence of two physically distinct flow regimes, known as laminar and turbulent flow.

Example 1.2

A plate moves at 2 m/s with a shearing force of 3.6 N/m^2. If the distance between moving and fixed plate is 0.08 mm, what is the viscosity of the fluid between plates?

Velocity of plate (u) = 2 m/s

$F/A = 3.6\ N/m^2$

Since shearing force is equal to $\mu\ du/dy$ and $du/dy \sim u/y$, y being small

Viscosity, $\mu = 3.6 \times (0.08/1000)/2 = 1.44\times10^{-4}\ Ns/m^2$

Example 1.3

Two rectangular flat plates of dimensions 850 mm by 600 mm are 0.02 mm apart. Oil of specific gravity 0.92 fills the space between them. The lower plate is fixed. The upper plate moves at 4.5 m/s and requires a force of 200 N to maintain this state. Determine dynamic and kinematic viscosity of oil.

Velocity of plate (u) = 4.5 m/s

Force (F) = 200 N

Area of plate (A) = 850 mm × 600 mm

Distance between plates (y) = 20 mm

Density of fluid = sp. gr × density of H_2O at 4°C = 0.92×1000 = 920 kg/m^3

Since shearing force, F/A is equal to $\mu\ du/dy$ and $du/dy \sim u/y$, y being small, we have

Dynamic viscosity, $\mu = (F\times y)/(A\times u)$

$200\times(0.02/1000)/4.5\times(850\times10^{-3}\times600\times10^{-3}) = 1.7429\times10^{-4}\ Ns/m^2$

Kinematic viscosity = Dynamic viscosity/fluid density = $1.7429\times10^{-4}/920 = 1.8945\times10^{-7}\ m^2/s$

Example 1.4

A liquid has a specific gravity of 1.9 and kinematic viscosity of 1.1 stoke. What is its dynamic viscosity?

Kinematic viscosity = 1.1 stokes = $1.1\times10^{-4}\ m^2/s$

Density of substance = sp. gr. × density of H_2O at 4°C = 1.9×1000 = 1900 kg/m^3

Dynamic viscosity = kinetic viscosity/density of substance = 1900/(1.1×10000) =0.1727 Ns/m^2

Vapour Pressure

Vapour pressure is a property exhibited by liquids as they have a tendency to vaporize or evaporate. This process takes place at free surface of liquid where liquid molecules continuously escape to the atmosphere. The liquid molecules, which escape from free surface of liquid, are in gaseous state. These exert partial pressure on surface of the liquid. Some molecules of the liquid in the vapour phase may convert back to the liquid phase after striking the walls of the containers or the surface of the liquid through the process known as condensation. As time passes, the numbers of molecules in vapour phase increase while the rate of condensation also increases. At some stage, rate of evaporation is equal to the rate of condensation, called the equilibrium stage. The vapour pressure thus, is the pressure exerted by the molecules ejected in phase equilibrium with its liquid phase at a given temperature.

In other words, vapour pressure is a measure of the tendency of a material to change into gaseous or vapour state. Higher the vapour pressure, higher the rate at which molecules will escape into the atmosphere. The surface area of the solid or liquid in contact with the gas has no effect on the vapour pressure. However, it does depend upon the types of molecules that make up a solid or liquid and its temperature.

If the intermolecular forces between molecules are relatively strong, the vapour pressure is relatively low. Contrarily, if these forces are weak, the vapour pressure is relatively high. Ethyl ether ($C_4H_{10}O$) exhibits a vapour pressure of 520 torr at 25°C (1 torr is equal to 1 mm of mercury or 101325/760 Pa or 133.32 Pa) because of the relatively weak dipole-dipole forces and London dispersion forces between molecules. Ethyl alcohol (C_2H_6O), on the other hand exhibits a pressure of only 75 torr at the same temperature. Although dipole-dipole forces and London dispersion forces also exist between ethyl alcohol molecules, the strong hydrogen bonding interactions results in much lower vapour pressure compared to ethyl ether. Mercury has a very low vapour pressure (0.002 mm Hg/21°C). It is why; mercury is used in barometers and thermometers for high accuracy. The pure liquids have higher vapour pressure compared to the liquid's solutions. Vapour pressure also increases with increasing temperature due to increase in the kinetic energy of its molecules (Table 1.3). At a higher temperature, more molecules have enough energy to escape from the liquid or solid compared to low temperature, where fewer molecules escape from the liquid or solid. A substance having high vapour pressure at normal temperatures is a volatile substance.

At saturation vapour pressure, the space above the liquid is saturated with vapours of the liquid and equilibrium exists between the liquid-vapour interface. It is the maximum attainable vapour pressure at a given temperature. Any change in the temperature will change the saturation vapour pressure. A liquid boils when its saturated vapour pressure becomes equal to the external pressure exerted on the liquid. The temperature at which it happens is the boiling point of the liquid. Besides, water boils at different temperatures if there is a change in pressure. It is why the boiling point of water decreases with increasing height above the mean sea level because the pressure decreases with increasing altitude. Similarly, as the temperature reduces water vapour in the air comes in the equilibrium with liquid water. The temperature at which it happens is the dew point. Below the dew point, liquid water begins to condense in droplets.

Even solids lose particles from their surface and form vapours. The process in this case is sublimation rather than evaporation. Thus, sublimation is the process of direct change from solid to vapour or vice-versa without going through the liquid stage.

An important implication of saturation vapour pressure is in turbo-machinery or in pipe networks. If the pressure, P at a point in the network of pipes is lower than the saturation vapour pressure, P_v, liquid begins to vaporize locally, creating cavities of vapour. These vapour cavities collapse when local P rises above P_v. Collapse of cavities is a violent process causing noise, and vibrations, which can damage pipe network or machinery. Rotodynamic pumps at its suction side often encounter such a phenomenon. This important aspect is a part of a forthcoming chapter on pumps (Chapter 13).

Surface Tension (σ)

In our daily life, we see that small liquid drops have spherical shapes. It is because of surface tension phenomenon where the liquid surface tries to have minimum surface area for a given volume (sphere has minimum surface area). Surface tension is the property of a liquid due to which its free surface tries to have minimum surface area and behaves as if it were under tension like a stretched elastic membrane. As a result, it is able to support comparatively heavier objects placed over it. One can also explain this process in a different manner. In the bulk of the liquid, neighbouring liquid molecules pull each molecule equally in every direction, resulting in a net force of zero (Figure 1.6 top). A molecule at the surface, on the other hand experiences a net attractive force pointing toward the liquid below, because there are no molecules of the liquid above the surface. Consequently, they cohere more strongly to those directly associated with them on the surface (Harvey, 1925). This forms a surface film, which makes it more difficult to move an object through the surface than to move it when it is completely submerged (Figure 1.6 bottom). It is the result of the force of surface tension. Thus, surface tension may be defined

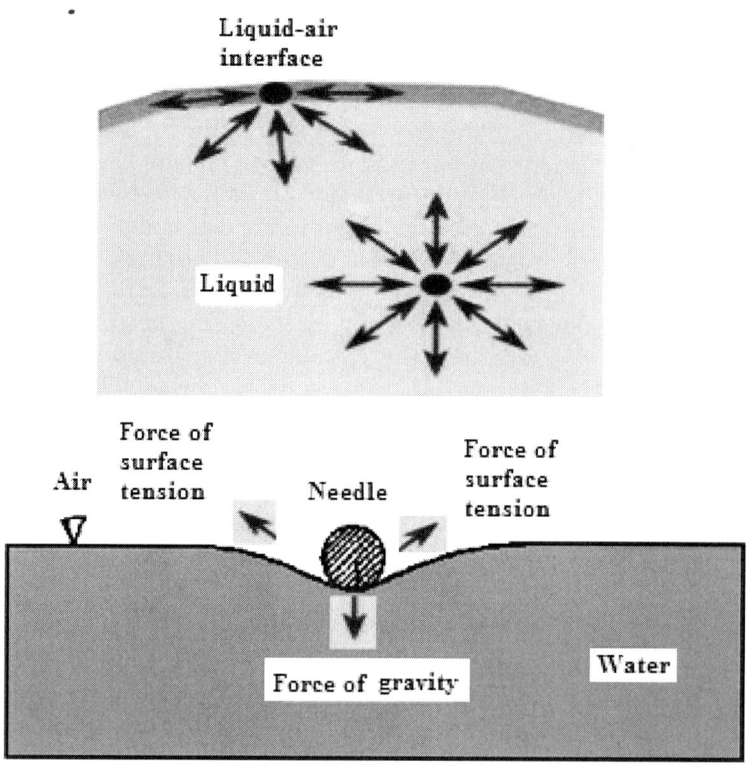

Figure 1.6. Surface Tension Phenomenon: Forces on an Object in Liquid and at an Interface (Top) and Floating Needle on Water Surface (Bottom).

as the tensile force acting on the surface of a liquid in contact with a gas or on the surface between two immiscible liquids such that contact surface behaves like a membrane under tension. In terms of energy, surface tension is the surface energy per unit area of the liquid surface. Surface tension results from cohesive forces. Thus, surface tension has the dimension of force per unit length or of energy per unit area. Some of the properties of surface tension are:

- ✰ It is a scalar quantity as it has a unique direction which is not to be specified
- ✰ It is temperature sensitive
- ✰ It is impurity sensitive
- ✰ It is a molecular phenomenon and its root cause is the electromagnetic forces
- ✰ Depends only on the nature of the liquid
- ✰ Unit of surface tension is N/m. In CGS system, it is Dyne/cm
- ✰ Dimension of surface tension, MT^{-2}.

Surface tension commonly denoted as σ of a liquid is the force acting per unit length on either side of an imaginary line on the free surface of liquid, the direction of this force being perpendicular to the line and tangential to the free surface of liquid. Surface tension of some of the liquids is reported in Table 1.4.

Table 1.4. Numerical Values of Surface Tensions of Select Liquids

Liquid	*Surface Tension, σ (N/m)*
Benzene (20 °C)	0.029
Blood (37 °C)	0.058
Glycerine (20 °C)	0.063
Mercury (20 °C)	0.47
Water (20 °C)	0.073
Water (100 °C)	0.059

If F is the force acting on one side of imaginary line of length L, then surface tension, σ = (F/L). For a floating body as in the case of Figure 1.6 (bottom), the horizontal components of the two forces of surface tension acting on the body point in opposite directions, so they cancel each other. On the other hand, vertical components point in the same direction and therefore, add up to balance the force of gravity or weight of the substance. When the weight of the substance is such that the surface tension is just able to support it then:

$$F_w = 2\,\sigma\,L\cos\theta \tag{1.9}$$

Here θ is the angle of contact and L is length of contact.

As already stated, liquids form bubbles and droplets due to surface tension. The inward surface tension force causes bubbles to be approximately spherical and raises the pressure of the gas trapped inside relative to atmospheric pressure outside. The surface tension balances the outward force due to the pressure difference between the inside and the outside air. For a bubble, we have two surfaces, the inner and the outer surface. The force from surface tension therefore is $F' = 2\sigma L = 2\sigma 2\pi r = 4\sigma\pi r = 2\sigma\pi d$ (Table 1.5). For the bubble to be stable and not collapse, the pressure inside the bubble must be higher than the pressure on the outside. Besides, the force due to the pressure difference, $(P_i - P_o)\pi r^2$ must balance the force from the surface tension. Thus

$$(P_i - P_o)\pi r^2 = 4\sigma\pi r$$

$$P = 4\sigma/r = 8\sigma/d \text{ (Table 1.5)}$$

Here r is the radius of the bubble, d the diameter of the bubble and P the gauge pressure inside the bubble *i.e.* difference between the pressure inside the spherical bubble and the atmospheric pressure outside $(P_i - P_o)$.

Table 1.5. Pressure and Surface Tension Forces in Water Droplet, Soap Bubble and Liquid Jet

Parameter	*Water droplet*	*Soap bubble*	*Liquid jet*
Pressure force, F	$F = P(\pi/4)(d)^2$	$F = P(\pi/4)(d)^2$	$F = PLd$
Surface tension force	$F' = \sigma(\pi d)$(around the circumference)	$F' = 2\sigma(\pi d)$(on both sides of the bubble)	$F' = \sigma 2L$(in jet)
At equilibrium	$F = F'$ or $P = 4\sigma/d$	$F = F'$ or $P = 8\sigma/d$	$F = F'$ or $P = 2\sigma/d$
Here P and d are the pressure inside the droplet and diameter of the drop or bubble or liquid jet and L the length of liquid jet			

Factors Affecting Surface Tension

Temperature

The surface tension of liquids decreases with rise in temperature (Table 1.3). The surface tension of liquid is zero at its boiling point and it vanishes at critical temperature. At critical temperature, intermolecular forces for liquid and gases become equal and liquid can expand without any restriction. For small temperature differences, the surface tension varies linearly with temperature as per the following relation:

$$\sigma_t = \sigma_0(1 - \alpha t) \tag{1.10}$$

Here σ_t, σ_0 are the surface tensions at t° C and 0°C respectively and α is the temperature coefficient of surface tension.

Impurities

The presence of impurities either on the liquid surface or dissolved in it, considerably affect the force of surface tension. Degree of contamination controls the actual effect. A highly soluble substance like sodium chloride when dissolved in water increases the surface tension of water. On the other hand, sparingly soluble substances like phenol dissolved in water, decreases the surface tension of water.

Practical applications of surface tension

If we place a needle of length L and weight W on the liquid surface with surface tension σ then the liquid can support the largest weight, W given by the relation:

$$W = 2\,L\,\sigma \tag{1.11}$$

(Since the surface tension acts on both sides of the needle, 2L)

Similarly, if F is the minimum force required to pull it away from the water, the following relation can calculate F.

$$F = 2L\,\sigma = W \tag{1.12}$$

Similarly, it is possible to derive relations for other bodies. For example, F for a hollow disc of inner and outer radius of r_1 and r_2 is:

$$F = 2\pi\,(r_1+r_2)\,\sigma \tag{1.13}$$

For a circular plate of radius r, following relation gives the value of F.

$$F = 2\pi\,r\,\sigma \tag{1.14}$$

For a ring

$$F = 2\times 2\pi\,r\,\sigma$$

Since the effective length is $2\times 2\pi\,r$ as the liquid wets it from inside as well as outside.

Example 1.5

A needle has a length of 3.2 cm. When placed gently on the surface of the water ($\sigma = 0.073$ N/m) in a glass, it floats. What is the weight of the heaviest needle that can float on this liquid?

$$W = 2\,L\,\sigma = 2\times 0.032\times 0.073 = 0.0047\text{ N}$$

Example 1.6

Surface tension of a liquid is 5.1 N/m. A small insect of about 1.6 cm long is sitting on it. If the liquid is just able to support this weight, what is the approximate mass of this insect?

From the surface tension, we calculate the force F, which in this case is the weight of the insect.

$$F = \sigma\,L = W$$

$$F = 5.1\times 0.016 = 0.0816\text{ N}$$

Mass of the insect is given as:

$$M = F/g$$

Note that $g = 9.8\text{ m/s}^2$ and therefore

$$M = 0.0816/9.8 = 0.00832\text{ kg or } 8.32\text{ g.}$$

Cohesion and Adhesion

We have seen that surface tension arises because of the intermolecular forces of attraction that molecules a liquid exert on one another. These forces, electromagnetic in nature, are of two types namely cohesion and adhesion. According to the definition, the force of cohesion is the force of attraction between molecules of the same substance enabling them to stick to each other due to mutual attraction. For example, solids having high cohesive properties do not stick to the surfaces they contact. On the other hand, gases have weak cohesion. A liquid coming in contact with a solid surface, such as glass, experiences additional forces of attraction. Known as adhesive forces, these occur between unlike molecules of the liquid and molecules of the solid surface. Thus, adhesive force is defined as the force of attraction between different substances to cling to each other. It may also be between two different liquids, which do not mix.

Water has both cohesive and adhesive properties. Water molecules stick to each other to form a spherical drop. It is the result of cohesive forces. When contained in a tube, the water molecules touching the surface of the container are at a higher level. It is due to the adhesive force between the water molecules and the molecules of the container. These forces are inversely proportional to the eighth power of distance between the molecules. The dewdrops on the leaves in the morning hours well illustrate the phenomena of cohesion, adhesion and surface tension. Cohesion causes water to form drops, surface tension causes them to be nearly spherical, and adhesion keeps the drops in place.

Wetting and non-wetting liquids

Wetting is the ability of the liquids to form interfaces with the solid surfaces when the two come in contact with each other. It is the result of intermolecular forces. The degree of wetting (wet ability) depends upon the force balance between the cohesive and adhesive forces. According to the surface wetting ability, liquids are classified as wetting and non-wetting liquids. As an illustration, mercury on a solid surface remains like a ball while water spreads well on the surface. In this illustration, mercury is non-wetting because adhesive forces are higher than the cohesive forces. Water on the other hand is a wetting liquid. A wetting fluid is able to maintain surface contact with the solid with contact angles being less than 90° (angle from the wetted surface to the surface of the fluid). The smaller the contact angle and the surface tension, the greater is the degree of wetting. A zero contact angle represents complete or perfect wetting (spreading liquid). On the contrary, non-wetting liquids are not able to keep the contact area as large as the wetting fluids. The contact angles in such cases are more than 90° (Figure 1.7).

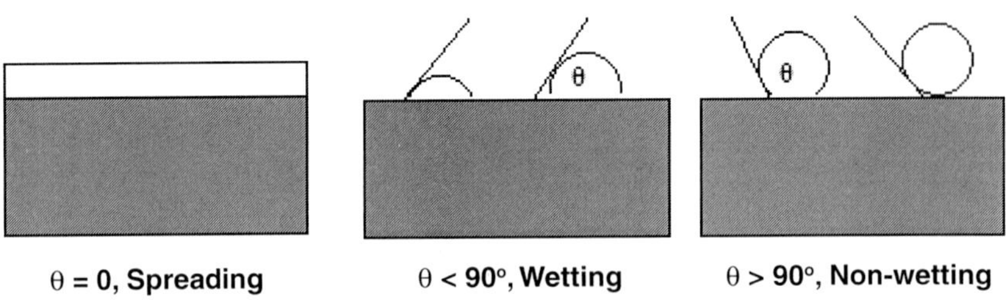

Figure 1.7. Description of Wetting and Non-wetting Fluids.

Capillarity

Capillarity is a phenomenon in which liquid in a capillary glass tube rises or falls from its original level due to the combined effect of cohesion and adhesion. The capillary rise in a capillary above the level of zero pressure occurs because of the net upward force exerted by the attraction of liquid molecules to a solid surface say glass or soil. It occurs when the adhesion to the walls is stronger than cohesion. If other conditions remain the same, the narrower capillary will show higher rise of fluid. If cohesion is greater than adhesion, a fall in capillarity is experienced (Figure 1.8 left). The meniscus in the former is concave, while in the latter it is convex. Angle of contact between liquid surface and the solid material is less than 90° in the former and greater than 90° in the later. As an example, capillary rise occurs in case of water, but a capillary depression occurs in case mercury (Figure 1.8). This phenomenon besides explaining the effect of cohesion, adhesion and surface tension, also explains whether the interaction by the fluid and the solid surface is wetted or non-wetted. If the adhesion of fluid molecules to the adjacent solid surface is stronger than the intermolecular cohesion, the fluid is said to wet (water) on the surface. Mercury, on the other hand, falls in a glass capillary due to the surface tension of the mercury (Figure 1.8 right). It also does not wet the glass surface. It is a non-wetted interaction (Mercury).

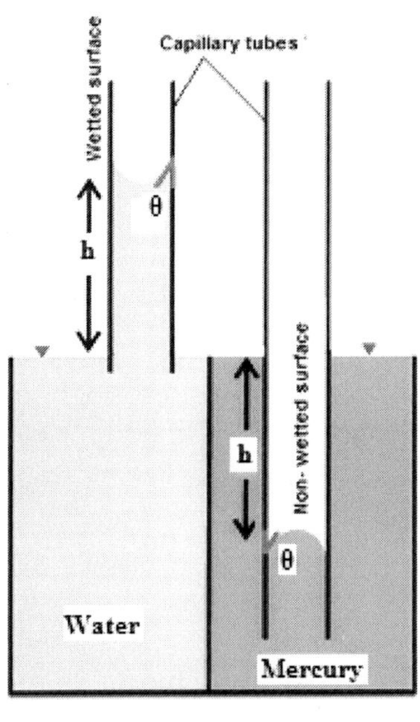

Figure 1.8. Capillary Phenomenon Illustrating the Wetting and Non-wetting Interactions.

Surface tension and gravity limits the height to which capillary action will take water in a uniform circular tube. The surface tension force acts along the tangent to the liquid surface at the point of contact, θ being the angle of contact (Figure 1.8). This force is resolved into two components namely horizontal, $\sigma \sin \theta$ and vertical, $\sigma \cos \theta$. While $\sigma \sin \theta$ cancel out each other, the weight of water in the capillary tube given as $V\rho g$ balances the force $\sigma \cos \theta$.

Let radius of glass tube = r (diameter, d = 2r)

Density of liquid = ρ

Rise in liquid level = h

Angle of contact between water surface and glass = θ

Force in upward direction = $\sigma \pi d \cos \theta$

Equating

$$\pi r^2 h \rho g = \sigma 2 \pi r \cos \theta \qquad (1.15)$$

$$h = 2 \sigma \cos \theta / r\rho g \qquad (1.16)$$

Here h is the height of liquid level in the capillary; r is the radius of bore of the capillary while all other parameters are as described earlier in this chapter. The capillary rise occurs in liquid solid interface, if contact angle is acute.

If the capillary solid material and fluid are same then one can write following equations for capillary rise in tubes of radius r_1 and r_2 respectively:

$$h_1 = 2 \sigma \cos \theta / r_1 \rho g \qquad (1.17)$$

$$h_2 = 2 \sigma \cos \theta / r_2 \rho g \qquad (1.18)$$

Here h_1 and h_2 are the respective capillary rise in the two tubes. Dividing eq. (1.17) by eq. (1.18), we get

$$h_1/h_2 = r_1/r_2$$

$$r_2 h_1 = r_1 h_2$$

This relation is known as the Jurin's law.

Example 1.7

A liquid of density 920 kg/m^3 rises to a height of 7.2 mm in a capillary of 1.8 mm diameter. If the contact angle is 25° find the surface tension of the liquid. Assume $g = 9.81\ m/s^2$.

From eq. (1.16), $\sigma = hr\rho g/(2 \cos\theta)$

$$\sigma = 7.2 \times 10^{-3} \times 0.9 \times 10^{-3} \times 920 \times 9.81/(2 \times 0.9063) = 0.0322\ N/m^2$$

Compressibility

In fluid mechanics, compressibility of the fluids is an important characteristic. Theoretically, all materials, solids or fluids (liquids/gases), change their shape under application of external force or stress. On the other hand, a fluid increases its pressure against compression, trying to retain its original volume. This characteristic is known as compressibility. In case of fluid, the force is generally termed as 'pressure' as it is uniform from all the directions. In simple words, compressibility is a measure of change in the volume of fluid under the effect of force/pressure or stress. Solids and liquids have very little free space between particles. Therefore, they are not easily compressible. For most applications, they including water are treated as incompressible. Gases, on the other hand, are easily compressible, having lot of free space between particles. Notwithstanding the above statements, compressible fluids exposed to smaller pressure difference can exhibit incompressible flow and incompressible fluids exposed to high-pressure difference can exhibit compressible flow. Thus, under pressurized situations it is not advisable to ignore the compressibility of liquids. On the other hand, whenever the change in pressure is small, one can neglect the compressibility of gases. Fluid flow is compressible if density of the fluid changes roughly 5 per cent of its original density during its flow. Since definition of compressibility involves change in volume due to change in pressure, hence compressibility can be isothermal, where volume change takes place at constant temperature or isentropic where volume change takes place at constant entropy.

In thermodynamics and fluid mechanics, compressibility also known as the coefficient of compressibility or isothermal compressibility is a measure of the fractional change in volume per unit increase in pressure.

$$\text{Compressibility, } \alpha = -(1/V)\,\partial V/\partial P = (1/\rho)\,\partial\rho/\partial P \tag{1.15}$$

Bulk modulus of elasticity

$$k = dP/(\Delta V/V) = \text{Change in pressure/volumetric strain} \tag{1.16}$$

Here V is volume and P is pressure. Units are the pressure units. Negative sign in the expression of compressibility takes care of volume decrease with increase in applied pressure. From eqs. (1.15) and (1.16), it may be seen that the compressibility, α is the reciprocal of the bulk modulus of elasticity, k of a fluid. The bulk modulus of elasticity, k for water and air at atmospheric pressure is approximately 2×10^6 kN/m^2 and 101 kN/m^2 (constant temperature bulk modulus) respectively. It indicates that air is about 20,000 times more compressible than water. For each atmosphere increase in pressure, the volume of water would decrease 46.4 parts per million.

Example 1.8

If the volume upon applying a pressure of 6 bars reduces to 0.30 per cent of its original volume then calculate the bulk modulus of elasticity

Volumetric stress = Pressure intensity = 6 bars = 6×10^6 N/m^2

Volumetric strain = 0.003

k = Volumetric stress/Volumetric strain = $6\times10^6/0.003 = 2\times10^3$ N/m^2

Example 1.9

A cylinder of 8-liter capacity is completely filled with oil. With application of pressure of 9.2 MN/m^2, volume of the oil decreases by 400 mL. Find the bulk modulus of oil.

Pressure (P) = 9.2 MN/m^2 = 9.2×10^6 N/m^2

Change in volume = ΔV = 400 mL = 0.4 L

Original volume = 8 L

$k = 9.2\times10^6/(0.4/8) = 9.2\times10^6/0.005) = 1.84\times108$ N/m^2.

Example 1.10

Pressure in the ocean at a depth of 7 km is 71.6 MPa. Calculate the following, if the specific weight of the water at the surface is 10.05 kN/m³ and bulk modulus of elasticity is 2.34×10^9 N/m².

a. Change in specific volume between the surface and at the depth of 7 km
b. Specific volume of water at the depth of 7 km
c. Specific weight of water at the depth of 7 km

$V_1 = 1/\rho_1 = g/\gamma_1 = 9.81/10050 = 0.000976$ m³/kg. Using eq. (1.16)

Change in specific volume, $\partial V = -\ 0.000976\ (71.6\times10^6 - 0)/2.34\times10^9 = -29.8\times10^{-6}$

$V_2 = V_1 + \partial V = 0.000976 - 0.0000298 = 0.000946$

$\gamma_2 = g/V_2 = 9.81/0.00946 = 10370$ N/m³

Temperature (T)

It is the measure of hotness and coldness of a substance. In thermodynamic sense, it is the measure of internal energy of a system. For temperature expressed in centigrade scale (°C), the freezing and boiling points of water are 0°C and 100°C, respectively. In SI system, the temperature is expressed in terms of absolute value in Kelvin scale, K = °C+ 273.15. Kelvin scale has an absolute zero at -273.15° K, below which temperature does not exist.

DIMENSIONS AND UNITS

Quantity is a generic term for all physical quantities that are measurable *e.g.* time, length, mass, force, and work done, *etc.* In physics, we study about physical quantities and their inter relationships. A set of physical quantities called Set of Fundamental Quantities are completely independent of each other. These fundamental quantities or their combinations can express all other physical quantities.

A physical quantity characterized by dimension, is a measure of the physical variable (without numerical values). Dimensions are of two types namely primary or fundamental and secondary or derived. Some basic dimensions such as mass (M), length (L), time (T) and temperature (θ) are base/primary/fundamental dimensions. While others such as velocity (L/T), pressure ($M/L\text{-}T^2$) *etc.* expressed in terms of primary dimensions are secondary/derived dimensions. In fluid mechanics, we can represent all the physical properties with three dimensions *i.e.* L, T and one of M or F. Thus, two combinations are either MLT or FLT system. Either mass ($M = FT^2L^{-1}$) or force ($F = MLT^{-2}$) can be used to represent the other. In this book, SI (Standard International) system of units and dimensions are used to express the properties of fluids. In the SI system (Système International), there are seven basic or fundamental quantities, namely length, time, mass, temperature, current, number of elementary particles, and luminous intensity (Table 1.6).

Measurement is the comparison of a quantity with a standard of the same physical quantity (Annexure I). All physical quantities are measured with respect to standard magnitude of the same physical quantity and these standards are called units *e.g.* second, meter, kilogram, *etc.* A unit is a way to assign a number or measurement to the dimension of a quantity. For example, length is a dimension, but measured in units of meters (m) or feet (ft). Similarly, the units of time, mass and temperature are the second (s), the kilogram (kg), and the temperature (K) respectively (Table 1.6). A quantity always has the same dimension but may have different units. For example, force has the dimension of Mass×Length/Time^2 but may have units of ergs, Newton *etc.* Solution to numerical and engineering problems becomes meaningless without units. Thus, while the dimension quantitatively expresses a physical variable, the unit is a particular way of attaching a number to the quantities of dimension. All the properties of fluid have some units and dimensions assigned to them.

Table 1.6. List of Complete Set of Basic Quantities

Quantity	*SI Name*	*SI Symbol*
Length, L	meter	m
Time, t	second	s
Mass, M	kilogram	kg
Temperature, T	Kelvin	K
Current, I	ampere	A
Number of elementary particles	mole	mol
Luminous intensity	candela	cd

Systems of Units

Absolute system of units

The system in which lengths, mass and time are the basic quantities, and from which the units of other quantities are derived, is called the absolute system of units.

MKS system of units

This is the system of units where the meter (m) is used for the unit of length, kilogram (kg) for the unit of mass, and second (s) for the unit of time as the base units.

CGS system of units

This is the system of units where the centimetre (cm) is used for length, gram (g) for mass, and second (s) for time as the base units.

FPS system of units

The foot-pound-second system or FPS system is a system of units built on the three fundamental units such as foot for length, pound for either mass or force and second for time.

International system of units (SI)

SI is the International System of Units abbreviated for French Système Internationale d'Unités. It is the system developed from the MKS system of units. It is a consistent and reasonable system of units, which makes it a rule to adopt only one unit for each of the various quantities used in such fields as science, education and industry. Table 1.7 reports the commonly used units for SI and other systems. It is possible to convert one system of units to another. Some important conversions are included in Annexure I.

Table 1.7. Commonly Used Units in CGS, MKS, FPS and SI

Dimension	*CGS Units*	*MKS Units*	*FPS Unit*	*SI Units*
Length (L)	Centimeter (cm)	Meter (m)	Foot (ft)	meter (m)
Mass (M)	Gram (g)	Kilogram (kg)	Pound (lb)	kilogram (kg)
Time (T)	Second (s)	Second (s)	Second (s)	Second (s)
Force (F)	Dyne (Dyn)	Kilogram-force (kgf)	Pound-force (lbF)	Newton (N) (kg m/s^2)
		Temperature (θ)		
Absolute	Rankine (R)	Celsius (°C)	Kelvin (K)	Kelvin (K)
Ordinary	Fahrenheit (°F)	Celsius (°C)	Kelvin (K)	Celsius (°C)

Quantities like force, pressure, energy, acceleration *etc.* are some combinations of fundamental quantities (dimensions) and are therefore, called derived quantities. For example force, F given by mass, m times the acceleration, a is written as

$$F = ma$$

$$[F] = MLT^{-2}$$

Square bracket means the dimensions of the quantity within the bracket. It also means that force has the dimensions of MLT^{-2}.

Dimensional formula is the expression of the physical quantity in terms of its basic units with proper dimensions. Dimensional formula of a physical quantity can be written only if its relationship with other physical quantities is known. Dimensional formula for force given above can be written as:

$$FT^2/ML = 1 \tag{1.17}$$

Thus, in fluid mechanics, as stated previously we can represent all physical quantities in terms of 3 dimensions, L, T and one of M and F. The two systems are then MLT and FLT systems. From eq. (1.17), we have $F = MLT^{-2}$ and $M = FT^2L^{-1}$. The density, ρ in MLT and FLT systems is:

$$[\rho] = ML^3$$

$$[\rho] = FT^2/L^4$$

Similarly, work in MLT system is ML^2T^{-2} (Work = Force × Distance), which in FLT system will be FL. Normally, we use MLT system in fluid mechanics. Table 1.8 lists some of the derived quantities along with their preferred units and dimensional formula. Table 1.9 reports commonly used pre-fixes for units. Some commonly used non-standard abbreviations in fluid mechanics are as follows:

fps = feet per second

gpm = gallons per minute

cfs or cusecs = cubic feet per second

cumecs = cubic meters per second

Table 1.8. Derived Quantities, their Preferred Units and Dimensional Formulae

Derived Quantity	*Preferred Units (SI)*	*Dimensional Formula (MLT)*
Area (A)	m^2	L^2
Volume (Vol)	m^3	L^3
Velocity (V)	m/s	LT^{-1}
Acceleration (a)	m/s^2	LT^{-2}
Discharge (Q)	m^3/s	$L^3 T^{-1}$
Force (F)	$N = kg\ m/s^2$	MLT^{-2}
Work	$J = kg\ m^2s^{-2}$	ML^2T^{-2}
Momentum	kg m/s	$M LT^{-1}$
Pressure (p)	Pa	$ML^{-1}T^{-2}$
Shear stress (τ)	N/m^2 (Force per unit area)	$ML^{-1}T^{-2}$
Density (ρ)	kg/m^3	ML^{-3}
Specific weight (ω)	N/m^3	$ML^{-2}T^{-2}$
Energy/Work/Heat (E)	J	ML^2T^{-2}
Power (P)	W	ML^2T^{-3}
Dynamic viscosity (μ)	$N\ s/m^2$ or Pa s	$ML^{-1}T^{1}$
Kinematic viscosity (õ)	m^2/s	$L^2 T^{-1}$ ($ML^{-1}T^1/ML^{-3}$) as õ= μ/ρ
Frequency	1/s	T^{-1}
Angle	Radian (Arc/radius)	$M^0L^0T^0$

Table 1.9. Unit Prefixes in SI System

Prefix	*Symbol*	*Factor*
yotta	Y	10^{24}
zetta	Z	10^{21}
exa	E	10^{18}
peta	P	10^{15}
tera	T	10^{12}
giga	G	10^9
mega	M	10^6
kilo	K, k	10^3
hecto	h	10°
deca	da	10
deci	d	0.1
centi	c	0.01
milli	m, k	10^{-3}

Prefix	*Symbol*	*Factor*
micro	μ, u	10^{-6}
nano	n	10^{-9}
pico	p	10^{-12}
femto	f	10^{-15}
atto	a	10^{-18}
zepto	z	10^{-21}
yocto	y	10^{-24}

A number of quantities are inherently non-dimensional, such as the numbers of counting. In addition, ratios of two quantities with the same dimension are dimensionless. For example, bulk strain, dV/V is the ratio of two quantities each with the dimension of volume. Therefore, it is non-dimensional. Angles are measured in radians. Being the ratio of an arc-length to a radius, an angle is the ratio of two lengths, and so it is non-dimensional. Non-dimensional quantities are independent of the system of units as long as the units are consistent, that is, if the same system of units is used throughout. Non-dimensional quantities have wide applications in fluid mechanics. This issue forms a part of chapter 10 of this book.

QUESTIONS (THEORY)

1. Define fluid mechanics and describe its various branches.
2. Define density, specific gravity, specific weight and specific volume of fluids.
3. Define viscosity and discuss the Newton law of viscosity.
4. Write short notes on: vapour pressure of liquids, compressibility of fluids and capillarity phenomenon.
5. Differentiate between: dimensions and units, Lagrangian and Eulerian approach, liquids and gases, solids and fluids, dynamic and kinematic viscosity, Newtonian and non-Newtonian fluids, ideal and real fluids, cohesion and adhesion, wetting and non-wetting fluids

Fluid Statics

Chapter 2

Pressure and Pressure Measurement

WHAT IS PRESSURE?

A fluid at rest exerts pressure perpendicular to any surface that it contacts. If we consider any point as a small cube, the water will exert equal pressure on all sides. There is no parallel component to cause the fluid to flow. Thus, the hydrostatic pressure is the pressure exerted by a fluid due to the force of gravity at equilibrium at a given point within the fluid. The pressure is defined as the amount of surface force per unit area exerted by the fluid on any boundary it is in contact with, and expressed as:

$$\text{Pressure} = \text{Force applied/area exposed} \tag{2.1}$$

The SI unit of pressure is Newton per square meter (N/m^2). It is equal to Pascal (Pa). Atmospheric pressure, on the other hand is the pressure exerted by the weight of air in the atmosphere of the earth. It decreases with increasing altitude and varies somewhat with changing weather conditions. At sea level, the values of standard atmospheric pressure are 760 mm of mercury column, 10.3 m of water column, 101.3 kN/m^2, 101.3 kPa or 1 bar. If the surface of the fluid is open to atmosphere, add the atmospheric pressure to work out the total pressure on the object.

HYDROSTATIC LAW

According to this law, rate of increase of pressure in a vertically downward direction in the fluid is equal to weight density of the fluid. It means pressure in a continuously distributed uniform static fluid varies only with vertical distance and is independent of the shape of the container.

Proof

Consider a fluid of density ρ, having fluid element of area ΔA having a height Δh at a depth h, m from the top surface of liquid (Figure 2.1).

Pressure acting on the left face along x-axis = Px

$$\text{Force} = Px\ \Delta y\ \Delta z \tag{2.2}$$

Pressure on right face along x-axis = $P(x+\Delta x) = Px+(\partial p/\partial x)\Delta x$ (2.3)

$$\text{Force} = (Px+(\partial p/\partial x)\Delta x)\ \Delta y\Delta z \tag{2.4}$$

For the element to be in equilibrium, net force should be zero. It means

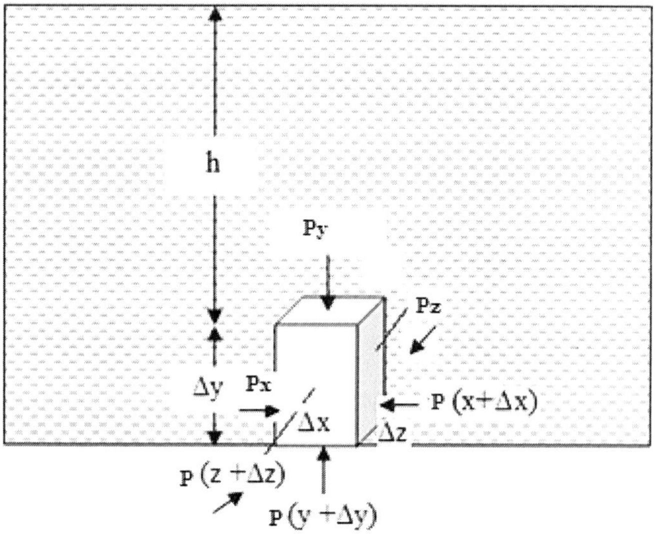

Figure 2.1. Forces Acting on a Fluid Element of Volume Δx Δy Δz.

$$Px\ \Delta y\Delta z = (Px+(\partial p/\partial x)\Delta x)\ \Delta y\Delta z \tag{2.5}$$

$$Px = Px+(\partial p/\partial x)\Delta x \tag{2.6}$$

$$(\partial p/\partial x)\Delta x = 0 \tag{2.7}$$

Since Δx is not zero

Therefore, $\partial p/\partial x = 0$

Similarly, it can be shown that $\partial p/\partial z = 0$

For equilibrium in y direction, forces acting in upward and downward direction should be equal

$$Py\ \Delta x\ \Delta z+ \gamma(\Delta x\ \Delta z\ \Delta y) = (Py+(\partial p/\partial y)\Delta y)\ \Delta x\ \Delta z \tag{2.8}$$

Here γ is the weight density or specific weight of the fluid and is equal to ρg such that ρ is the density of the fluid and g is acceleration due to gravity. Cancelling out $\Delta x\ \Delta z$ on both sides of eq. (2.8), we have

$$Py + \gamma\,\Delta y = Py+(\partial p/\partial y)\Delta y \tag{2.9}$$

$$\gamma\,\Delta y = (\partial p/\partial y)\ \Delta y \tag{2.10}$$

Since P is a function of y only, we replace partial differentiation with perfect differentiation in the above relation such that:

$$dP/dy = \gamma = \rho g \tag{2.11}$$

For an incompressible fluid ρ being constant, integration yields

$$P_2 - P_1 = \rho g\,(y_2 - y_1) \tag{2.12}$$

Here P_1 is taken at y_1 *i.e.* at the water surface (1 atm) and P_2 at the particle of very small thickness at h cm below the water surface. Incorporate $(y_2 - y_1)$ at y_2 as h, the height of the water column above the top of the element. Thus, pressure at the element due to the column of water h will be $P_2 - P_1$ because the atmospheric pressure will cancel out with the atmospheric pressure at the water surface. Writing P for $P_2 - P_1$, we have:

$$P = \rho g h \tag{2.13}$$

Here h is the pressure head, m. Above basic pressure height relationship describes the pressure variation in any static fluid. Restating the same, 'pressure in a continuously distributed uniform static fluid varies only with vertical distance. It is independent of the shape of the container'. It is why; a common practice is to state all kinds of pressures in terms of column height units such as mm of Hg or mm of water. As an illustration, we measure pressure in manometers in terms of liquid column height. Further note that pressures are additive in nature such that

Pressure at depth, h = Pressure acting on the fluid surface + Pressure due to height of liquid

$$P_h = P_0 + \rho g h \tag{2.14}$$

Here P_h and P_0 represent pressures at h cm below the surface and at the surface respectively. The pressure diagram in static water is in the form of a prism (Figure 2.2), the pressure increasing with depth. At a depth of 10 meters under water, pressure is nearly twice the atmospheric pressure at sea level, and increases by about 100 kPa for each increase in 10 m depth (See Example 2.2).

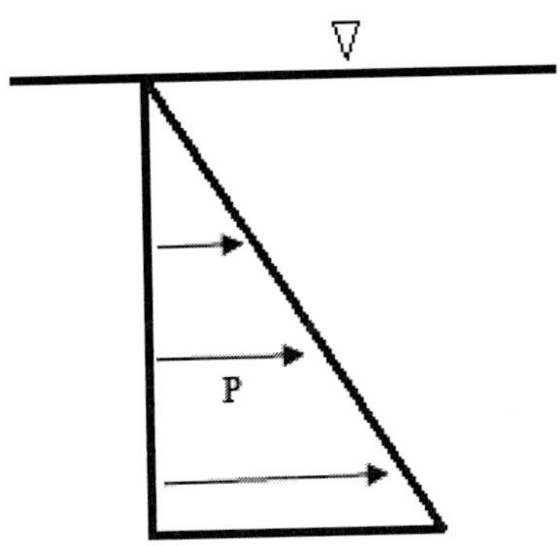

Figure 2.2. Variation of Hydrostatic Pressure in a Static Fluid.

It is also possible to derive eq. (2.13) following a different approach. Let us consider a small cylindrical element of fluid of cross-sectional area A, and height

($h = Z_2 - Z_1$), surrounded by the fluid of mass density, ρ (Figure 2.3). The pressure at the bottom of the cylinder is P_2 at level Z_2, and is P_1 at level Z_1. Since the fluid is in equilibrium and at rest, all the forces in the vertical direction should sum to zero.

$$\text{Force due to } P_1 \text{ (downward)} = P_1 A \tag{2.15}$$

$$\text{Force due to } P_2 \text{ (upward)} = P_2 A \tag{2.16}$$

$$\text{Force due to weight of element} = mg = \rho g A(Z_2 - Z_1) \tag{2.17}$$

Taking the summation of forces (upward as positive), $(\uparrow)\Sigma F=0$

$$P_1 A - P_2 A + \rho g A(Z_2 - Z_1) = 0 \tag{2.18}$$

$$P_1 - P_2 = -\rho g (Z_2 - Z_1) = -\rho g h \tag{2.19}$$

$$P_2 - P_1 = \rho g h \tag{2.20}$$

Thus, for any fluid at rest, the pressure increases with increasing depth and vice-versa *i.e.* pressure decreases with decreasing depth.

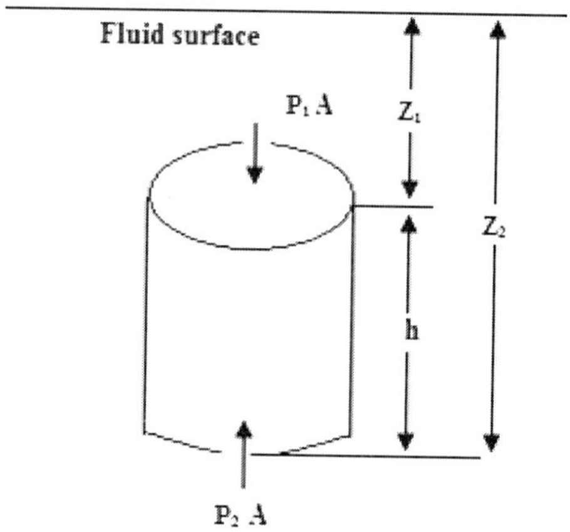

Figure 2.3. Schematics of Forces Acting on a Submerged Element in the Fluid.

Example 2.1

Calculate the pressure at depths of 100 cm, 300 cm and 10 m in a tank filled with water. Take the density of water as 1000 kg/m^3.

P_1 at 100 cm $= \rho g h = 1000 \times 9.8 \times 100/100 = 980$ Pa

P_2 at 300 cm $= \rho g h = 1000 \times 9.8 \times 300/100 = 2940$ Pa

P_3 at 10 m $= \rho g h = 1000 \times 9.8 \times 10 = 9800$ Pa

Example 2.2

Show that a scuba diver experiences nearly double the total pressure than given by the fluid pressure exerted by seawater at 10 m below the sea surface. Assume seawater density as 1.03×10^3 kg/m^3. Also, assume the standard atmospheric pressure as 1.01×10^5 N/m^2.

P due to fluid = $1.03\times10^3 \times9.81\times10 = 1.01\times10^5$ N/m^2

P due to air = 1.01×10^5 N/m^2

Total pressure = $1.01\times10^5 + 1.01\times10^5 = 2.02\times10^5$ N/m^2

Apparently, the pressure at 10 m depth is double the pressure exerted by the seawater.

GAUGE PRESSURE AND ABSOLUTE PRESSURE

Differential pressure is defined as the pressure difference between measurements at two points in a system. This measurement is quite significant in applications that have pressure functionality, such as weather instrumentation, airplanes and many other applications. Differential pressure is measured and reported as absolute or gauge pressure (Anonymous, 2018). Therefore, it is important to understand the difference between the two pressures since it can make lot of difference in reporting or in making calculations of pressures.

Gauge Pressure

It is often convenient to measure pressure taking atmospheric pressure as reference (Figure 2.4). The pressure measured taking the atmospheric pressure as reference is known as gauge pressure. The atmospheric pressure on the scale is marked as zero.

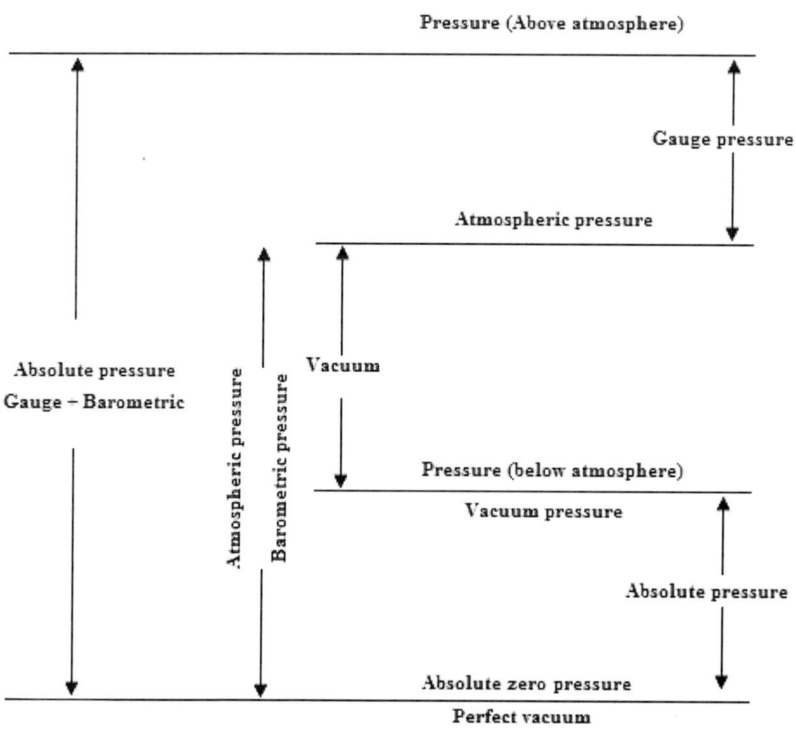

Figure 2.4. Concept of Gauge and Absolute Pressure.

Gauge pressure = Absolute pressure - atmospheric pressure

Gauge pressure can be positive or negative. Conventionally, one avoids the use of negative sign with the pressure. Therefore, to distinguish a negative pressure from the positive ones, either the word vacuum is appended with the pressure value or the gauge itself is labelled as a vacuum gauge.

Absolute Pressure

Since, atmospheric pressure changes with altitude and other atmospheric conditions, in the absolute standard, the pressure in a perfect vacuum is taken as zero. Absolute pressure is thus, referenced to perfect vacuum (Figure 2.4) so that it is always a positive quantity. Thus

$$\text{Absolute pressure} = \text{atmospheric pressure} + \text{gauge pressure} \qquad (2.21)$$

Example 2.3

Convert 2 bar gauge pressure to absolute pressure.

Since, 1 atmosphere is nearly equal to 1 bar, we have

Absolute pressure = Gauge pressure + Atmospheric pressure = 2+1 = 3 bars.

Example 2.4

A manometer indicates a negative gauge pressure of 60 mm of mercury. What is the absolute pressure in the pipe in N/m^2?

Manometer pressure = - 60 mm of mercury

In terms of $kN/m^2 = -\rho gh = -13600 \times 9.81 \times 0.06 = -8.004\ kN/m^2$

$P_{abs} = 101.213 + (-8.004) = 93.209\ kN/m^2$

PASCAL'S LAW

Blaise Pascal, a French mathematician and physicist proposed the Pascal's law or Pascal's principle. According to this principle, any increase in the liquid pressure because of an external force applied at the liquid surface is transmitted equally in all directions and to the walls of the container in which it is filled. In general, it can be described as 'Pressure applied at any point of a liquid enclosed in a container, is transmitted without loss to all other parts of the liquid and to the walls of the container. The pressure acts perpendicular to the boundaries of the container walls (Figure 2.5). It has a magnitude but no specific direction. It is therefore, a scalar quantity (Shrotri, 2017). The Pascal's law is valid only for fluid flows where shear stresses do not exist. Thus, it applies to cases where

- Fluid is at rest
- No relative motion exists between different fluid layers. For example, fluid at a constant linear acceleration in a container
- Ideal fluid flows where viscous force is negligible.

Pascal's law finds many applications in our daily life. Few important among them are: hydraulic brake systems, hydraulic jack, hydraulic press and several hydraulic machines. Besides, a siphon works because gravity pulling down on the taller column of liquid causes reduced pressure at the top of the siphon. It also finds applications in artesian wells, water towers, and dams.

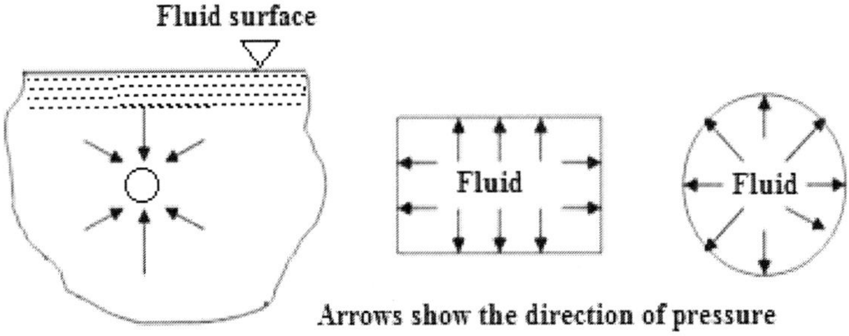

Figure 2.5. Pressure within Fluid and along Boundaries of the Containers.

Theorem

Let us consider a fluid element of linear dimensions Δx, Δy and Δz in the x, y and z coordinates as shown in the Figure 2.6. If Ps is the pressure exerted on the inclined surface, then equation for horizontal forces is given as:

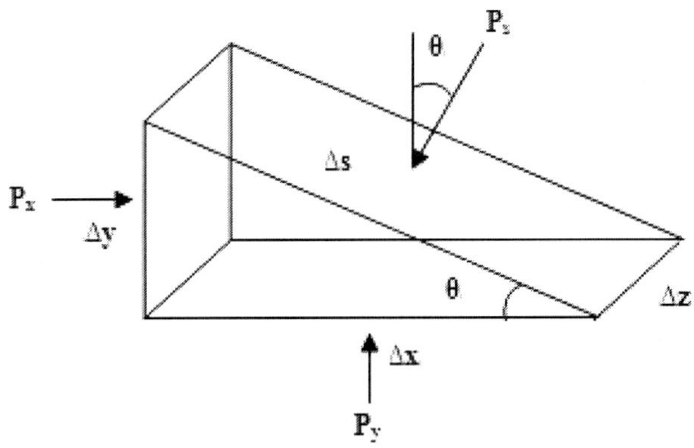

Figure 2.6. Pressure on an Inclined Fluid Element in Static Fluid.

$$P_x \Delta y \Delta z = P_s \sin\theta \, \Delta z \Delta s \tag{2.22}$$

Since $\sin\theta = \Delta y/\Delta s$, we have

$$P_x = P_s \tag{2.23}$$

Equation for vertical forces:

$$P_y \Delta x \Delta z = P_s \cos\theta \, \Delta z \Delta s + 0.5\, \rho g \, \Delta x \Delta z \Delta y \tag{2.24}$$

Since $\cos\theta = \Delta x/\Delta s$

$$P_y \Delta x \Delta z = P_s \, \Delta z \Delta x + 0.5 \rho g \, \Delta x \Delta z \Delta y \tag{2.25}$$

Since Δ is very small (In the limit when the wedge shrinks to a point), one can neglect the second term on the right hand side of eq. (2.25), such that

$$P_y = P_s \tag{2.26}$$

From eq. (2.23) and eq. (2.26), we can write that

$$P_x = P_y = P_s \tag{2.27}$$

It means the pressure exerted on all the sides is same. This proves the Pascal's principle.

Applications

Let us consider the schematic diagram of a hydraulic press shown in Figure 2.7. Let us also assume that pressure P_1 is applied on the left side on the smaller piston. It is transmitted to the right hand side since in a closed system the pressure is transmitted to all parts. Thus, we have

$$F_1 = P_1\, A_1 \text{ and } F_2 = P_1\, A_2$$

$$F_2/F_1 = A_2/A_1 \tag{2.28}$$

$$F_2 = F_1\,(A_2/A_1) \tag{2.29}$$

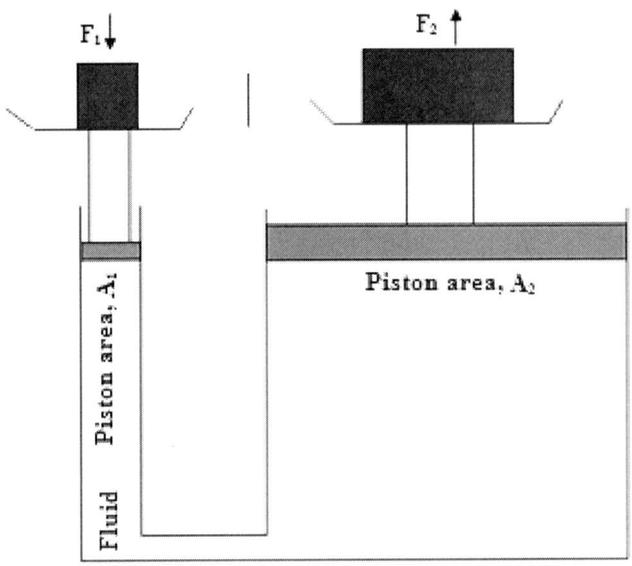

Figure 2.7. Schematic Diagram of a Hydraulic Press Illustrating the Application of Pascal's Law.

Here A_1 and A_2 are the areas of the small and big pistons and P_1 is the pressure applied on the smaller piston. Since A_2/A_1 is more than 1, a large force is transmitted in response to the pressure applied on the smaller piston. In terms of diameters of the pistons, we have

$$F_2 = F_1\,(d_2/d_1)^2 \tag{2.30}$$

Here d_2 and d_1 are the diameters of larger and smaller pistons respectively.

It may not be out of place to mention that this law does not violate the law of conservation of energy. There will be no gain in the work output. Since we can write

Work done = Force × Distance

$$F_1 D_1 = F_2 D_2 \tag{2.31}$$

Where D_1 and D_2 are the distances travelled by the smaller and bigger pistons in one stroke of the smaller piston. The distance travelled by the larger piston will be in the same proportion less than the increase in the force.

Example 2.5

In a hydraulic system a piston has a cross sectional area of 15 cm^2. It pushes an incompressible fluid with a force of 30 Newton. Calculate the force on the second piston connected to the other end having a cross-sectional area of 90 cm^2.

Using eq. (2.29), we have

$F_2 = 30\ (90/15) = 180$ Newton

Example 2.6

A hydraulic jack lifts a car at the back end (half the weight) of a car having mass of 2200 kg. What is the minimum force applied if the areas ratios of the cylinders at the two ends is 0.1 (The area of the large cylinder is 10 times the area of the small cylinder).

According to the Newton's Second Law, the force acting on the large cylinder is:

$F_1 = m\ a$

Such that m is mass (kg) and a is the acceleration due to gravity (m/s^2)

$F_1 = 0.5 \times 2200 \times 9.81 = 10791$ (N)

We use eq. (2.29) to calculate the force acting on the small cylinder, F_s

$F_s = 10791 \times 0.1 = 1079.1$ (N)

Example 2.7

The piston of a hydraulic jack weighs 100 N and supports a mass of 52 kg. They are both at same height and made of the same material. Calculate the force required at the plunger to support the piston and mass to carry it at a level 15 cm above that of plunger. The diameter of plunger and piston are 300 and 1200 mm respectively. The relative density of oil in the jack is 0.9.

Area A_1 of piston = $(\pi/4)\ d^2 = 1.131$ m^2

Area A_2 of plunger = 0.0314 m^2

Pressure at the plunger = Due to piston and weight of the oil because of difference in level

$(52\times9.81 + 100)/1.131 + 900\times9.81\times15/100 = 539.45 + 1324.35 = 1863.8$ N/m^2

Pressure at the piston = $F_2/0.0314$ N/m^2

Equating

$F_2/0.0314 = 1863.8$ such that $F_2 = 58.52$ N

HYDROSTATIC FORCES ON SUBMERGED SURFACES

There is no relative motion between adjacent fluid layers and both the velocity gradient and shear stress are zero for fluids in static equilibrium. The forces acting on fluid particles are only due to pressure of fluid normal to the surface and due to gravity. Assessment of these forces is quite important in the design of storage tanks, ships, dams and other hydraulic structures.

Plane Horizontal Surface

For a horizontal plane submerged in a liquid (Bottom of the tank in Figure 2.8 left), the pressure, P is equal at all the points of the surface. Thus, the resultant force on horizontal surface is the simple product of pressure times the area of interest, *i.e.*

Force = Pressure × Area of plane

$$F = PA = \gamma hA \tag{2.32}$$

Here A is the area and h is the depth from the water surface to the center of gravity of the surface, the centroid. If free surface is at some pressure, P_1, then eq. (2.32) modifies to

$$F = \gamma hA + P_1/\gamma \tag{2.33}$$

This force acts vertically downward through the center of pressure. Since pressure in such cases is constant and uniformly distributed over the surface, center of pressure and centroid of the area coincide such that resultant force also acts through the centroid of the area.

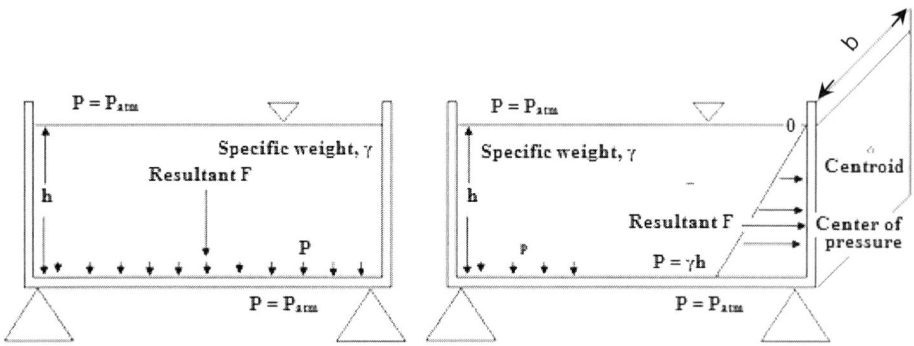

Figure 2.8. Hydrostatic Forces on a Horizontal (Left) and a Vertical Plane (Right).

Plane Vertical Surface

Let us consider the vertical wall of the tank with a width b and height h (Figure 2.8 right). The pressure exerted at the water surface is zero and maximum at the bottom. Pressure distribution is linearly distributed and is like a prism. Therefore, for an incompressible fluid, it is possible to take an average value. The resultant force, F_r on the rectangular wall is the volume of the pressure prism:

$$F_r = (0 + \gamma h)/2 \times b\,h = (\gamma h/2) \times bh = 0.5\,\gamma h\,A \quad (2.34)$$

(Since area, $A = bh$)

Since $h/2$ is the location of the centroid of the area, eq. (2.34) states that the magnitude of the resultant force is equal to the pressure at the centroid of the area multiplied by the total area. Centroid of the area as per definition is the point at which the area balances itself if suspended from that point. It is equivalent to the center of gravity of a solid body. The resultant force in this case acts on the center of pressure, which is located at $h/3$ from the base. It is the centroid of the prism. Thus, center of pressure is located below the centroid of the area, which is at $h/2$.

For a more general case, consider a plane surface of arbitrary shape immersed vertically in a static mass of fluid (Figure 2.9). Let us define C as the center of gravity so that h_c is the depth of center of gravity from free liquid surface and E, the center of pressure with h_p as the depth of center of pressure from free surface of liquid.

Consider an elementary strip of area A having width and thickness as b and dh located at a distance h from the free surface of liquid. If pressure acting on the strip is P then total pressure force (F) acting on the elementary strip is

$$F = P\,b\,dh = \rho g\,h\,b\,dh \quad (2.35)$$

We can determine the total pressure force acting on whole area by integrating the right hand side as

$$\rho g \int h\,b\,dh \quad (2.36)$$

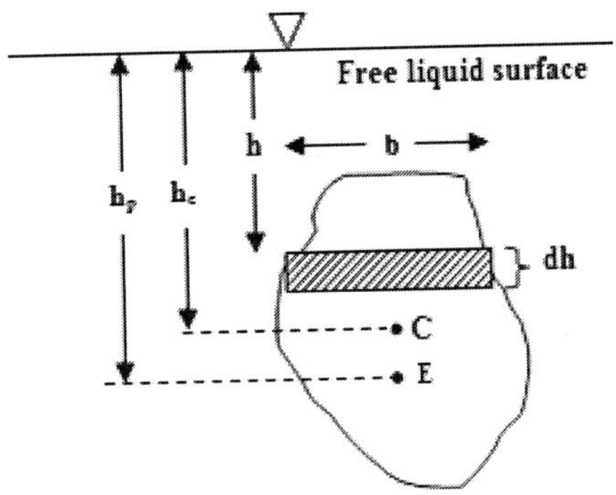

Figure 2.9. Hydrostatic Forces on a Vertically Immersed Plane Surface of an Arbitrary Shape.

Since $\int h\,b\,dh$ is the moment of surface area about the free liquid surface, we can take its value, h_c A. Incorporating, we get

$$\text{Total pressure force, } F = \rho g\,h_c\,A \quad (2.37)$$

Here A is total area of plate. Since $\rho g\,h_c$ is the pressure exerted by the fluid at its center of gravity, we can state that the resultant force acting on a vertical object is the force exerted by the fluid at its center of gravity. The resultant force acts through the center of pressure.

Example 2.8

Calculate the hydrostatic thrust acting on the bottom and side of a container filled with water. The container dimensions are 1 m wide, 2 m length and 2 m deep.

The force at the bottom, F_b as per eq. (2.32) will be

$F_b = \gamma h A$

$F_b = 1000\times9.81\times2\times1\times2 = 39240\ N/m^2$

On the vertical wall, the pressure at top is zero and maximum being at the bottom. Since it is linearly distributed, take an average value as:

$F_s = \gamma A (h_t+h_b)/2 = 1000\times9.81\ \times1\times2\ (0+2)/2= 19620\ N/m^2$

Plane Inclined Surface

For a general case, let us assume that the plate is inclined at an angle θ to the free surface (Figure 2.10). Considering one side only, there will be a force due to fluid pressure acting on each element of area dA, the magnitude of the pressure being dependent on the vertical depth h of the element below the free surface. Taking the pressure at the free surface as 0 (atmospheric), the pressure at a distance h below the free surface is:

$$P = \rho gh \tag{2.38}$$

Force on elemental area dA

$$dF = \rho gh\ dA \tag{2.39}$$

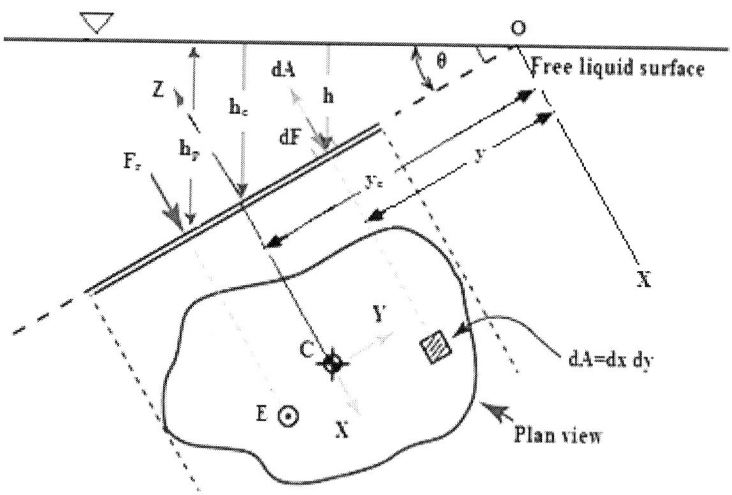

Figure 2.10. Hydrostatic Forces on an Inclined Plane Surface Immersed in a Static Fluid.

The resultant force, F_r acting on the plane is found by integrating all the forces on the small element:

$$F_r = \int P dA = \int \rho g h \, dA \tag{2.40}$$

Assuming that ρ and g are constant, and putting $h = y \sin\theta$

$$F_r = \rho g \int y \sin\theta \, dA \tag{2.41}$$

$$F_r = \rho g \sin\theta \int y \, dA \tag{2.42}$$

The integral $\int y \, dA$ is the first moment of area about the x-axis and is equal to Ay_c, where A is the area of the whole immersed surface and y_c is the vertical distance from the free surface to the centroid of the area, C, of the immersed surface.

Substituting into eq. (2.42) will give

$$F_r = \rho g y_c \sin\theta A \tag{2.43}$$

From trigonometry, $h_c = y_c \sin\theta$

$$F_r = \rho g h_c A = P_c A \tag{2.44}$$

Total pressure force is equal to total area multiplied by the pressure at the center of gravity, P_c of the plate surface immersed in the liquid. Clearly, the resultant force, F_r is independent of the angle of inclination θ so long as the depth of the centroid y_c remains unchanged. The point of application of the resultant force on the submerged area is termed as the center of pressure. The resultant force acts perpendicular to the immersed surface at the center of pressure, E.

Coordinates of the center of pressure

Referring to Figure 2.8 (right), one can state that the pressure on an immersed surface increases with depth and the resultant force acts at center of pressure rather than centroid of the plane. Suppose the coordinates of center of pressure, C_p at E are x_p and y_p. To find the co-ordinates of C_p, we know that the moment of the resultant force about any axis must be equal to the moment of distributed force about the same axis. We have

$$x_p F_r = \int x \, dF \text{ and } y_p F_r = \int y \, dF \tag{2.45}$$

Taking the moments about x-axis and incorporating the equation for F (eq. 2.43) and dF (eq. 2.39), we get

$$y_p \rho g \sin\theta \, y_c A = \int y \rho g \sin\theta \, y \, dA \tag{2.46}$$

$$y_p y_c A = \int y^2 dA \tag{2.47}$$

The term $\int y^2 dA$ is second moment of the area about the x-axis denoted as I_{xx}. Therefore

$$y_p = I_{xx}/(y_c A) \tag{2.48}$$

Using parallel axis theorem, I_{xx} is related to the x-axis that passes through the centroid of the area, I_{xc} as follows:

$$I_{xx} = I_{xc} + Ay_c^2$$

Such that eq. (2.48) can be written as

$$y_p = y_c + I_{xc}/(y_c\,A) \tag{2.49}$$

Using eq. (2.49) to find the point of action of F_r on a rectangular plate forming a pressure prism, we have

$$y_p = y_c + I_{xc}/(y_c A) = 0.5\,h + I_{xc}/(0.5\,h\times b\times h)$$

In case of rectangular plate, $I_{xc} = (1/12)\,A\,h^2 = (1/12)\,b\,h^3$ (Table 2.1)

$$y_p = (1/12)\,b\,h^3/(0.5\,h\times b\times h) + 0.5\,h = h\,(1/6 + ½) = (2/3)\,h \text{ from the top}$$

Similarly taking moments about the y-axis, we have

$$x_p\,F = \int x\,\gamma\sin\theta\,y\,dA$$

Following similar arguments as for y_p, x_p is,

$$x_p = I_{xy}/(y_c\,A)$$

Here I_{xy} is the product of inertia with respect to x and y-axis. Again using parallel axis theorem, we have

$$x_p = I_{xyc}/(y_c\,A) + x_c \tag{2.50}$$

Values of parameters used in these derivations for various shapes are given in Figure 2.11 (centroid) and Table 2.1. Since the surface is often symmetrically loaded, I_{xx} is zero and hence $x_c = 0$ or the center of pressure is located directly below the centroid on the line of symmetry. Following these arguments, note that I_{xyc} is zero for a circle, a rectangle and a semi-circle. In these cases, center of pressure is just below the center of gravity and only eq. (2.49) is sufficient to calculate the location of the center of pressure below the center of gravity. As the surface becomes more deeply submerged, the center of pressure approaches the centroid and x_{cp} and y_{cp} approaches 0,0.

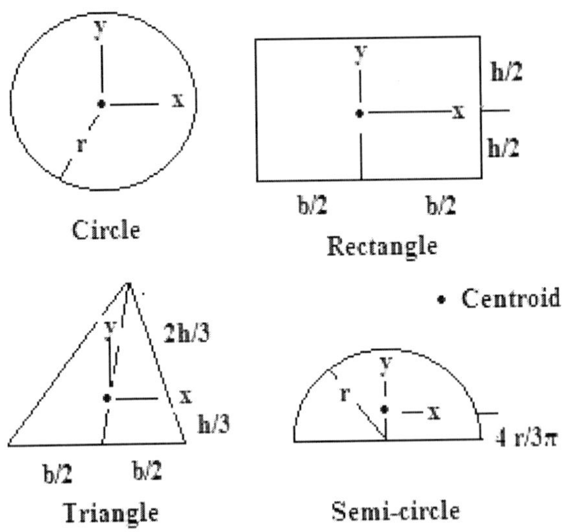

Figure 2.11. Location of the Centroid for Elements of Various Geometrical Shapes.

Table 2.1. Products of Inertia for Various Geometrical Shapes

Shape	*Area*	I_{xc}	I_{yc}	I_{xyc}
Circle	$\pi r^2/4$	$\pi r^4/4$	$\pi r^4/4$	0
Rectangle	bh	(1/12) b h^3	(1/12) b^3 h	0
Triangle	0.5bh	(1/36) b h^3	–	(1/72) b h^2 (b-2h)
Semi-circle	$\pi r^2/2$	0.1098 r^4	0.3927 r^4	0

Example 2.9

A rectangular plate 3 m × 5 m vertically immersed in water has its 5 m side parallel to free surface. If the top of rectangular plate touches the free liquid surface, calculate the total pressure force and the location of the center of pressure. Also, calculate the total pressure force and the location of the center of pressure if the plate immerses 2 m below the surface of water.

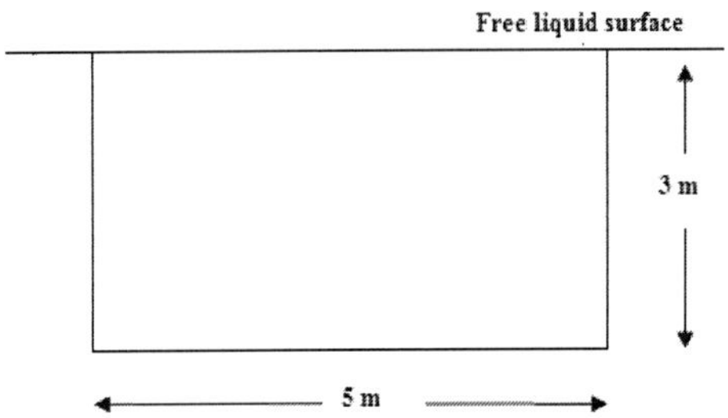

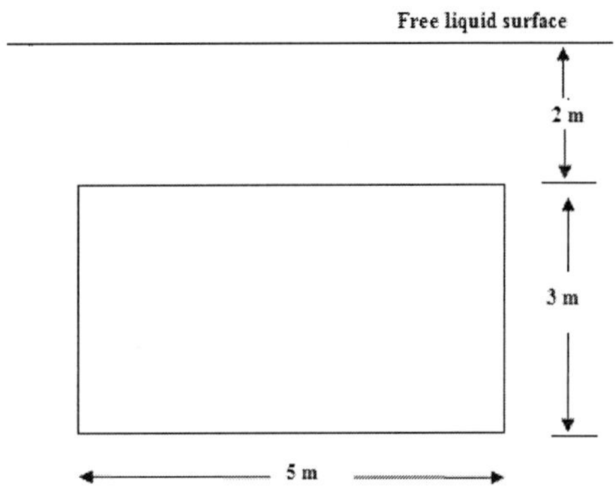

First case

Total pressure force = $\rho g h_c \times A$ (Note h_c = 1.5 m)

Total pressure force = 1000 × 9.81 × 1.5× 15 = 220725 N

The y coordinate of center of pressure, h_p is given as (Refer Table 2.1 for rectangle)

$h_p = I_{xc}/(Ah_c) + h_c = bh^3/(12\ A\ h_c) + h_c$

$h_p = 135/270 + 1.5 = 2.0$ m

Or it is 0.5 m below the centroid (2 m-1.5 m)

Second case

Total pressure force = $\rho g h_c \times A$ (Note h_c = 3.5 m)

Total pressure force = 1000 × 9.81 × 3.5× 15 = 515025 N

The y coordinate of center of pressure, h_p is

$h_p = I_{xc}/(Ah_c) + h_c = bh^3/(12\ A\ h_c) + h_c$

$h_p = 135/630 + 3.5 = 0.214 + 3.5 = 3.714$ m

Or 0.214 m below the centroid.

Example 2.10

A vertical gate 3 m wide and 5 m high closes a tunnel running full of water. The pressure at the bottom of the gate is 165 kN/m^2. Determine the total pressure on the gate and position of the center of the pressure.

Since the total pressure at the bottom of the gate is given. We can determine the head of the water as:

$h = P/\gamma = 165000/(1000\times9.81) = 16.81$ m

$y_c = 16.81 - 5/2 = 14.31$ m

Area of the gate = 5×3 = 15 m^2

$F = 1000\times9.81\times14.31\times15 = 2105715$ N or 2.10 MN

Since the gate is rectangular, $I_{xc} = bh^3/12$

$h_p = I_{xc}/(Ay_c) + yc = 3\times5^3/(12\times15\times14.31) + 14.31$

$h_p = 0.146 + 14.31 = 14.456$ m

Curved Surfaces

We encounter many arbitrarily shaped bodies immersed in liquids such as pipes and walls of containers. Forces acting on these bodies can be determined in the same manner as in the previous section. More often, an alternate approach is used since the integration terms involved are quite tedious to resolve. Most common approach is to first determine horizontal and vertical components of the forces. Once known, these are vectorially combined to give the resultant force. For a curved surface, one can anticipate the following two situations.

Surface containing liquid

Let us consider the surface AB, which contains liquid on its left hand side as shown in Figure 2.12. Using the imaginary plane ACD, we can see that the horizontal component of the force on the surface must balance the horizontal force F_{ac} given as:

$$F_{ac} = F_x = \rho g\, h_c\, A \qquad (2.51)$$

Here h_c is the center of gravity of the projected plane CA. The force F_x is equal and opposite this force and will act at center of pressure of the projected surface.

The vertical component of force on the surface is equal to the weight of liquid above the surface. Hence:

$$F_y = \rho g V \qquad (2.52)$$

Here V is the volume of water over the curved surface including DEBC. This component must act through the center of gravity of the area ABED, shown as G on the diagram.

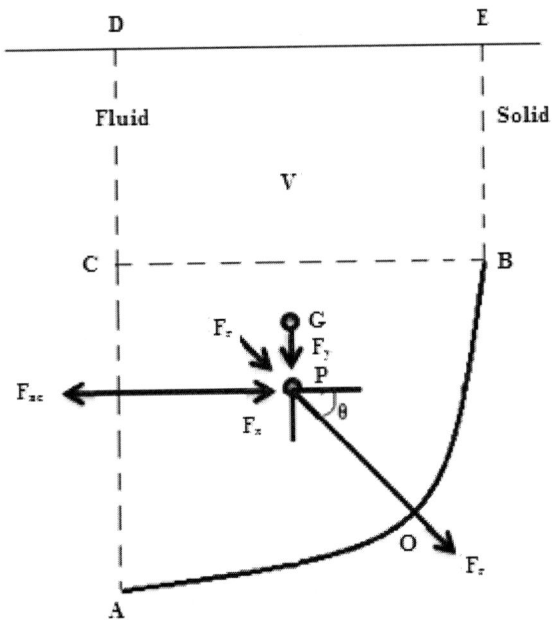

Figure 2.12. Hydrostatic Forces on a Curved Surface Containing Liquid.

The resultant force will act through a point where lines of forces of F_x and F_y intersect, say point P. The resultant force is then:

$$F = \sqrt{(F_x^2 + F_y^2)} \tag{2.53}$$

This force acts at an angle θ with the horizontal:

$$\tan \theta = F_y/F_x \tag{2.54}$$

If we extend the line of the resultant force, it meets the curved surface at some point say O. It is the point where the resultant force will be operative on the curved surface.

Surface displacing liquid

Consider the surface AB, which displaces liquid as shown in Figure 2.13. Similar to the previous case, the horizontal component of force on the surface must balance with the horizontal force F_{bh} such that

$$F_x = \rho g h_c A$$

Note that h_c is the vertical distance to the centroid of the plane projected area BH. This force also acts at the same level as F_{bh} as before. In this case, we imagine that the area ABED is filled with the same liquid as on the right. Thus, F_y will be equal to the imaginary weight of the water, ABEDA. This force will act upward. Hence:

$$F_y = \rho g V \tag{2.55}$$

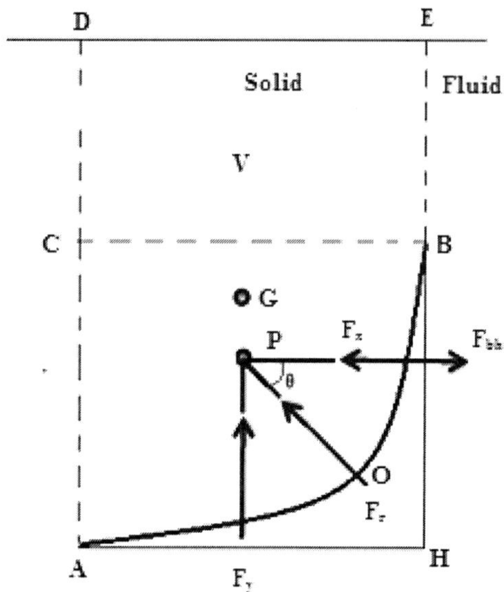

Figure 2.13. Hydrostatic Forces on a Curved Surface Displacing Liquid.

This component acts through the center of gravity of the imaginary liquid in area ABED, shown as G on the diagram.

Further discussion and analysis is similar to what is included in the previous case.

Example 2.11

Determine the resultant force and its direction on the gate (quadrant circle) shown in figure below:

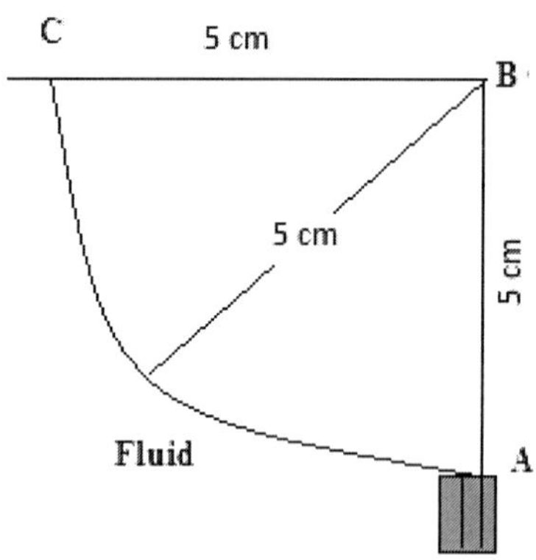

The horizontal force, per meter run of the gate, is that of the surface projected onto a vertical plane of length CB:

$$F_x = \rho\, g\, A_{ab}\, h_{ac} = 1000 \times 9.81 \times 5 \times 1 \times 5/2 = 122.625 \text{ kN/m}^2$$

This force acts at $2\times5/3 = 3.33$ m from the surface.

$F_y = \rho\, g\, V = 1000\times9.81\times\pi\times5^2\times1/4 = 192.643\ kN/m^2$

The vertical force is located at $4r/3\pi = 2.12$ m to the left of line AB.

The resultant force = $\sqrt{(122.625)^2 + (192.643)^2} = \sqrt{(15036.9 + 37111.3)}$

The resultant force acts at an angle of θ to the horizontal given by eq. (2.54).

$\theta = \tan^{-1}(192.643/122.625) = 57.5^\circ$

The resultant passes through point B. There should be no net moment about B since the force on each infinitesimal length of the surface passes through B. We can check this as follows.

$122.625\times3.33 - 192.643\times2.12 = 408.34 - 408.40 \sim 0$

PRESSURE MEASUREMENT

Manometers and pressure gauges are commonly used to measure pressure and vacuum. Amongst the two, application of manometers is common to measure low pressures and vacuum especially in hydraulic laboratories.

Manometers

Manometer, a simple and precise device that measures pressure, balances a column of liquid of known specific gravity against the pressure to be measured. Commonly used manometers classified in four groups are as follows:

1. **Simple manometers**
   a. Piezometer
   b. U-tube manometer
2. **Differential manometers**
   a. U-tube upright differential manometer
   b. U-tube inverted differential manometer
3. **Well type manometers**
   a. Vertical column manometer
   b. Inclined column manometer
4. **Micro-manometers**

Advantages and limitations

Manometers offer the following advantages:

- Simple and time proven
- Reasonable cost
- Reasonably accurate and sensitive
- Availability of a wide range of filling fluids of varying specific gravities (Low density fluids can be used to amplify the deflection of liquid column manometer)

Major limitations of manometers are:

- These are only suitable for low pressure or low differential pressure applications
- Being fragile, their portability is a problem
- Accurate levelling is must for measurement accuracy.

Simple Manometer: Piezometer

Simplest amongst all the manometers is a piezometer. It is a vertical transparent glass tube used to measure very low pressures. The lower end of the piezometer is connected to the point in the vessel or pipe at which static pressure is to be measured (Figure 2.14). The upper end is open to the atmosphere. Measured value of h is used to calculate the pressure P at point A (N/m^2) as per the following equation.

$$P = \rho g h \tag{2.56}$$

Here, ρ is the density of liquid, kg/m^3, h is the height of liquid in the piezometer from the center of the pipe, m and g is the acceleration due to gravity, m/s^2.

Since the tube is open to the atmosphere, the pressure measured is the gauge pressure.

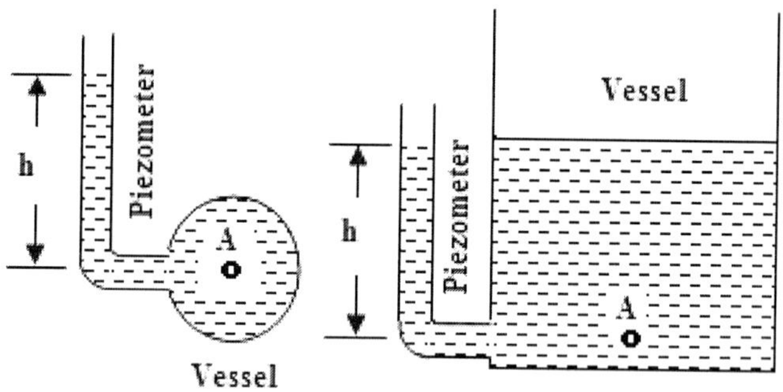

Figure 2.14. Set-up of a Piezometer in a Pipe and in a Vessel.

Example 2.12

Calculate the height of liquid columns in the three piezometers shown below if the liquids A, B and C have specific gravities of 0.8, 0.9 and 0.95 respectively. Assume datum line at bottom of each reservoir for the calculation of height of the liquid columns.

Pressure at 3, P_3 is calculated with the help of eq. (2.56) given by sum of the pressures due to columns of liquids A, B and C as follows:

$P_3 = (0.8 \times 1000 \times 9.81 \times 1.5) + (0.9 \times 1000 \times 9.81 \times 2.0) + (0.95 \times 1000 \times 9.81 \times 3.0) = 57388.5$ N/m^2

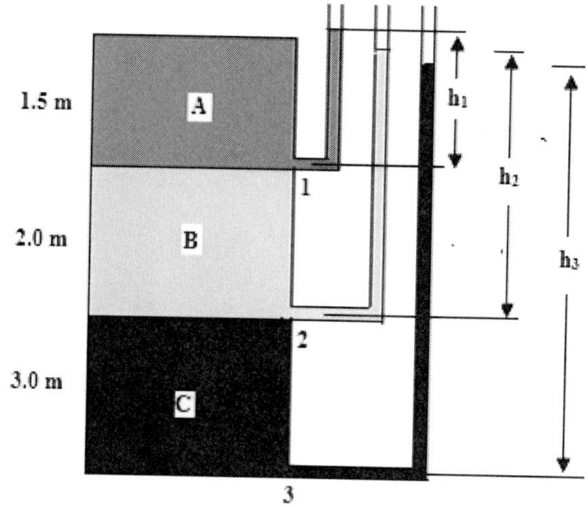

The height to which the liquid with a specific gravity of 0.95 will rise is calculated as:

$h_3 = P_3/(0.95\times1000\times9.8) = 57388.5/9319.5 = 6.16$ m of water at 4°C

Since the density of water 1000 kg/m³ is at a temperature of 4°C, it is conventional to mention the temperature.

Similarly, pressure at 2, P_2 is calculated with the help of eq. (2.56) given by sum of the pressures due to columns of liquid B and C as follows:

$P_2 = (0.8\times1000\times9.81\times1.5) + (0.9\times1000\times9.81\times2.0) = 29430\ N/m^2$

$h_2 = 29430/(0.90\times1000\times9.81) = 29430/8829 = 3.33$ m of water at 4°C

And pressure at 1, P_1 is calculated with the help of same equation given by the pressure due to column of liquid A as follows:

$P_1 = 0.8\times1000\times9.81\times1.5 = 11772\ N/m^2$

$h_1 = 11772/(0.80\times1000\times9.81) = 11772/7848 = 1.5$ m of water at 4°C

Example 2.13

A pipe is carrying oil at a pressure of 19 kN/m². Find the rise of liquid level in the piezometer connected to the pipe. Specific gravity of oil is 0.9.

$h = P/(S_o\rho g) = 19000/(0.9\times1000\times9.81) = 2.15$ m of water at 4°C

U-tube Manometer

A U-tube manometer consists of a glass tube having the shape of alphabet 'U'. It is why; the manometer is named as U tube manometer. While one end of the manometer is connected to the point at which pressure is to be measured, the other end remains open to the atmosphere (Figure 2.15).

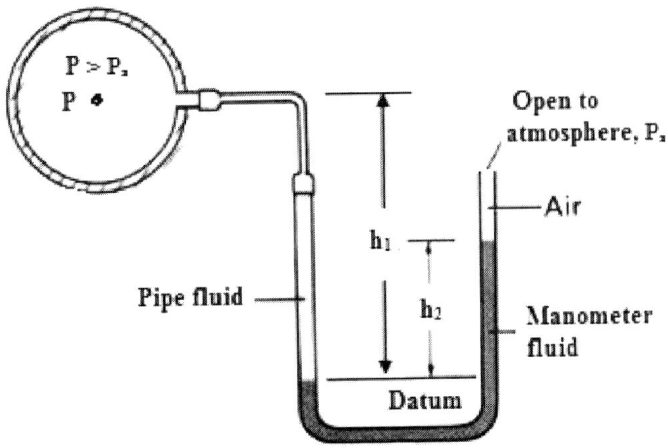

Figure 2.15. Schematic Diagram of an U Tube Manometer (P > P_a).

Let P be the gauge pressure of the fluid in the pipe at the desired point of measurement. Let h_1 be the height of the fluid in the left limb from the datum line to the center of pipe containing the fluid and h_2 the height of manometer liquid from the datum line in the right limb. Then we can write

$$\text{Pressure in the left limb at datum} = P + \rho_1 g h_1 \quad (2.57)$$

$$\text{Pressure in the right limb at datum} = \rho_2 g h_2 \quad (2.58)$$

Here ρ_1 and ρ_2 are the densities of fluid in the vessel and the fluid of the manometer. As per Pascal's law, the pressure at datum line will be equal. Therefore,

$$P + \rho_1 g h_1 = \rho_2 g h_2 \quad (2.59)$$

$$P = \rho_2 g h_2 - \rho_1 g h_1 = \gamma_2 h_2 - \gamma_1 h_1 \quad (2.60)$$

Here γ_1 and γ_2 are the specific weights of fluid in the pipe and the fluid of the manometer respectively. If we divide, both sides of the above equation with ρg, such that ρ is the density of water, we get:

$$P/\rho g = \rho_2 g h_2/\rho g - \rho_1 g h_1/\rho g \quad (2.61)$$

Since $P/\rho g$ is h, head of water and ρ_2/ρ and ρ_1/ρ are the specific gravity of fluids of manometer and in the vessel respectively, we have

$$h = S_2 h_2 - S_1 h_1 \quad (2.62)$$

Here S_2 and S_1 are the specific gravities of fluids of manometer and in the vessel respectively. One of the major limitations of a U-tube manometer is that we need to measure two values of h. It is likely to increase the errors in measurements.

The preceding case depicts a situation where pressure in the vessel is more than the atmospheric pressure, P_a. If the pressure in the pipe is less than the atmospheric pressure (vacuum), then the manometer will appear as shown in Figure 2.16. Following the same procedure, we have

$$P + \rho_1 g h_1 + \rho_2 g h_2 = P_a = 0 \quad (2.63)$$

$$h = -(S_2 h_2 + S_1 h_1) \quad (2.64)$$

Use the following guidelines to balance the known and unknown pressures in a manometer.

- ☆ Begin from a point usually from the side with unknown pressure. In a differential manometer, one can begin from either side. Assign a zero value to atmospheric pressure
- ☆ Add/subtract $\gamma\Delta h$ as moving down/up from the reference point
- ☆ Jump across U-tube at any two points at the same elevation having the same liquid as the two points will be at the same pressure
- ☆ Equate with the pressure on the other side.

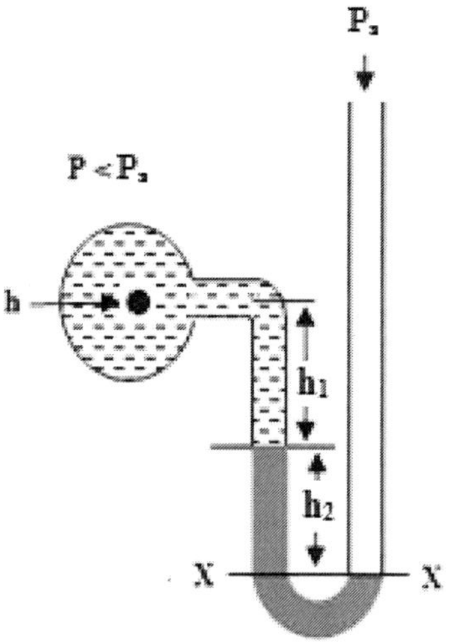

Figure 2.16. Schematic Diagram of an U Tube Manometer (P< P_a).

Example 2.14

A U-tube manometer is used to measure the pressure of oil (specific gravity 0.80) flowing in a pipeline. The center of pipe is 100 mm below the level of mercury in the right limb. Determine the head and pressure, if the difference of mercury level in the two limbs is 60 mm.

As per statements made in the question, following diagram depicts the problem.

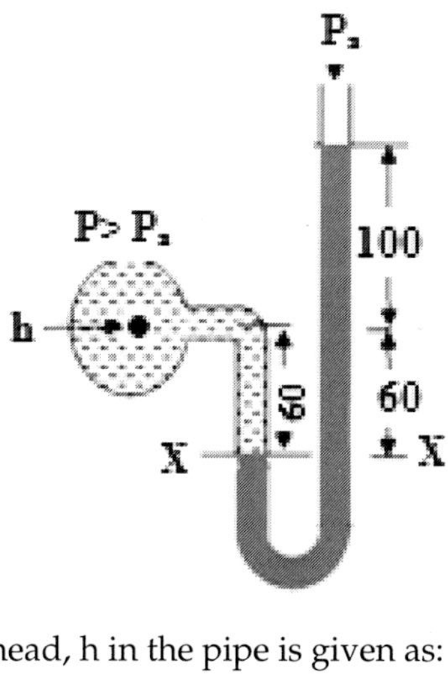

Using eq. (2.62), the head, h in the pipe is given as:

$h = 13.6 \times 0.16 - 0.8 \times 0.06 = 2.176 - 0.048 = 2.128$ m (Specific gravity of mercury is 13.6)

$P = \rho gh = 1000 \times 9.81 \times 2.128 = 20875.7\ N/m^2 = 20.88\ kN/m^2$.

Example 2.15

An U-tube manometer containing mercury (Specific gravity 13.6) was connected to the water pipe and its right limb was exposed to the atmosphere. What is the negative pressure in the pipe if height of water up to center of pipe in the left limb was 40 mm. Difference between the mercury levels in the two limbs was also 40 mm?

Using eq. (2.64), we have

$h = -(s_1 h_1 + s_2 h_2) = -(13.6\times0.04 + 1.0\times0.04) = -0.584$ m

$P = -0.584\times1000\times9.81 = -5.729$ kN/m^2

Differential Manometers

In many cases, one is interested to know the difference between the pressures at two points rather than the actual values of the pressure at each of these points. Differential manometers are used to measure the pressure difference between two points in a pipe or in two different pipes at the same or different levels. There are two types of differential manometers namely U-tube upright differential manometer and U-tube inverted differential manometer.

U-tube upright differential manometer

Let us consider a U-tube upright differential manometer connected at two points in a pipe at same level (Figure 2.17).

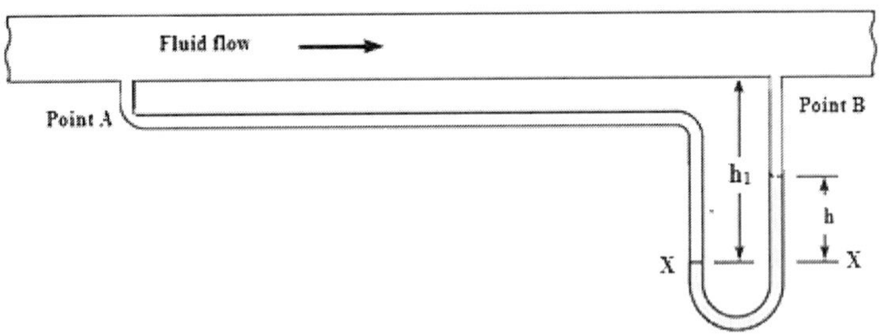

Figure 2.17. U-tube Upright differential Manometer.

To calculate the pressure difference at two points in the pipeline, proceed as follows:

Left limb: $h_a + h_1 S_p$ (2.65)

Right limb: $h_b + (h_1 - h) S_p + h S_m$ (2.66)

Equating the two expressions at x-x, we have

$$h_a + h_1 S_p = h_b + (h_1 - h) S_p + h S_m \quad (2.67)$$

$$h_a - h_b = -h_1 S_p + (h_1 - h) S_p + h S_m = h (S_m - S_p) \quad (2.68)$$

Here h_a and h_b are the fluid heights at points A and B respectively, S_p and S_m are the specific gravity of fluids flowing in the pipe and manometer fluid respectively. To determine the pressure difference, multiply both sides of eq. (2.68) by ρ_p, the density of the fluid flowing in the pipe such that

$P_a - P_b = \rho_p g h (S_m - S_p)$

Here P_a and P_b are pressures at points A and B respectively.

Let us now consider the case of a U-tube upright differential manometer connected between two pipes at different levels and carrying different fluids (Figure 2.18). Following a similar procedure as in the previous case, one can write the following equation

$h_a + h_1 S_{p1} = h S_m + h_2 S_{p2} + h_b$

Rearranging, we get

$$h_a - h_b = - h_1 S_{p1} + h_2 S_{p2} + h S_m \quad (2.69)$$

Here S_{p1}, S_{p2} and S_m are the specific gravity of fluids flowing in the left side pipe, in the right side pipe and manometer fluid respectively.

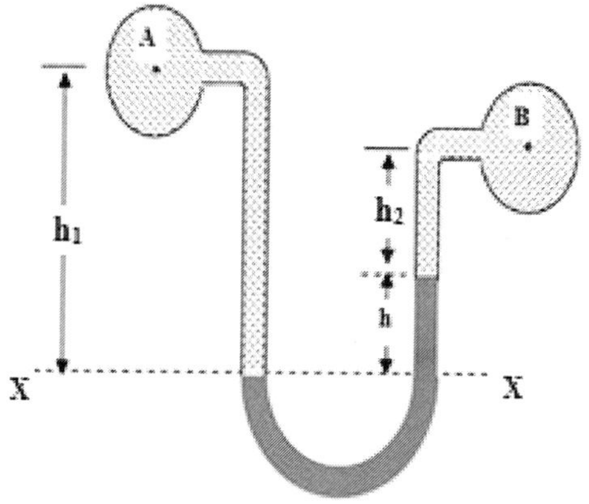

Figure 2.18. Upright (Vertical) Differential Manometer to Measure Pressure difference between Two Pipes.

Example 2.16

The difference in the mercury level in the two limbs of a differential manometer used to measure the difference in pressure at two points A and B in a pipeline is 250 mm. Calculate the pressure difference, if the specific gravity of water flowing in the pipe is 0.99.

Using eq. (2.68)

$h_a - h_b = h (S_m - S_p) = h (13.6 - 0.99) = 0.25\times 12.61 = 3.15$ m

$P_{diff} = 3.15\times0.99\times1000\times9.81 = 30592.5$ N/m^2

U-tube inverted differential manometer

The manometer fluid used in this type of manometer is lighter than the working fluids. Normally, air, fed through the air entry valve provided at the top, is used as the manometer fluid (Figure 2.19). These manometers are quite sensitive to measure small pressure differences because the lighter fluid enhances the height difference between the two limbs. For the configurations shown in Figure 2.19, we have

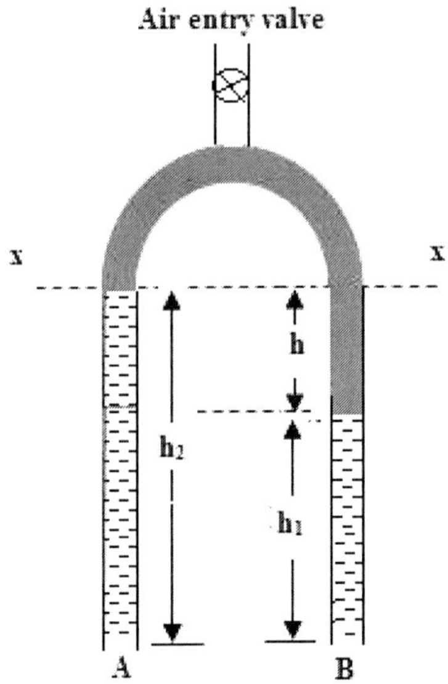

Figure 2.19. Inverted Differential Manometer.

The pressure at x-x

Left limb: $P_a - \rho_w g\, h_2$ (2.70)

Right limb: $P_b - \rho_w g h_1 - \rho_m g h$ (2.71)

Comparing the pressure on both ends on x-x, and rearranging, we have

$$P_a - P_b = \rho_w g h_2 - \rho_w g h_1 - \rho_m g h \quad (2.72)$$

Since $h_2 = h_1 + h$, we have

$$P_a - P_b = \rho_w g\,(h_1+h) - \rho_w g h_1 - \rho_m g = gh\,(\rho_w - \rho_m) \quad (2.73)$$

Here ρ_w and ρ_m are the densities of fluids in the pipe and manometer respectively. Besides, h, h_1 and h_2 are respective heights as shown in Figure 2.19. If ρ_m of the manometric fluid is much less than ρ_w, then

$$P_a - P_b = \rho_w g h \quad (2.74)$$

Example 2.17

Two pipes on the same elevation convey water and oil (specific gravity 0.88). A U-tube manometer connects the two pipes with the manometric liquid having a specific gravity of 1.25. Find the pressure difference in two pipes, if the manometric liquid in the limb connecting the water pipe is 2 m higher than the other pipe. The height of the oil column above the manometric fluid up to the pipe center is 5 m.

With the given information, following diagram describes the problem

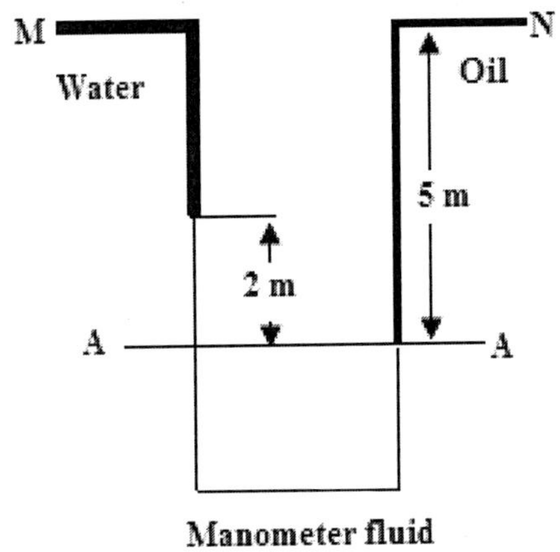

Equating the pressures at A-A, we have

$P_m + 2\times1.25\times\rho_w\times g + (5\text{-}2)\times\rho_w\times g = P_n + 5\times 0.88\times\rho_w\times g$

$P_m - P_n = 5\times 0.88\times\rho_w\times g - 2\times1.25\times\rho_w\times g - 3\times\rho_w\times g$

$P_m - P_n = 43164 - 24525 - 29430 = -10791\ \text{N/m}^2$

Example 2.18

Determine the pressure in the pipe B if the pressure in pipe A is 2 m of water. The values of h, h1 and h2 in Figure 2.19 are 100, 200 and 400 mm. The inverted tube differential manometer has an oil of specific gravity 0.9.

We use the full equation (eq. 2.73)

$P_a - P_b = \rho_w g h_2 - \rho_w g h_1 - \rho_m g h$

$2 - P_b = 1\times0.4 - 1\times0.2 - 0.9\times0.1$

$P_b = 2 - 0.11 = 1.89$ m

For water using the relation $P = \rho_w gh$, we have

$P_b = 1000\times9.81\times1.89 = 18540\ \text{N/m}^2$

Well Type Manometers

The main difference between a U-tube and a well type manometer is that one of limbs of the U-tube is made like a large well. Its cross sectional area is about 100 times the cross sectional area of the U-tube. This well acts as a reservoir to hold the manometer fluid. The change in the level of fluid in the tank is negligible with any change in the pressure. Neglecting this change for low-density fluids, the height of liquid in the other column is a measure of the pressure. It is why; it is categorized as single column manometer. Two kinds of such manometers are vertical column manometer and inclined column manometer.

Vertical column manometer

Initially, when no fluid is flowing through the pipe, the level of manometer liquid in the well is at XX. As soon as the tank is connected to some unknown pressure, the level of manometer liquid moves down from XX to YY in the tank and rises in the right limb to the distance h_2 from XX (Figure 2.20). Since it is difficult to measure Δh, we calculate it from volumetric balance. The volume change in the manometer fluid in the well is equal to the volume change in the right limb resulting in rise of level by h_2. Therefore:

$\Delta h A = a\, h_2$

$$\Delta h = a\, h_2 / A \tag{2.75}$$

Here a is the area of the manometric tube and A the area of the well.

$$\text{Pressure in the left limb at YY} = P_x + (h_1 + \Delta h)\,\gamma_1 \tag{2.76}$$

$$\text{Pressure in the right limb at YY} = (h_2 + \Delta h)\,\gamma_2 \tag{2.77}$$

Here γ_1 and γ_2 are the specific weights of fluid in the pipe and manometric fluid respectively.

As per the Pascal's law,

$$P_x + (h_1 + \Delta h)\,\gamma_1 = (h_2 + \Delta h)\,\gamma_2 \tag{2.78}$$

$$P_x = h_2\gamma_2 - h_1\gamma_1 + \Delta h(\gamma_2 - \gamma_1) \tag{2.79}$$

Incorporating Δh as calculated previously (eq. 2.75), we have

$$P_x = h_2\gamma_2 - h_1\gamma_1 + h_2(a/A)\,(\gamma_2 - \gamma_1) \tag{2.80}$$

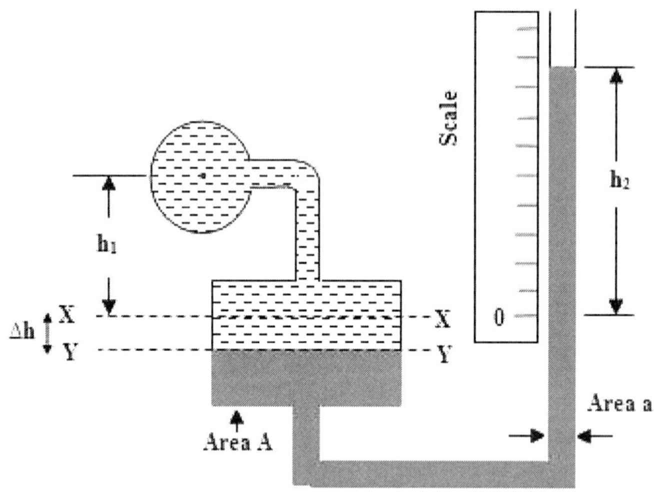

Figure 2.20. Well Type Vertical Column Manometer.

$$P_x = h_2\gamma_2\,(1 + a/A) - \gamma_1\,(h_1 + h_2\, a/A) \tag{2.81}$$

Since a is quite small than A, a/A approaches zero, so that

$$P_x = h_2\gamma_2 - h_1\gamma_1 \tag{2.82}$$

Now, if the specific weight of the fluid in the pipe is small such as that of the gases, even the second term on the right hand side can be neglected. Eq. (2.82) in that case reduces to

$$P_x = h_2\gamma_2 \tag{2.83}$$

For such cases, the manometer has a scale to measure h_2. The scale may be in terms of pressure so that one can directly read the pressure (Figure 2.20).

Example 2.19

A vertical column manometer is connected to a pipe containing oil of specific gravity 0.92. The ratio of area of reservoir to that of manometer tube is 180. Calculate the oil pressure in the pipe, if the pipe center is 240 mm above the manometer fluid in the well and the mercury level in the tube is 340 mm above the center line of the pipe (Assume specific gravity of mercury as 13.6).

Based on the data, following diagram is constructed.

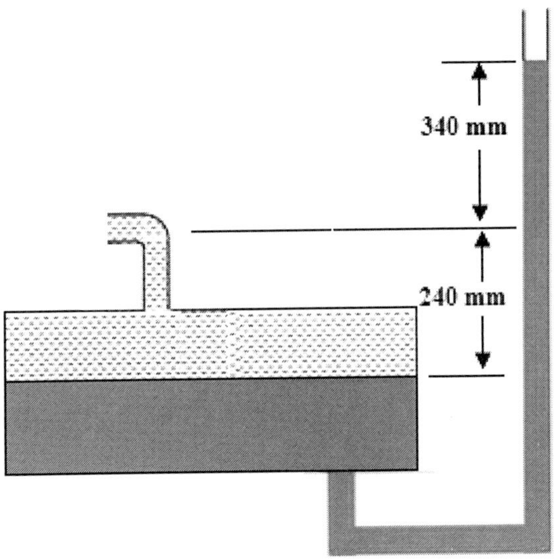

Since a/A = 1/180 =.0055, it is neglected and the simplified eq. (2.82) is used.

$P_x = h_2\gamma_2 - h_1\gamma_1 = (0.34+0.24) \times 13600 \times 9.81 - 0.24 \times 920 \times 9.81$

$77381.3 - 2166.0 = 75215.3$ N/m^2 or 75.22 kN/m^2

Inclined column manometer

A modified version of vertical column manometers is an inclined manometers (Figure 2.21). The main advantage of inclined angle manometer is that even a small or low amount of pressure will produce a large liquid movement relative to the tube's graduations. As a result, these manometers are highly sensitive as graduation scale can be very precise.

Equations derived for vertical manometer are applicable for this except for the value of h_2, which in inclined angle manometer is:

$$h_2 = L \sin \theta \tag{2.84}$$

Here L is the length and θ the angle of the inclined tube with the horizontal (Figure 2.21). Incorporating h_2 in the simplified equation of vertical column manometer, we get

$$P_x = L \sin \theta \, \gamma_2 - h_1 \gamma_1 \tag{2.85}$$

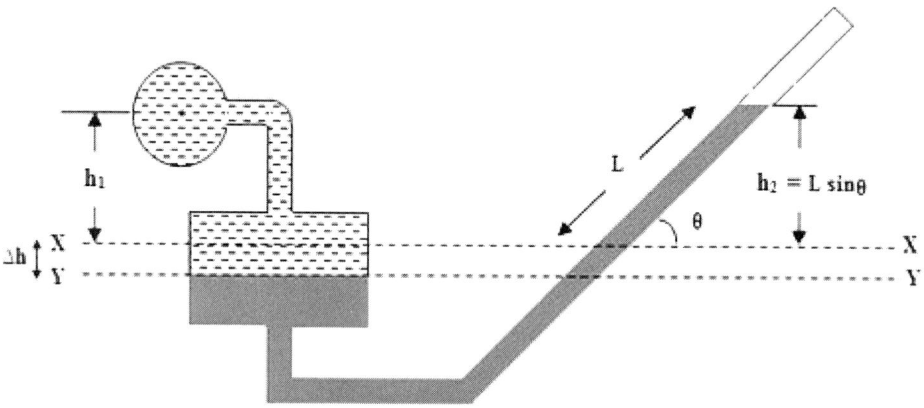

Figure 2.21. Inclined Column Micro-manometer.

In terms of head h and specific gravities of the fluids, equation takes the form:

$$h = S_2 L \sin \theta - S_1 h_1 \tag{2.86}$$

At the same pressure difference, equations for vertical and inclined manometers indicate that inclined tube manometer amplifies the length of measurement by 1/sinθ, which is the primary advantage of such a manometer. Even a U-tube manometer can be inclined to improve its accuracy (Figure 2.22).

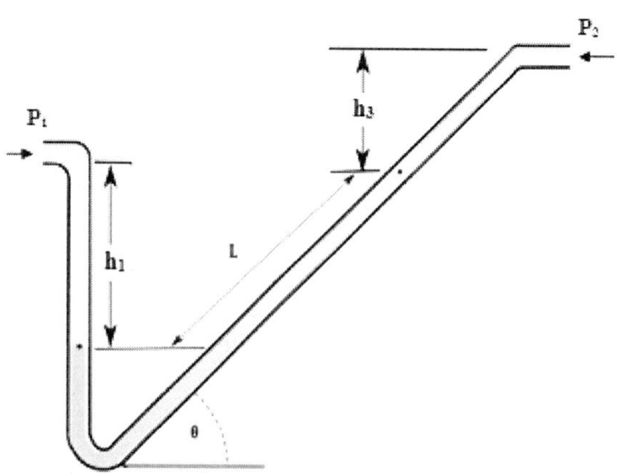

Figure 2.22. Modified Inclined U-tube Manometer.

Micro-manometer

A typical micro-manometer has a well (reservoir), which moves up and down with the help of a micrometer screw (Figure 2.23). The other leg is an inclined tube connected to the reservoir with the help of a flexible tube. A magnifier helps to view the meniscus in the inclined tube located at a reference level. Before the application of pressure, move the reservoir up and down so that fluid level in the reservoir coincides with the reference mark. Apply the pressure on the well side. It disturbs the liquid level. Using the micrometer, restore the reservoir level once again to the reference. The extent of the micrometer movement gives the pressure difference between the two points. Micro-manometers are commonly used in the laboratories to calibrate manometers used in wind tunnels.

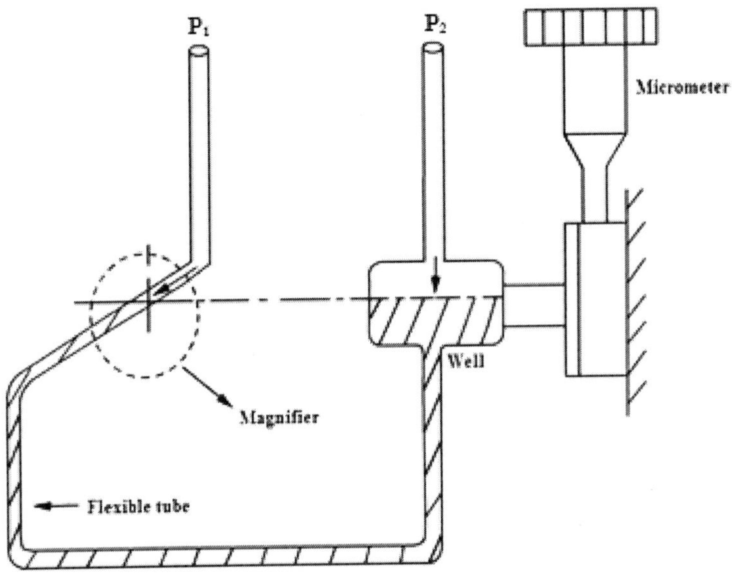

Figure 2.23. Schematic Diagram of a Micro-manometer.

Multi-fluid Manometer

A multi-fluid manometer, also known as compound or multiple tube manometers, is a kind of differential manometer used to measure the pressure difference of the fluid flowing in two pipes. All principles related to manometers discussed so far are applicable. The following illustration explains the principle of a multi-fluid manometer (Figure 2.24). Let us consider that oil is flowing in the pipe on the right of the manometer. Specific gravity of the oil is 0.90. While the specific gravity of mercury in the U-tubes is 13.6, the specific gravity of water in pipe on the left is 1.0. Thus, there are 3 fluids in the manometer instead of 2 discussed so far.

Following the earlier guidelines, we begin with the centerline of pipe on the far left with pressure P_b. Then

$$P_b + \rho_w g h_1 + \rho_m g h_2 - \rho_w g h_3 + \rho_m g h_4 - \rho_o g h_5 = P_o \qquad (2.87)$$

$$P_o - P_b = \rho_w g h_1 + \rho_m g h_2 - \rho_w g h_3 + \rho_m g h_4 - \rho_o g h_5 \quad (2.88)$$

If the values of h_1, h_2, h_3, h_4 and h_5 in Figure (2.24) are 150, 150, 250, 200 and 150 mm respectively, then eq. (2.88) is solved as follows.

$$P_o - P_b = (1000\times9.81)\ (1\times0.15 + 13.6\times0.15 - 1\times0.25 + 13.6\times0.2 - 0.9\times0.15)$$

$$P_o - P_b = 9810\times4.525 = 44390\ N/m^2$$

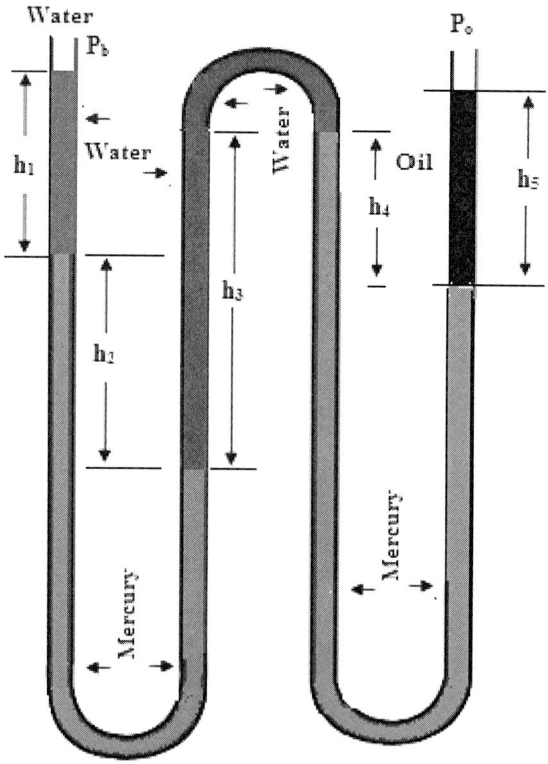

Figure 2.24. Schematics of a Multi-fluid Manometer.

Manometric Fluid

To end the section on manometers, it may be useful to discuss the importance of selecting a manometric fluid. The selection of manometric fluid normally depends on the operating range. Compared to water, mercury moves only about 1/13.6 whereas oil may move about 1.2 times farther than water in response to a given pressure. Besides, one should look for the following characteristics.

- ☆ The manometric fluid and the fluid whose pressure is to be measured should not mix with each other
- ☆ The specific gravity of the fluid must be known with sufficient accuracy. Besides, one must be able to decipher changes in specific gravity with local changes in the atmospheric conditions

- ☆ It should have good wetting characteristics
- ☆ It should be able to form a well shaped meniscus

SIMPLE MERCURY BAROMETER

A simple barometer, known as Torricellian barometer, named after its inventor an Italian mathematician Evangelista Torricelli, measures the atmospheric pressure. It consists of a tube more than 760 mm long, closed at the upper and open at the bottom end. The tube is inserted in an open container full of mercury. Mercury extends from the container up into the tube (Figure 2.25 left) because of atmospheric pressure on the mercury surface. The void produced at the top of the tube contains mercury vapours. Since the saturation vapour pressure of mercury is quite small, the void space is assumed a perfect vacuum. At the sea level, mercury rises in the tube to a height of approximately 760 mm. The level of mercury will rise and fall with changes in atmospheric pressure. The instrument is calibrated in such a way that the mercury level gives the direct reading of prevailing atmospheric pressure, P_a as a pressure head (of mercury). The following relation can determine the pressure or conversion table can be used for change to the desired units (Annexure I).

$$P_a = \rho g h \tag{2.89}$$

Here ρ is the density of mercury, kg/m^3 (13534 kg/m^3), and h the height of mercury in the barometer, m.

A barometer measures the absolute atmospheric pressure. A major disadvantage of the instrument is that it is quite fragile. Moreover, spilling of mercury, which is toxic in nature, cannot be ruled out. It can create environmental problems. It is why aneroid barometers are fast replacing the mercury barometer (Figure 2.25 right).

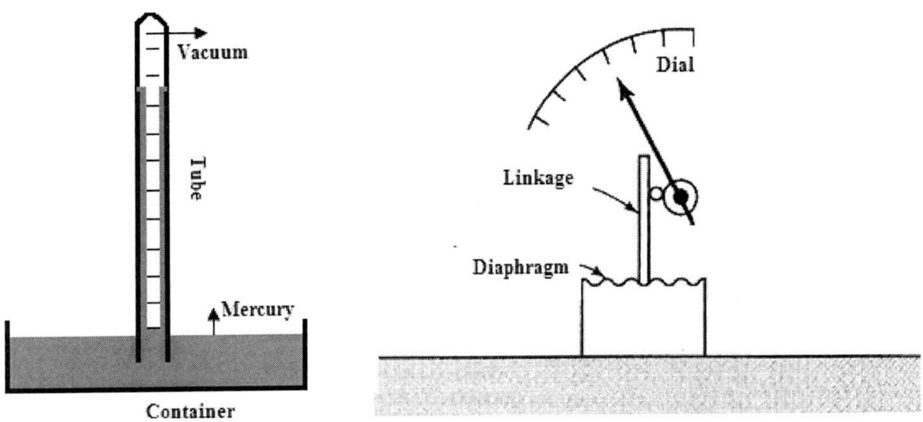

Figure 2.25. Construction Details of a Mercury (Left) and an Aneroid Barometer (Right).

Aneroid Barometer

Aneroid barometer dispenses with the use of toxic mercury. Instead, it uses elastic diaphragm to measure atmospheric pressure (Figure 2.25 right).

Example 2.20

A gauge measured pressure is 85 kPa above the atmospheric pressure. The barometer measured atmospheric pressure is 760 mm of Hg. What is the absolute pressure of the gauge? Assume specific gravity of mercury as 13.54.

Atmospheric pressure in kPa = $13.54 \times 1000 \times 9.81 \times 760/(1000 \times 1000) = 100.95$ kPa

Absolute pressure = Gauge pressure + Atmospheric pressure = 85 +100.95 = 185.95 kPa.

ELECTROMECHANICAL GAUGES

Manometers are capable of measuring low-pressures near to the atmospheric pressure. For medium and the high pressures, one uses elastic pressure gauges such as tubes, diaphragms, bellows *etc.* Elastic deformation of the elements used in these instruments helps to measure the pressure. Since, the elements deform within elastic limits of the materials, these gauges are also known as elastic gauges. These instruments are also known as secondary instruments because we calibrate these gauges using primary instruments such as manometers.

Bourdon Tube Pressure Gauge

Bourdon pressure gauge is the oldest pressure-sensing element patented in 1849 by Eugene Bourdon. It consists of metal (Phosphor bronze, Beryllium bronze or Beryllium copper) tube of elliptical cross section and shaped in the form of English alphabet letter 'C' (Figure 2.26) or coiled into a spiral shape or helix. While one end is free, the other fixed end connects to the pressure source. When connected to the pressure source, it responds by bending inward due to increase in pressure. The maximum displacement occurs at the free end. A simple pinion and sector arrangement converts the linear movement of the tube into angular movement of the pointer. The pointer over dial graduated on a suitable scale indicates the pressure. The ratio of the major and the minor axes of the elliptical cross-section govern the sensitivity of the Bourdon tube. The larger is the ratio, higher the sensitivity. It has wide range, being capable of measuring high and low pressures. Moreover, it has a very fast response. A major limitation of the instrument is its sensitivity to temperature variations.

Diaphragm Gauges

Diaphragm gauges are either flat or corrugated or capsules/bellows types. The pressure sensors in the flat diaphragm gauge make use of flat elastic diaphragms fixed in a tubular member (Figure 2.27 left). The diaphragm deflects upwards upon application of the pressure. A strain gauge or transducer senses this movement and converts the pressure signal into an electrical signal. In analogue devices, a rack and pinion system moves the pointer. Transmission lag in these instruments is much less because these can be installed flush with the surface of the body.

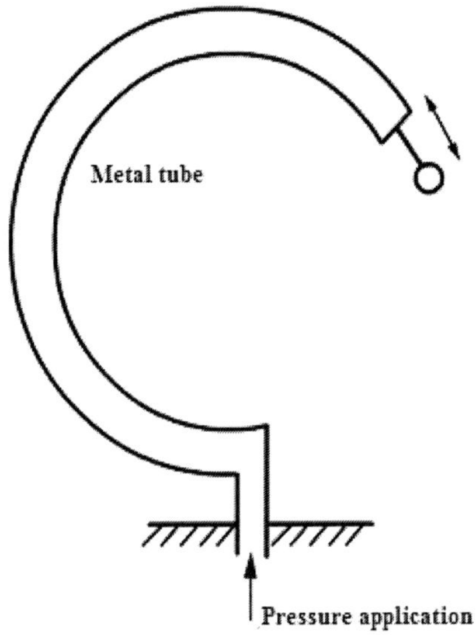

Figure 2.26. Bourdon Tube Pressure Gauge.

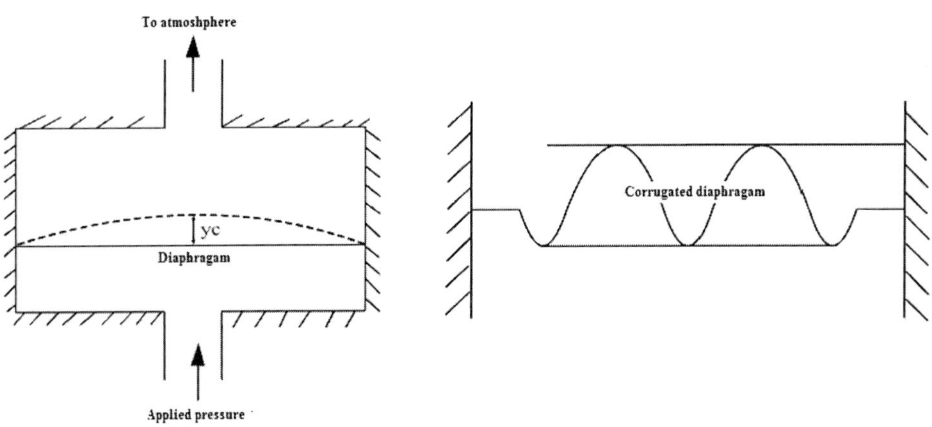

Figure 2.27. Diaphragm Pressure Gauges: Flat (Left) and Corrugated (Right).

Another version of these instruments uses corrugated diaphragms that permit considerably larger deflections than flat diaphragms (Figure 2.27 right). Numbers of corrugations and their depth governs the response and sensitivity. The greater the number and depth, the more linear is its deflection and greater the sensitivity. In a still improved version for larger deflections, corrugated diaphragms are made into boxes or bellows. These bellows are capable of measuring small steady pressures.

Pressure Transducers

The basic principle of a transducer is to convert one form of energy to another form. A pressure transducer often called a pressure transmitter or electronic pressure transducer senses the signals and converts them into electronic/electrical signals, which are further processed within the instrument. Some of these are piston, strain, piezoelectric, capacitive and magnetic gauges *etc.*

Piston type pressure transducer

In this instrument, a piston and a spring arrangement measures the pressure. The pressure port connected to the pressure to be measured moves the piston (Figure 2.28 left). This movement accentuates the spring. The movement is either measured through a meter placed outside or is converted to an electrical signal, which is used to measure the pressure.

Diaphragm type pressure transducers

These instruments are similar to ones discussed in the previous section. Only difference is that the pressure in this case is converted into electrical signals. Depending upon the kind of pressure transducer used, it measure changes in inductive, capacitive or ohmic resistances caused by the deformation of an elastic element. Upon application of pressure, the diaphragm deforms with its center deflecting the most. This initiates changes in the electrical circuit. These changes are calibrated against pressure to be measured (Figure 2.28 right).

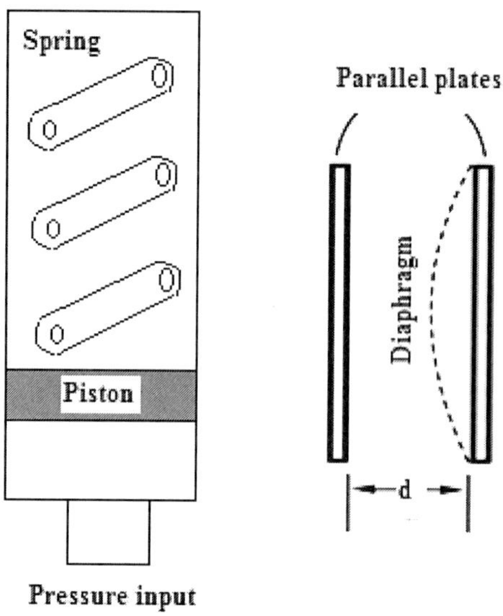

Figure 2.28. Pressure Transducers: Piston (Left) and Elastic Diaphragm in a Parallel Plate Capacitor (Right).

Piezo-electric pressure transducers

The word piezo-electric derived from the Greek word "piezein" means squeeze or press. Certain materials possess the ability to generate electrical potential when subjected to mechanical strain. Conversely, they change dimensions when supplied with electrical voltage. Materials exhibiting piezo-electric properties are quartz, tourmaline, barium titanate, and lead zirconate. Quartz is the preferred material, as it possesses good mechanical properties as well. Moreover, it is a good insulator least influenced by the moisture. Some of the principles used in these gauges are as follows.

Capacitive: It uses a diaphragm and pressure cavity to create a variable capacitor to detect strain due to applied pressure.

Magnetic: It measures the displacement of a diaphragm by means of changes in inductance, LVDT, Hall effect, or by eddy current principle.

Piezoelectric: It uses the piezoelectric effect in certain materials such as quartz to measure the strain upon the sensing mechanism due to pressure.

Optical: It uses the physical change of an optical fiber to detect strain due to applied pressure.

Potentiometric: It uses the motion of a wiper along a resistive mechanism to detect the strain caused by applied pressure.

Resonant: It uses the changes in resonant frequency in a sensing mechanism to measure stress, or changes in gas density, caused by applied pressure.

Piezo-electric transducers are capable of measuring dynamic pressures in the range of frequencies from kHz to 100MHz.

Pressure Sensitive Paints (PSP)

PSPs are used to measure air pressure or local oxygen concentrations. PSPs are paint like coating, composed of luminescent molecules dispersed in oxygen permeable polymer binder. After curing in an oven, the PSP is placed in an appropriate test environment. With PSP exposed to a blue or an ultraviolet light, the luminescent molecules are excited to a higher energy level. LED lamps of specific wavelengths excite the pressure sensitive probe within the paint. Fluorescence is dimmer under high and brighter at low pressure.

Electrical Resistance Gauges

Electrical resistance gauges, also known as Bridgeman gauges, are capable of measuring very high pressures (say above 1000 bars). They make use of resistance change brought about by direct application of pressure to the conductor itself. The sensing element is the thin wire of magnanin (84 per cent Cu + 12 per cent Mn + 4 per cent Ni) or an alloy of gold and chromium, which is loosely wound (Figure 2.29). When pressure is applied, bulk compression effects produce a change in resistance, which is calibrated against pressure.

Vacuum Gauges

Bourdon gauge is capable to measure vacuum. The only difference is that in this case tube has an inward instead of the outward bend. The instrument is graduated in mm of Hg below atmospheric pressure. Similarly, we have diaphragm vacuum gauge, capsule vacuum gauge, capacitance vacuum gauge, thermal conductivity gauge and hot and cold cathode ionization vacuum gauge. Detailed discussion on these kinds of gauges is left out being beyond the scope of this book.

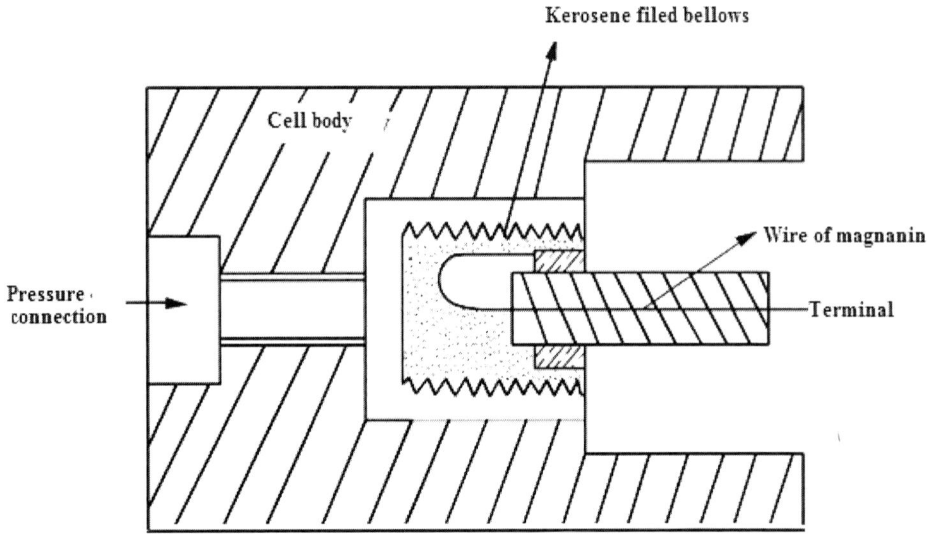

Figure 2.29. A Schematic Diagram of a Bridgeman's Gauge.

Table 2.2 lists the range of measurement for some selected equipment.

Table 2.2. Measurement Range of Various Pressure Measuring Instruments

Instrument	*Range*
Liquid column manometers	Vacuum to 1.5 bars
Bourdon tube	Vacuum to few thousands bars
Diaphragm	Vacuum to few hundred bars
Bellows	Vacuum to a hundred bars
Electrical resistance gauges	>1000 bars

QUESTIONS (THEORY)

1. Define pressure and hydrostatic law. Prove the hydrostatic law.
2. Explain and prove the Pascal's law.
3. Discuss hydrostatic force on plane horizontal and vertical surfaces. Illustrate your answer using a neat diagram.
4. Discuss hydrostatic force on an inclines surface. Illustrate your answer using a neat diagram.
5. Discuss hydrostatic force on curved surfaces. Use a neat diagram to illustrate your response.

6. Explain the working of a manometer using a neat diagram of a U-tube upright differential manometer connected between two pipes at different levels and carrying different fluids.

7. Explain the working of a well type manometer using a neat diagram.

8. Draw a neat diagram and explain the working of an inclined tube manometer.

9. List the characteristics of a good manometer fluid.

10. Explain the principle and working of a simple mercury barometer and list its limitations.

11. What is pressure transducer? Briefly explain the various kinds of pressure transducers.

12. Differentiate between gauge pressure and absolute pressure and flat and corrugated diaphragm pressure gauges.

Chapter 3

Archimedes Law and Stability of Floating Bodies

If a solid body (or any other body) is submerged or floating in a liquid, the surfaces in contact with water experience hydrostatic pressure forces. In the previous chapter, we have seen that the pressure in a fluid column increases with depth. The pressure at the bottom of the object submerged or floating in the fluid is more than that at the top. This pressure difference results in a net upward force on the object known as the buoyant force. Fishes, whales, submarines, balloons and airships all owe their ability to float to buoyancy, the lifting power of water and air. Similarly, stability of floating bodies is of vital importance in the design of ships, submarines and various other floating structures. The shape of the floating body and its mass distribution determine whether it will float stably or capsize. This chapter focuses on the Archimedes's principle that explains the force of buoyancy. Further, it dwells upon the conditions necessary to ensure stability of floating and submerged bodies.

BUOYANCY

An important extension of the operation of vertical forces on curved surfaces discussed in the previous chapter is the concept of buoyancy. The buoyancy of a body wholly or partly immersed in a fluid at rest, situated in a gravitational field or other field of force, is defined as the upward thrust of the fluid on the body. In other words, buoyancy is the tendency of a body to lift upward in a fluid against the force of gravity. Buoyancy force on a body is equal and oppositely directed to the weight of the fluid that the body displaces. This also explains the loss of weight in a body immersed in the fluid, which is equal to the weight of fluid displaced by it. The unit of buoyant force is Newton, N. Center of Buoyancy is the point at which all the forces of buoyancy act in a vertically upward direction (Figure 3.1). The center of buoyancy will be the center of gravity of the fluid displaced.

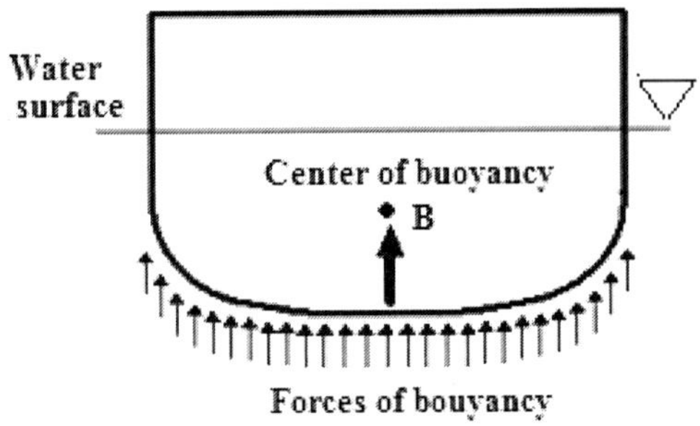

Figure 3.1. Depiction of Forces of Buoyancy on a Floating Object.

If a floating body displaces a weight of fluid equal to its own weight; then the volume of fluid displaced by the floating body is just enough to balance its weight. The floating body is then in equilibrium. If a solid object with a uniform density exactly equal to that of water is immersed in water, the force of buoyancy will be exactly equal to the weight of the body. The center of buoyancy will be the center of gravity of the body. The object will be in equilibrium with the surrounding fluid (Anonymous, 2018). The principle applies to both floating and submerged bodies and to all fluids *i.e.* to gases and liquids.

Some important characteristics of the buoyancy force are as follows.

- The buoyancy of any body is vectorially equal and opposite to the weight of the fluid displaced by the body and has the same line of action
- The buoyancy force acts through the center of buoyancy. It is the point where the center of gravity of the volume of displaced fluid is located. In Figure (3.1), point B is the center of buoyancy. The center of buoyancy is not the same as the center of gravity, which relates to the distribution of weight within the object
- In the case of a body lying at the interface of two immiscible fluids, the buoyant force equals the sum of the weights of the fluids displaced by the body
- Buoyancy does not depend upon the shape of the object but only on the volume of the object.

ARCHIMEDES LAW

The basic principle discovered by Archimedes, a Greek scientist, is an important scientific law of buoyancy. Archimedes' law states that a body immersed in a fluid either wholly or partially, displaces its own weight in the fluid in which it floats and a body immersed in a fluid experiences a vertical buoyant force equal to the weight of the fluid it displaces. Some common phenomena in daily life based on Archimedes law are:

- Use of hydrometer to test the purity of milk
- A balloon filled with a gas lighter than air rises to such a height that the weight of the air displaced is equal to the weight of the gas in the balloon

- ☆ A ship weighing tons easily floats in sea where as a hammer sinks in water
- ☆ A person keeps floating without sinking in the Dead Sea
- ☆ A life jacket helps to float in water.

Application of the principles of Archimedes resolves all problems relating to buoyant force.

Fully Submerged Body

For a fully submerged body we can find out the buoyant force by referring to Figure (3.2). A fully submerged body with top and bottom areas as A is located at a distance h_1 and h_2 from the surface of the fluid (Figure 3.2 left), then

$$F_1 = h_1 \rho\, g\, A \tag{3.1}$$

$$F_2 = h_2 \rho\, g\, A \tag{3.2}$$

Here g is the acceleration due to gravity and ρ is the density of the fluid.

Since $h_2 > h_1$, $F_2 > F_1$ and buoyancy force F_b is given as

$$F_b = F_2 - F_1 = h_2 \rho\, g\, A - h_1 \rho\, g\, A = (h_2 - h_1)\, \rho\, g\, A \tag{3.3}$$

Since $(h_2 - h_1)$ A is the volume, V of the fully submerged object, we get

$$F_b = \rho\, g\, V \tag{3.4}$$

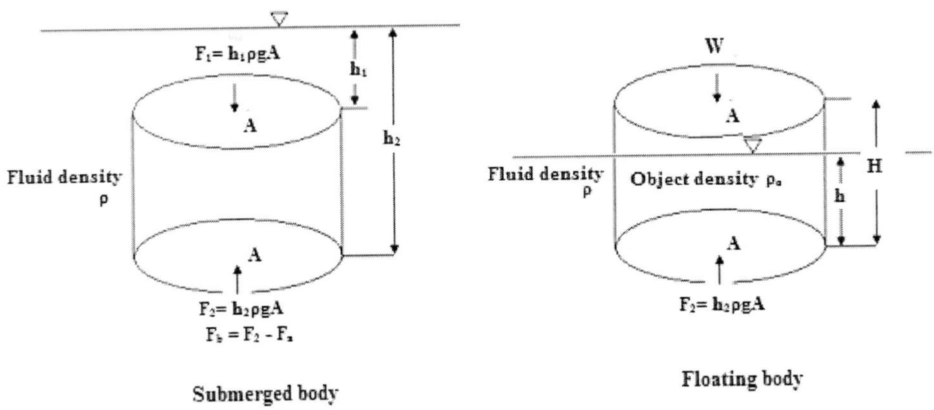

Figure 3.2. Illustration of Buoyant Force on a Fully Submerged (Left) and a Floating Body (Right).

Example 3.1

A balloon weighs 80 kg has a capacity of 1200 m³. The balloon is filled with helium having density of 0.18 kg/m³. How much payload this balloon will be able to take? Express the payload in Newton. Assume the density of air as 1.3 kg/m³.

Total mass that can be lifted by the balloon= 1200× 1.3 = 1560 kg

Since it includes the weight of the balloon, helium and payload, the mass of the payload is

Payload = 1430 – 80 – 1200× 0.18 = 1134 kg

Payload = 1134 × 9.81 = 11125 N

Floating Body

For a floating body, only a portion of the body submerges in the fluid, while the remaining portion remains out of the free surface. We can explain this issue with the principle of flotation (Poso, 2015), which states that the body will float if the weight of the body is equal to the weight of the liquid displaced by it *i.e.* the buoyant force. In this case, weight of the liquid displaced by the submerged portion of the object is equal to the total weight of the object. Principle of Archimedes can be used to determine the fraction of the body submerged in the fluid. According to this principle, the buoyant force of a partially immersed body is equal to the weight of the displaced liquid. For a floating body in static equilibrium and in the absence of any other external force, the buoyant force must balance the weight of the body, W (Figure 3.2 right).

$$\text{Fraction of the body submerged} = V_s/V_o = V_f/V_o \quad (3.5)$$

Here V_s and V_o are the volume submerged and total volume of the object respectively. Since the volume submerged equals the volume of fluid displaced, V_f, we can also write V_f in place of V_s. In terms of mass and densities, we write

$$V_f/V_o = (m_f/\rho_f)/(m_o/\rho_o) \quad (3.6)$$

Here m_f and m_o are the mass of the fluid displaced and mass of the object respectively, ρ_f and ρ_o are the densities of the fluid and the average density of the object. Note that ρ_o is average density of the object and not of the material of its construction. Since body floats, its mass and that of displaced fluid are equal. Thus

$$\text{Fraction submerged} = V_f/V_o = \rho_o/\rho_f \quad (3.7)$$

Above equation makes sense only if $\rho_o < \rho_f$. Otherwise, it becomes more than one and the object will sink. For the object shown in Figure (3.2 right), it is also possible to write

$$AH\rho_o = Ah\rho_f \quad \text{(Since volume = Area × height)} \quad (3.8)$$

$$\text{Or } h/H = \rho_o/\rho_f \quad (3.9)$$

Such that H is the height of the body and h the height of the body submerged in the fluid.

Example 3.2

A wooden block 3 m × 2 m × 1.5 m (L ×B× D) is floating in the water. If the density of block is 750 kg/m^3, determine the position of center of buoyancy.

W = (3 × 2 × 1.5) × 750 × 9.81= 66217.5 N

Force of buoyancy, $F_b = V_f \times 1000 \times 9.81$

According to Archimedes Principle $W = F_b$

$V_f \times 1000 \times 9.81 = 66217.5$

$V_f = 6.75\ m^3$

$h \times 2 \times 3 = 6.75$

$h = 1.125$ m

We can directly use eq. (3.9) to obtain h as

$h/1.5 = 750/1000 = 0.75$

$h = 1.125$ m

Since the block is of uniform density, the position of center of buoyancy = h/2 = 0.562 m

Example 3.3

A wooden block floating in the water is 3.25 cm in diameter, 8 cm long having a density of 720 kg/m^3. What is the submerged depth of the block? How much weight can be placed on this block so that it is completely immersed in water?

Area = $(\pi/4)(3.25\times3.25) = 8.3$ cm^3

Volume = $8.3\times8 = 66.4$ cm^3

W = $66.4\times720\times9.81 = 0.469$ N

Volume displaced = $0.469/(1000 \times 9.81) = 0.000048$ m^3 = 48 cm^3

Submergence depth = $48/8.3 = 5.78$ cm

Additional volume to be displaced = $66.4 - 48 = 18.4$ cm^3

Weight of water to be displaced = $18.4\times1000\times9.81\times10^{-6} = 0.181$ N

A weight of 0.181 N should be placed on the wooden block.

Example 3.4

A hydrometer of mass 27 g floats in water with 4 cm of its stem being out of the water. Now the hydrometer is immersed in another liquid having relative density of 1.25. Find out the total volume of the hydrometer and the length of the stem outside the liquid. The cross section of the hydrometer of the stem is 0.70 cm^3. Assume that cross section of the stem is flat.

Volume of the water displaced = 27 cm^3 (Density of water is taken as 1 g/cm^3)

Volume of the stem out of water = $4\times0.70 = 2.8$ cm^3

Total volume = $27 + 2.8 = 29.8$ cm^3

Volume of the second liquid displaced = $27/1.25 = 21.6$ cm^3

Length of the stem out of the second liquid = $(29.8 - 21.6)/0.7 = 11.7$ cm

FUNDAMENTALS OF STABILITY

To understand the basics of stability, let us consider a cone resting on a table in three different positions (Figure 3.3). The stability of the cone depends on how close the center of gravity is to the plane on which it is resting.

Position A is stable. In this case, cone on slight displacement comes back to its original position.

Position B is unstable. In this case, the cone will fall even with slight disturbance.

Position C is neutral. The cone when slightly displaced remains in the new position. It neither comes to its original position nor falls.

Similarly, in the case of immersed and floating bodies, the position of center of buoyancy determines the stability or equilibrium.

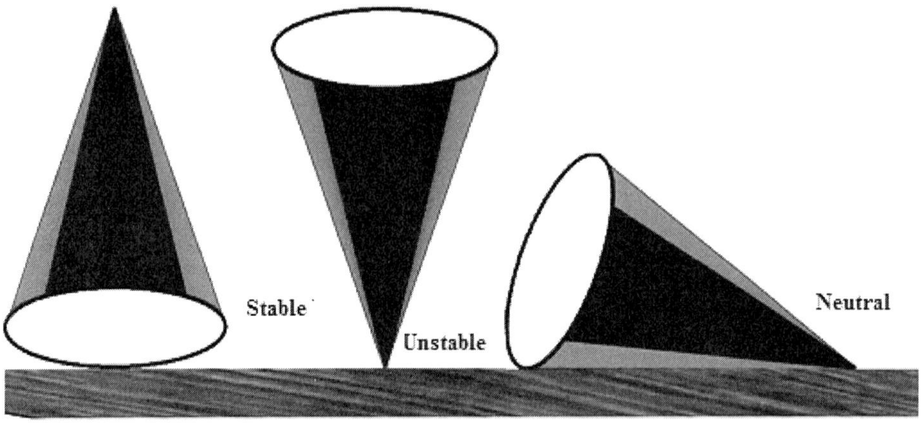

Figure 3.3. Illustration of Principle of Stability.

Before we consider the issue of stability, let us look at few important reference points that describe the stability of an object. For an object floating in a fluid, three important points are G, the center of gravity, M, the metacenter, and B, the center of buoyancy (Figure 3.4 left).

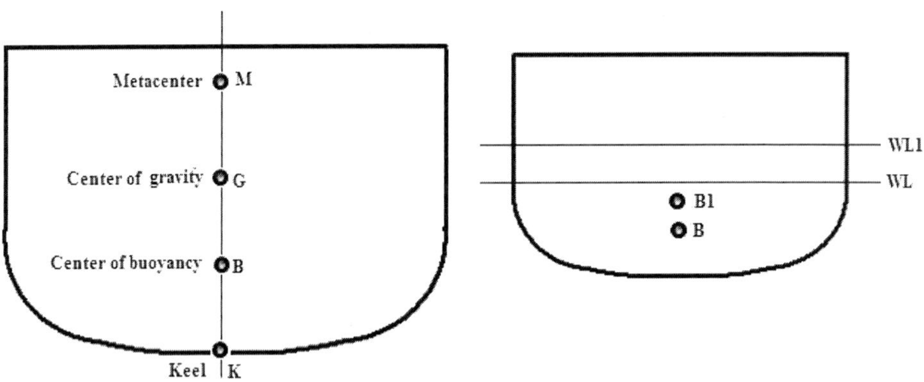

Figure 3.4. Reference Points for the Stability of Floating or Submerged Bodies (Left) and Movement of Center of Buoyancy with Increasing Fraction of Submergence (Right).

Center of Gravity

G, the center of mass of the body, is the point at which all forces of gravity acting on the body act. The position of G is dependent upon the distribution of weights within the body. In the case of a ship, any change in the distribution of weights will alter the position of G as follows.

- ☆ G moves towards a weight addition
- ☆ G moves away from a weight removal
- ☆ G moves in the same direction as a weight shift

Center of Buoyancy

As already stated previously in this chapter, a body made heavier will sit deeper in the water and center of buoyancy, B moves up, from B to B1 (Figure 3.4 right). Contrarily, if the body lightens, it sits shallower in the water and B moves down.

Metacenter

If a small angular displacement is given to a body floating in a liquid, the body oscillates about some point M. The point about which the body oscillates is the metacentre, M. It is a theoretical point at which an imaginary line passing through the center of buoyancy and center of gravity intersects the imaginary vertical line through a new center of buoyancy created as the body tips in the fluid (Figure 3.5).

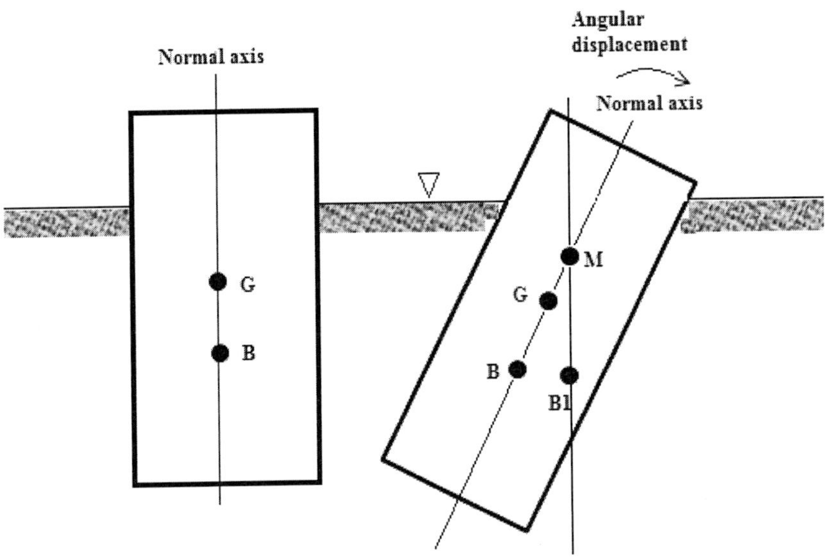

Figure 3.5. Explanatory Diagram Showing the Location of Metacenter.

In ships, another point K is the keel (Figure 3.4 left). It is the base line reference point used to measure all other reference points. We have the following linear measurements in the case of ships.

KG: Height of the ship's center of gravity above the Keel

KM: Height of metacenter above the Keel

GM: Metacentric height calculated by subtracting KG from KM (GM = KM - KG). GM is a measure of the ship's initial stability.

BM: BM, known as metacentric radius, is the distance between the center of buoyancy and the metacenter. It is actually the radius of the circle for the movements of "B" at small angles of heel.

Some of these linear measurements will find application in the stability analysis of floating and submerged bodies.

Stability

The equilibrium of a body submerged in a liquid requires that the weight of the body acting through its center of gravity should be collinear with equal hydrostatic lift acting through the center of buoyancy. If the distribution of mass over the entire volume of the body is not homogeneous, then the location of center of gravity, G does not coincide with the center of volume *i.e.* the center of buoyancy, B. Relative location of the points G and B determine the stability of the body, which is the tendency of a body to rotate one way or the other when forcibly inclined. Depending upon the relative locations of G and B, a floating or submerged body attains three different states of equilibrium.

- ✰ Stable equilibrium
- ✰ Unstable equilibrium
- ✰ Neutral equilibrium

Stable equilibrium

A body is said to be in stable equilibrium, if it returns to its original position by retaining the original vertical axis as vertical due to internal forces after it was slightly tilted (small angular displacement) by some external force. In the case of a floating body metacenter (M) should be above the center of gravity (G). Since the metacenter is located above the body's center of gravity, the buoyant force and the weight always produce a restoring couple trying to return the body to its original position upon its slight displacement (Figure 3.6 left). Another terminology in the ship industry for this is the creation of righting arms, which tend to return the ship to its original vertical position (Figure 3.6 right). As shown in this figure, the

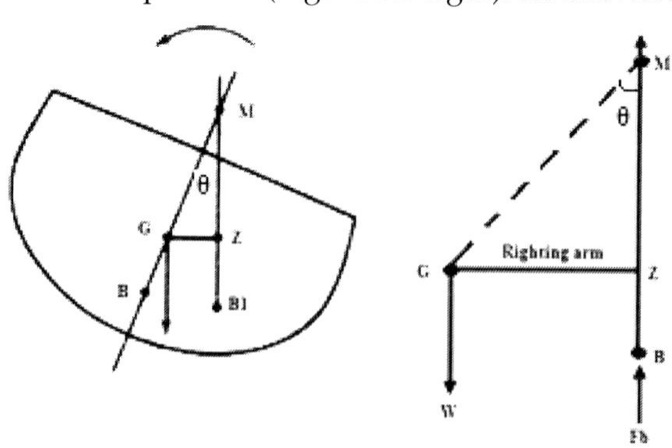

Figure 3.6. Illustration of Stable Equilibrium of a Floating Body.

righting arm is a perpendicular line drawn from the center of gravity to the point of intersection on the force of buoyancy line. It results in the formation of a stability triangle. The righting moment is equal to the righting arm multiplied by body's displacement. This rotation returns the ship to where the forces of buoyancy and gravity balance out. For small angles of heel (0° through 7° to 10°, metacenter does not move), the value of the ship's righting arm (GZ) is (using trigonometry):

$$GZ = GM \sin\theta \tag{3.10}$$

Unstable equilibrium

Given a slight angular displacement, a body instead of returning to its original position moves farther away is said to be in unstable equilibrium. In a floating body, such a situation exists when the metacenter (M) is below the center of gravity (G). As the body is inclined, creation of negative righting arms (called upsetting arms) tends the body to sink or capsize the ship (Figure 3.7).

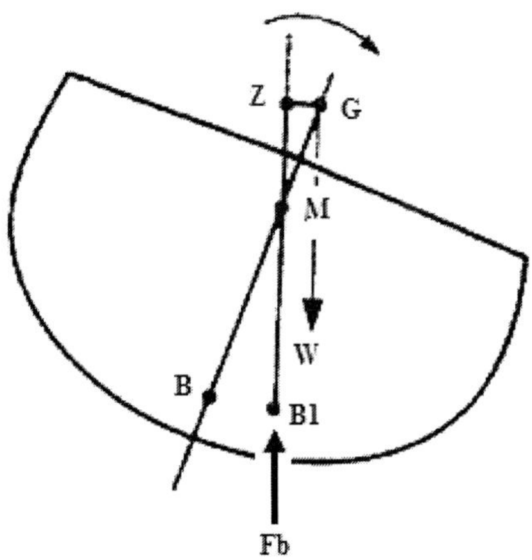

Figure 3.7. Illustration of Unstable Equilibrium of a Floating Body.

Neutral equilibrium

A body is in neutral equilibrium, if upon slight angular displacement it occupies a new position and remains at rest in that new position. It neither returns to its original position nor there is further displacement. For a floating body to be in neutral equilibrium, metacenter (M) coincides with center of gravity (G). Upon slight inclination of the body, neither a restoring couple nor negative righting arm is created (Figure 3.8).

To summarize the conditions of equilibrium for a floating body in terms of metacentric height, we can write:

GM > 0 (M is above G) stable equilibrium

GM = 0 (M coinciding with G) neutral equilibrium

GM < 0 (M is below G) unstable equilibrium

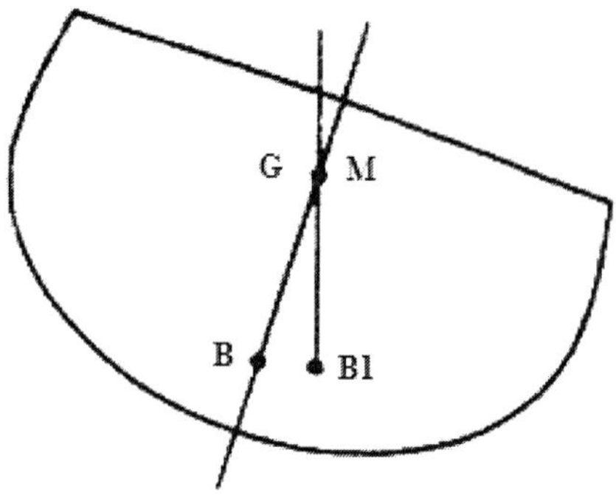

Figure 3.8. Illustration of Neutral Equilibrium of a Floating Body.

Submerged Bodies

For stable equilibrium of submerged bodies, the center of gravity should be below the center of buoyancy (B). In that case, a restoring couple is produced that tends the body to return to its equilibrium position (Figure 3.9 left).

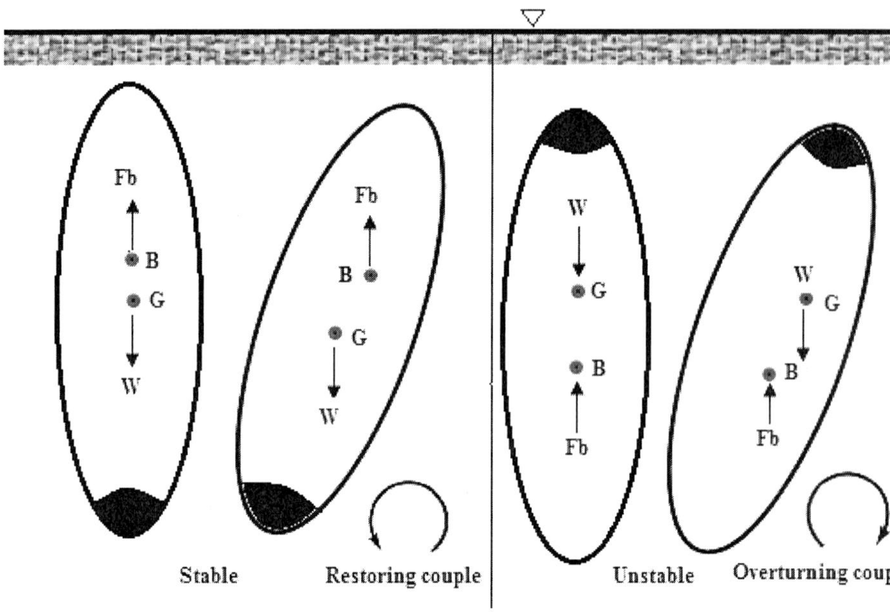

Figure 3.9. Stable and Unstable Equilibrium of Submerged Bodies.

Note: As the body is totally submerged, the shape of displaced fluid remains unaltered with tilting of the body. As such, the center of buoyancy relative to the body remains unchanged.

For a submerged body, the center of buoyancy (B) below the center of gravity (G) produces an over turning moment that makes the body in unstable equilibrium (Figure 3.9 right).

A submerged body is in neutral equilibrium if the center of buoyancy (B) coincides with center of gravity (G).

It is possible to change the center of gravity, G by adding weights at appropriate places (Figure 3.9). The addition of weight at the bottom end lowers G while adding weight at the top, raises G. The exact quantum of change in G depends upon the weight added. In ships, often there is a change in both the point of buoyancy and center of gravity with angular displacement, because there is some adjustment in weights at various places in the ship.

Metacentric Height

Consider an immersed rectangular body (Figure 3.10a) with its center of buoyancy at B, the center of gravity at G and the water line at P-N. Upon slight angular displacement of the body in the clockwise direction, a new water line P_1-N_1 emerges. The center of buoyancy now moves to B_1 (Figure 3.10b). Now an upthrust due to buoyancy acts at B_1 while the weight of the body (W) acts at G in the downward direction. A couple begins to act to restore the body to its original position. We need to find GM, the distance between the center of gravity, G and the metacentre, M also known as the metacentric height (Figure 3.10b). Metacentric

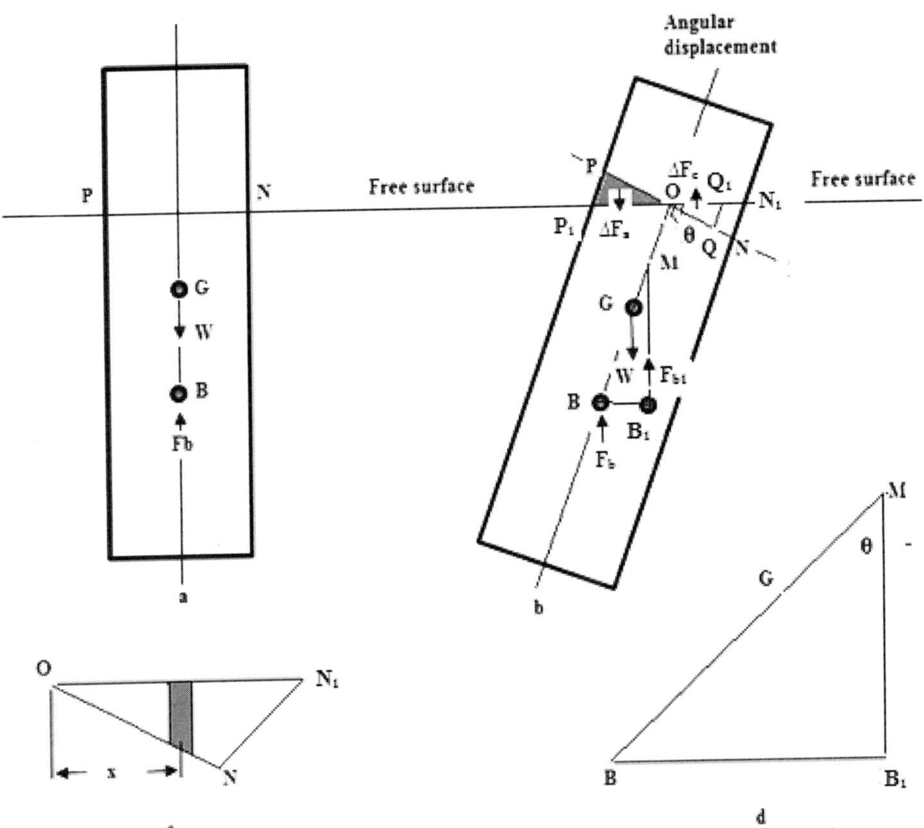

Figure 3.10. Schematic Diagram to Determine the Height of Metacenter.

height is a good measure of the body's initial stability. Either analytical techniques or experimental methods are applied to determine the metacentric height. Herein, an analytical technique is described to determine GM.

Analytical method

For a small angular displacement of the body in the clockwise direction, the wedge shaped prism on the right side of the axis (region NON_1) submerges into water. The identical wedge shaped prism on the left side of the axis (region POP_1) emerges out of water. The wedge on the right side shows the gain of buoyancy force on the right side represented by a vertical upward force ΔFc. Similarly, the wedge on the left side shows the loss of buoyancy force (which is equal and opposite of gained buoyancy force) represented by a downward force ΔF_a. The couple formed due to these two opposite forces tends to rotate the ship in counter clockwise direction.

Consider a small elementary prism QQ_1 on the right of the axis at a distance x (not shown) from point O. (join Q and Q_1 to form a triangle QOQ_1)

Height of the prism h, $QQ_1 = x \times \theta$ (distance from Q to Q_1 is h)

(Since $\tan\theta = h/x$. if θ is small then $\theta = h/x$)

If we take thickness of prism as dx and length of the floating body as L

Area of the prism A = Height × Thickness = $x \times \theta \times dx$

Volume of prism V = Area × Length = $x \times \theta \times L \times dx$

Weight of the prism W = Density × Gravitational force × Volume

$$W = \rho \times g \times x \times \theta \times L \times dx \quad (3.11)$$

Similarly, if we consider a small prism of thickness dx at a distance x from point O towards the left of the axis, the weight of the prism will be

$$W = \rho \times g \times x \times \theta \times L \times dx$$

These two weights are acting in the opposite direction and hence constitute a couple.

Moment of couple = weight of one prism W times perpendicular distance between these two weights (forces) = $[\rho \times g \times x \times \theta \times L \times dx] \times [x]$

This produces the couple (due to symmetrically located element on P_1N_1) = $2\rho g x^2 \theta L\, dx$

$$\text{Moment of the couple of whole wedge} = \int 2\rho g x^2 \theta L\, dx \quad (3.12)$$

In the above figure we can see L × dx is the elemental area on the water line. So $L \times dx = dA$.

Moment of the couple for the whole surface (See chapter 2)

$$M = \rho g \theta \int x^2 dA = \rho g \theta I_{yy} \quad (3.13)$$

(Here I_{yy} = moment of inertia of the sectional area of the plane of body at water surface about the axis Y – Y)

Moment of couple, M due to shifting of center of buoyancy from B to B_1

$M = F_b \times BB_1$ (join B and B_1 to form triangle MBB_1 Figure 3.10 d)

Note that due to symmetry of the situation, displaced volume remains unaltered and hence the buoyancy force $F_{b1} = F_b$.

$$M = F_b \times BM \times \theta \tag{3.14}$$

Since $\tan\theta = (BB_1/BM)$, if θ is small then $\tan\theta = \theta$

Since F_b = Weight of the object, W

$$M = W \times BM \times \theta \tag{3.15}$$

The above two couples are same. Equating eqs. (3.13) and (3.15), we get

$$W \times BM \times \theta = \rho\, g\, \theta\, I_{yy} \tag{3.14}$$

$$W \times BM = \rho\, g\, I_{yy}$$

$$BM = \rho\, g\, I_{yy}/W \tag{3.15}$$

W = Weight of the body = Weight of the fluid displaced by the body

$= \rho\, g \times$ volume of the fluid displaced by the body $= \rho\, g \times$ volume of the body submerged in water

$= \rho\, g \times V$

$$BM = (\rho\, g\, I_{yy})/(\rho\, g\, V) = I_{yy}/V \tag{3.16}$$

From triangle (Figure 3.10d)

$$GM = BM \pm BG \tag{3.17}$$

Use +ve sign if G is lower than B, and -ve sign if G is higher than B.

Steps to solve buoyancy problems

- ☆ From geometry of body and densities of fluid and body equate: Weight of displaced fluid = Total weight of body. This gives the depth of immersion of the body or the weight of the body, whichever is unknown.
- ☆ To assess stability, first find the location of the center of gravity G of the body.
- ☆ Then, find the location of the center of buoyancy B (centroid of displaced volume). For a regularly shaped body this will be at half the height of the immersed portion of the body.
- ☆ Calculate the distance GB.
- ☆ Calculate BM, using $BM = I_{yy}/V$. Note $I_{yy} = \pi D^4/64$ for a circular section body and $bd^3/12$ for a rectangular section body (D is diameter; b and d are the sides of the rectangle).
- ☆ Calculate metacentric height using eq. (3.17). If $GM > 0$ then body is stable. If $GM < 0$ then body is unstable.

Example 3.5

A hollow cylinder open at both ends has an internal diameter of 25 cm, a wall thickness of 5 cm, and a length of 80 cm. Given that the weight of the cylinder is 80 kgf, comment on the stability of the cylinder.

As per the statements, it is possible to construct the following diagram. Let h be the depth of submersion in water.

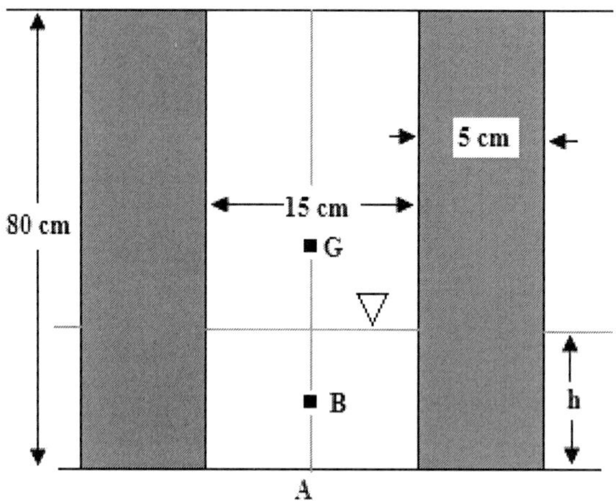

Assuming density of water to be 1000 kg/m^3, for equilibrium of the cylinder weight of the cylinder = weight of the displaced water

$80 = 1000\times\pi\ (0.25^2 - 0.15^2)\ h$

$h = 2/\pi = 0.636$

$AB = h/2 = 1/\pi = 0.318$ m

$AG = 0.8/2 = 0.4$ m

$BG = AG - AB = 0.4 - 0.318 = 0.082$ m

I_{yy} for a hollow cylinder is given as

$I_{yy} = \pi/4\ (do^4 - di^4)$ such that do and di are the outer and inner diameters of the hollow cylinder.

$I_{yy} = (\pi/4)\ (0.25^4 - 0.15^4) = 0.002673\ m^4$

$MG = (\rho\ g\ I_{yy}/W) - BG = (1000\times9.81\times0.002673)/(80\times9.81) - 0.082 = 0.033 - 0.082 = -0.049$ m

Therefore, MG < 0 implies cylinder is unstable

Example 3.6

A wooden block of size 1.2 m long, 1.0 m wide and 0.5 m deep floats on water. Find its metacentric height, if the density of the wooden block is 720 kg/m^3. Comment upon the stability of the wooden block.

Following diagram represents the problem.

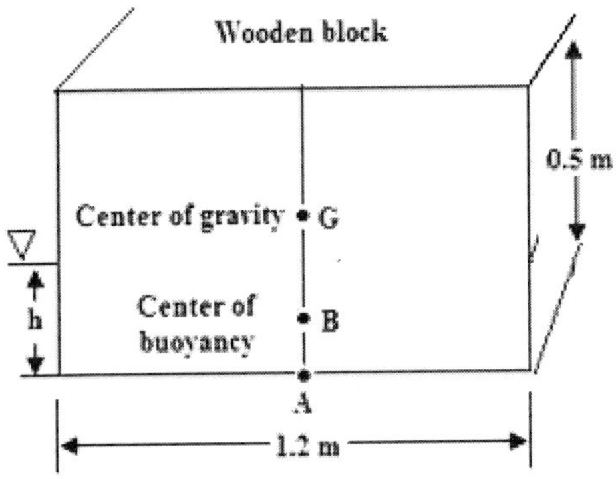

AG = 0.5/2 = 0.25 m

The volume of wooden block, V in the water is

$V \times 1000 \times 9.81 = (1.2\times 1.0 \times 0.5) \times 720 \times 9.81$

$V = 0.43\ m^3$

$h \times 1.0 \times 1.2 = 0.43$

$h = 0.36$

AB = h/2 = 0.18 m

BG = AG - AB = 0.25 - 0.18 = 0.07 m

$BM = I_{yy}/V$

Such that $I_{yy} = LB^3/12 = 1.2\ (1.0)^3/12 = 0.1$

BM = 0.1/0.43 =0.23

GM = BM – BG = 0.23 - 0.07 = 0.16 m

Since GM is positive, the wooden float is in stable equilibrium.

In addition, the moment caused by the displacement of center of buoyancy from B to B1 is also in the counter clockwise direction. Thus, both these couples must be equal.

QUESTIONS (THEORY)

1. Define buoyancy, list its characteristics and give two examples of this phenomenon in daily life.

2. State the Archimedes Law and describe it with the help of a neat diagram.

3. Discuss the state of stability for floating bodies.

4. Discuss the state of stability for submerged bodies.

5. Derive the equation to determine the metacentric height.

Fundamentals of Fluid Flow

Chapter 4

Fluid Kinematics and Visualization

WHAT IS FLUID KINEMATICS?

Fluid kinematics, one of the branches of fluid mechanics, deals with the motion of fluids without considering the forces and moments, which create the motion. Three main aspects of fluid kinematics are:

- ☆ Development of methods and techniques to describe and specify the motion of fluids
- ☆ Determination of the conditions for the kinematic possibility of fluid motion
- ☆ To characterize kinds of motion and associated deformation rates of fluid element

The fluid kinematic studies help to understand the fluid motion from kinematic point of view involving position, velocity and acceleration of the fluid. Following tools are used to visualize or describe the fluid motion.

- ☆ Material derivative and its relationship to Lagrangian and Eulerian descriptions of fluid flow
- ☆ Flow visualization
- ☆ Plots of flow data
- ☆ Fundamental kinematic properties of fluid motion and deformation
- ☆ Reynolds transport theorem.

This chapter discusses some of these issues in a detailed manner, while others are included in brief.

METHOD OF DESCRIBING FLUID MOTION

Lagrangian Description

The description, named after Italian mathematician Joseph Louis Lagrange (1736-1813), comprises marking the individual fluid particles, and track their positions, velocities, *etc.* as a function of time. Figure (4.1 left) shows the position and velocity vectors of particles A and B at an instant of time. As the particles move in the flow field, their positions and velocities change with time. The physical laws, such as Newton's laws and conservation of mass and energy apply to each particle. Understandably, if there were only a few particles, the Lagrangian description will be desirable. Since the fluid flow is a continuum phenomenon down to the molecular level, it may not be possible to track each particle and interaction between molecules in a complex flow field. Thus, Use of the Lagrangian description is rare in fluid mechanics. However, the approach is useful for specialized applications such as in sprays, bubble dynamics and rarefied gases.

Eulerian Description

The description named after Swiss mathematician Leonhard Euler (1707-1783) comprises defining a control volume instead of identifying the individual fluid particles (Figure 4.1 right). Field variables are described as functions of space and time. The kinematic behaviour or properties of fluid particle passing a given point in space (x, y, z) with time, t are recorded. For example, a pressure field as a function of space and time may be described as:

$$\text{Pressure field, } P = P(x, y, z, t) \tag{4.1}$$

Similarly, velocity and accelration fields are written as:

$$\text{Velocity field, } \bar{V} = \bar{V}(x, y, z, t) \tag{4.2}$$

$$\bar{V} = (u, v, w) = u(x, y, z, t)\bar{i} + v(x, y, z, t)\bar{j} + w(x, y, z, t)\bar{k} \tag{4.3}$$

$$\text{Acceleration field, } \bar{a} = \bar{a}(x, y, z, t) \tag{4.4}$$

$$\bar{a} = (a_x, a_y, a_z) = a_x(x, y, z, t)\bar{i} + a_y(x, y, z, t)\bar{j} + a_z(x, y, z, t)\bar{k} \tag{4.5}$$

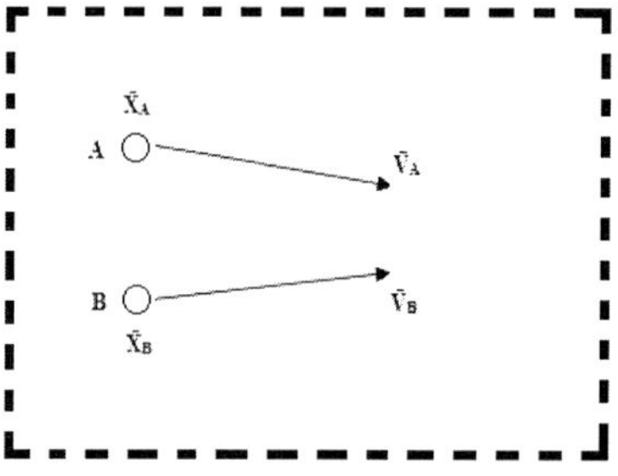

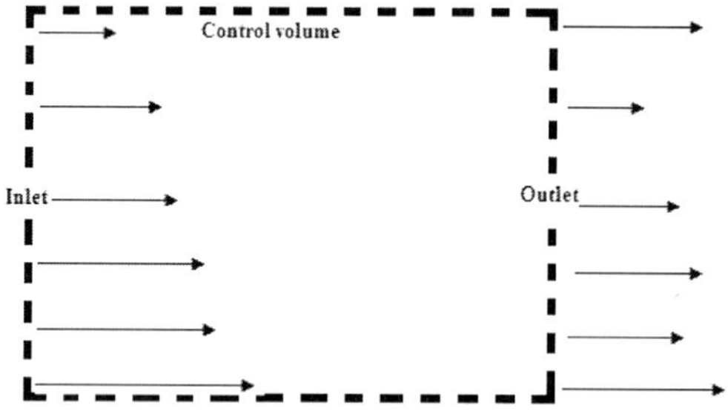

Figure 4.1. Description of Lagrangian (Left) and Eulerian (Right) Flow Fields.

These (and other) field variables define the flow field. Since fluid flow is a continuum phenomenon, people prefer to use the Eulerian description in fluid mechanics. Most physical laws such as Newton's laws and the laws of conservation of mass and energy apply directly to particles in a Lagrangian description, but some translation or reformulation of these laws is required for use with an Eulerian description. This description suits well to formulate initial boundary-value problems.

Coupled Eulerian-Lagrangian Description

It may not be easy to resolve fast transient problems involving fluid-solid interactions using any of the two approaches. In such cases, coupling of the Eulerian capability with traditional Lagrangian approach helps to model interactions between highly deformable materials and relatively stiff bodies. The need for such coupling arises when either specific solution methods or accuracy considerations necessitate that different and disjoint sub-domains should be treated with different (Eulerian or Lagrangian) schemes. For example; forensic analysis of Columbia accident was made through simulation of shuttle debris trajectory using Eulerian computational fluid dynamics (CFD) for flow field and Lagrangian method for the debris.

THE MATERIAL DERIVATIVE

There are many common names for the material derivative. They include total derivative, convective derivative, substantial derivative, substantive derivative, and others. The material derivative computes the time rate of change of any quantity such as discharge, temperature or velocity (which gives acceleration) for a portion of a material moving with a velocity, V. One can arrive at this by applying the chain rule to Eulerian quantities that are dependent on time and position. The material derivative of some quantity defines the rate of change of that quantity following a fluid particle. Let us consider a fluid flowing through a converging nozzle (Figure 4.2). Apparently, any particle of fluid will speed up as it flows along the path of decreasing cross-section. But if one only focuses on a single (x, y) point in space, he will not be able to notice this acceleration because the fluid velocity at that point appears constant. The problem with this interpretation is that one is not following the same particle to see it speeding up over time. The material derivative effectively corrects this confusing effect to give a true rate of change of a quantity. Moreover, material derivative also serves as a bridge between Lagrangian and Eulerian descriptions.

Let us consider the case of an arbitrary fluid property, Q.

$$DQ/Dt = dQ/dt = \partial Q/\partial t\,(dt/dt) + \partial Q/\partial x\,(dx/dt) + \partial Q/\partial y\,(dy/dt) + \partial Q/\partial z\,(dz/dt) \tag{4.6}$$

$$dQ/dt = \partial Q/\partial t\,(1) + \partial Q/\partial x\,(u) + \partial Q/\partial y\,(v) + \partial Q/\partial z\,(w) \tag{4.7}$$

$$dQ/dt = \partial Q/\partial t + u\,\partial Q/\partial x + v\,\partial Q/\partial y + w\,\partial Q/\partial z \tag{4.8}$$

Note that the notation, DQ/DT, used by some authors to emphasize that this is a material or total derivative, as opposed to some partial derivative, is identical to dQ/dt. In the above derivation, dt/dt = 1 by definition. Since a fluid particle is being followed, dx/dt = u, *i.e.* the x-component of the velocity of the fluid particle. Similarly, dy/dt = v, and dz/dt = w following a fluid particle. Thus, at some given spatial location (x,y,z) and at some given time (t), DQ/Dt = dQ/dt = the material derivative of Q, and is defined as the total rate of change of Q with respect to time

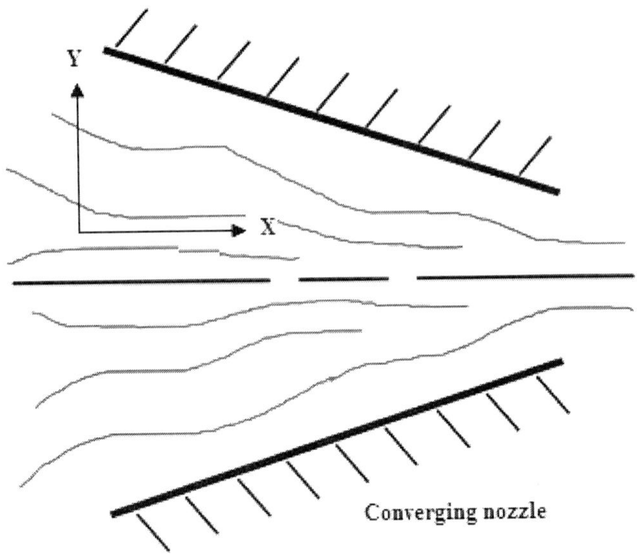

Figure 4.2. Flow Field in a Converging Nozzle.

as one follows whatever fluid particle happens to be at that location at that instant of time. Q changes for two reasons: First, if the flow is unsteady, Q changes directly with respect to time. It is the local or unsteady rate of change of Q. Second, Q changes as the fluid particle migrates or convects to a new location in the flow field. It is the convective or advective rate of change of Q.

Q in the above derivation can be any fluid property, scalar or vector. For example, Q can be a scalar like the pressure, in which case one gets the material derivative or substantial derivative of the pressure. Q can also be the velocity vector, such that one gets the material acceleration through the material derivative of the velocity *i.e.* the rate of change of velocity following a fluid particle. The material acceleration is:

$$d\bar{V}/dt = \bar{a} = \partial\bar{V}/\partial t + u\,\partial\bar{V}/\partial x + v\,\partial\bar{V}/\partial y + w\,\partial\bar{V}/\partial z \tag{4.9}$$

The term on the left hand side is the total acceleration. It represents the actual acceleration vector experienced by whatever fluid particle happens to reside at the location and time of interest. The first term on the right hand side is the local acceleration or the unsteady acceleration. It is only non-zero in an unsteady flow. The last three terms make up the convective acceleration, which is the acceleration due to convection or movement of the fluid particle to a different part of the flow field, where the velocity is different. The convective acceleration can be non-zero even in a steady flow as is shown with a physical and mathematical example in the following cases. In other words, even when the velocity field is not a function of time (*i.e.* a steady flow), a fluid particle is still accelerated from one location to another. Advective acceleration is nonlinear and is a source of many phenomena, being a primary challenge in solving fluid flow problems. It also explains transformation between Lagragian and Eulerian frames.

Case 1

Consider steady flow in a converging duct (Figure 4.3). In order to conserve mass, the flow must accelerate through the nozzle *i.e.* the u-component of velocity must increase as a fluid particle passes through the nozzle. Thus, the acceleration vector is non-zero, even though it is a steady-state flow.

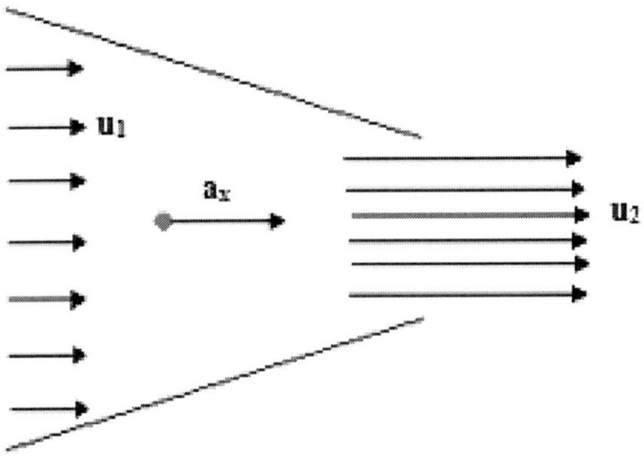

Figure 4.3. Flow Field in a Converging Duct.

As an illustration of the material acceleration, let us assume that 2-D velocity field for a steady state is:

$$\bar{V} = 3x\bar{i} - 3y\bar{j} \tag{4.10}$$

To find the acceleration field, plug this velocity into the acceleration equation:

$$d\bar{V}/dt = a = \partial\bar{V}/\partial t + u\,\partial\bar{V}/\partial x + v\,\partial\bar{V}/\partial y + w\,\partial\bar{V}/\partial z \tag{4.11}$$

$$d\bar{V}/dt = a = 0 + 3x\,(3\bar{i}) - 3y\,(-3\bar{j}) + 0 = 9x\bar{i} + 9y\bar{j} \tag{4.12}$$

Note that the acceleration is not zero. Thus, the acceleration vector can be non-zero, even for a steady-state flow.

Case 2

The description of the position, y, of a ball being thrown straight up into the air at any time, t, is given by the following relation.

$$y = y_0 + v_0 t - \tfrac{1}{2} g t^2 \tag{4.13}$$

Such that y is the position of the ball at any time, y_0 is the position at t=0, v_0 is the velocity at the initial state and g is the acceleration due to gravity (negative sign for ball being thrown up). Equation for the velocity of the ball is then:

$$dy/dt = v = v_0 - g\,t \tag{4.14}$$

The acceleration, a then is

$$a = dv/dt = -g \tag{4.15}$$

Thus, the acceleration is constant and equal to the acceleration of gravity.

If instead of position as a function of time, the velocity field is given as a function of position, we have $v^2 = v^2_0 + 2aS$, such that a is the acceleration and S the distance, which can be written as $y - y_o$. Then the velocity field is (a is equal to –g):

$$v_y = \pm\sqrt{[v^2_o - 2\,g\,(y - y_o)]} \tag{4.16}$$

Here positive root applies to the ball rising, and the negative to the ball falling. It is not readily apparent that the velocity equation, as a function of position, actually represents a constant acceleration equal to gravity. To see this, apply the material derivative. It is 1-D problem and so we have

$$a = dv_y/dt = \partial v_y/\partial t + v_y\,\partial v_y/\partial y \tag{4.17}$$

$\partial v_y/\partial t$ is zero because the velocity, as described here, is not transient. It does not change with time at any given location. We can obtain $\partial v_y/\partial y$ as

$$\partial v_y/\partial y = \mp/[\sqrt{v^2_0} - 2g\,(y - y_o)] \tag{4.18}$$

Multiplying v_y (eq. 4.16) and $\partial v_y/\partial y$ (eq. 4.18) gives

$$v_y\,(\partial v_y/\partial y) = [\pm\sqrt{[v^2_o - 2g\,(y - y_o)]}][\mp/\sqrt{[v^2_o - 2g(y - y_o)]}] \tag{4.19}$$

$$v_y\,\partial v_y/\partial y = -\,g \tag{4.20}$$

Since $v_y\,\partial v_y/\partial y$ is equal to a (eq. 4.17), we arrive at the same value derived before (eq. 4.15).

FLOW VISUALIZATION

Flow visualization is the visual examination of features of a flow-field. It is an art of making flow patterns visible. It is an important element of both physical experiments and numerical solutions. A number of methods are available for flow visualization. Some of these are:

- ☆ Streamlines and stream tubes
- ☆ Path lines
- ☆ Streak lines
- ☆ Timelines
- ☆ Refractive techniques
- ☆ Surface flow techniques

Streamline

Streamlines, the geometrical representation of the flow velocity, are drawn to visualize the flow field in fluid mechanics. A streamline is an imaginary line that is tangential to the instantaneous velocity direction (velocity is a vector having the magnitude and direction, Figure 4.4). In other words, a streamline at any instant is an imaginary curve or line in the flow field so that the tangent to the curve at any point represents the direction of the instantaneous velocity at that point. Consider a simple shear flow between parallel plates (Figure 4.5). At some instant of time, one can draw a streamline by connecting the velocity vector lines such that the streamline is everywhere parallel to the local velocity vector. In this example, streamlines are simply horizontal lines. In a steady flow, the orientation or the pattern of streamlines is fixed. On the other hand, in an unsteady flow where the velocity vector changes with time, the pattern of streamlines also changes from one instant to another.

Consider an arc length, dr (Figure 4.4) such that

$d\bar{r} = dx\bar{i} + dy\bar{j} + dz\bar{w}$

Since $d\bar{r}$ must be parallel to the local velocity vector

$\bar{V} = u\bar{i} + v\bar{j} + w\bar{k}$

Geometric arguments result in the equation of a streamline in the 3D as:

$$dr/V = dx/V_x = dy/V_y = dz/V_z \tag{4.21}$$

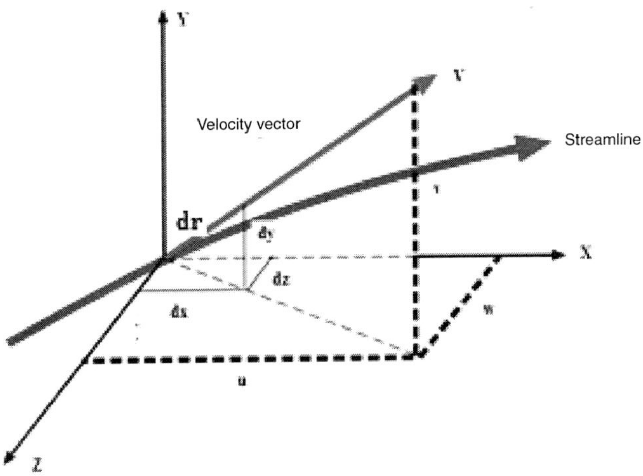

Figure 4.4. Definition Sketch of a Streamline.

From the above equation, writing u for V_x and v for V_y equation of a streamline in a 2-D flow field is

$$u\,dy - v\,dx = 0 \tag{4.22}$$

In a 2-D flow, lines of constant stream function are streamlines.

Stream tube

The Eulerian method gives a series of instantaneous streamlines of the state of motion. It is easy to imagine that a bundle of neighbouring streamlines form a passage through which the fluid flows (Figure 4.6). This passage is known as a stream-tube. Properties of stream tube are:

- ☆ The stream tube is bounded on all sides by streamlines
- ☆ Since fluid velocity does not exist across a streamline; no fluid may enter or leave a stream tube except through its ends
- ☆ The entire flow in the flow field is imagined comprising flows through stream tubes arranged in some arbitrary positions.

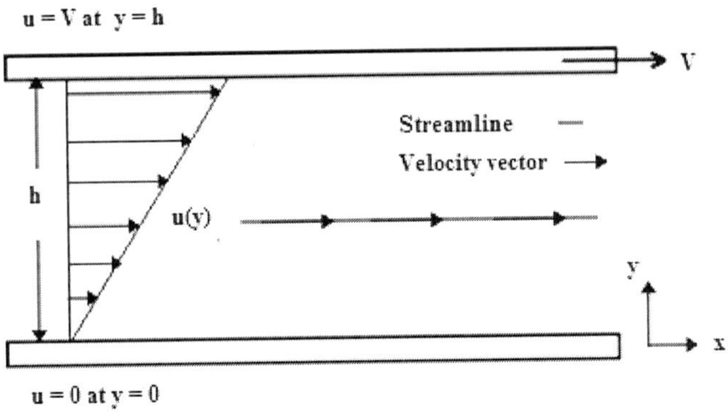

Figure 4.5. Streamline in a Flow Field between Two Parallel Plates where Upper Plate moves with Velocity V.

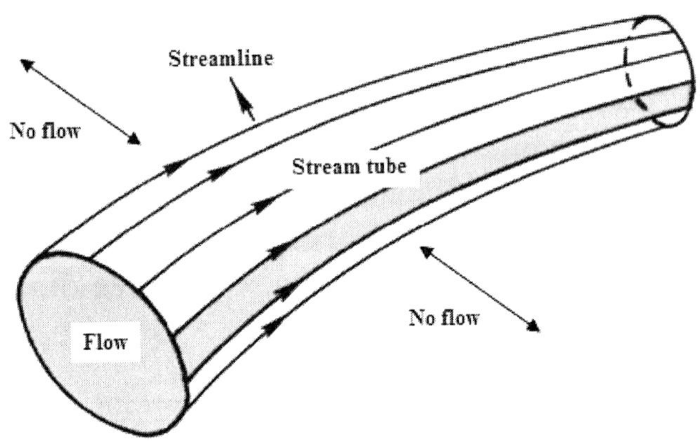

Figure 4.6. Illustration of a Stream Tube.

Streak Line

Streak line, a fluid-material line is the locus of all particles, which have earlier passed a specified (tagging) location in some interval of time. It represents multiple particles at one snapshot in time. Dye, smoke or hydrogen bubbles are commonly used to generate streak lines in experiments to visualize the flow field (Figure 4.7 top). The streak line in Figure (4.7 top) shows a turbulent flow. As time elapses, the streak line gets longer and longer, and represents an integrated history of the dye streak.

Path Line

Path line is the actual path traversed by a given (marked) fluid particle (Figure 4.7 bottom). The path line depends on the location where the particle is injected in the flow field and in case of unsteady flow field also at the time when it was injected. A path line represents an integrated history of where the fluid particle has been. In other words, a path line represents one particle in multiple snapshots in time. It

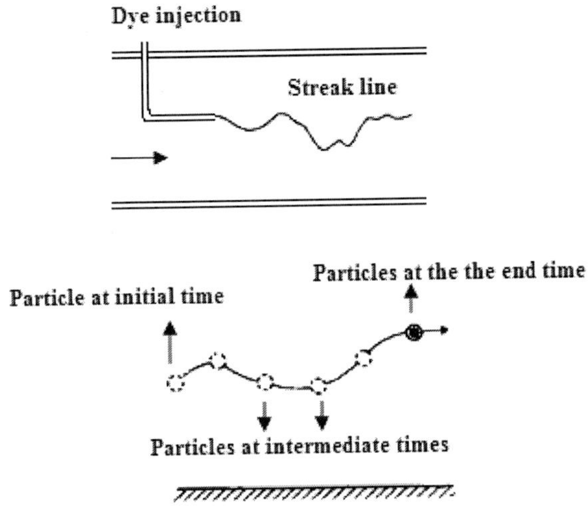

Figure 4.7. Illustration of a Streak Line (Top) and a Path Line (Bottom).

is the Lagrangian viewpoint of the flow field. A family of path lines represents the trajectories of different particles (Figure 4.8), say, P_1, P_2, P_3, *etc.* Path lines of two particles may intersect each other.

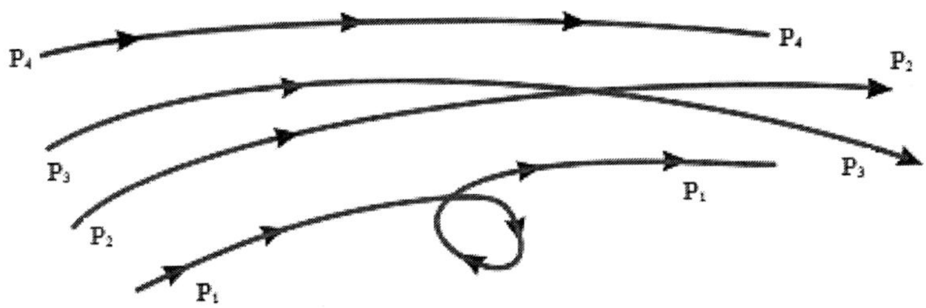

Figure 4.8 Trajectories of path lines of different particles

Experimentally, path lines can be visualized by tagging a fluid particle and tracing its motion throughout the flow field (Figure 4.7 bottom). In unsteady flows, path lines are difficult to follow and are not easy to create experimentally. While streak lines and path lines coincide in steady flows, they may be very different in unsteady flows.

For a known flow field, and specified initial location of the particle, the trajectory can be calculated by integrating the advection equation from t_o to t:

$$X = X_o + \int \overline{V}\, dt \tag{4.23}$$

Comment on path line, streamline and streak line

The characteristics of path line and streamline reported in Table 4.1 helps to understand the differences between the two.

Table 4.1 Characteristics of path lines and streamlines

Path Line	*Streamline*
It refers to a path followed by a fluid particle over a period of time	It is an imaginary curve in a flow field for a fixed instant of time, tangent to which gives the instantaneous velocity at that point
Two path lines can intersect each other, a single path line can form a loop as different particles, or even same particle can arrive at the same point at different instants of time	Two streamlines can never intersect each other, as the instantaneous velocity vector at any given point is unique

It is possible to draw a streak line from known path lines (Figure 4.9). Assume P be a fixed point in space through which particles of different identities pass at different times. In an unsteady flow, the velocity vector at P will change with time and hence the particles arriving at P at different times will traverse different paths like PAQ, PBR and PCS, which represent the path lines of the particle (Figure 4.9). Now, let us say that at an instant of time, these particles arrive at points Q, R and S. Thus, Q, R and S represent the ends of the trajectories of these three particles at that instant. The curve joining the points S, R, Q and the fixed point P will define the streak line at that instant. Even the fixed point P will also lie on the streak line, since at any instant; there will always be a particle of some identity at that point.

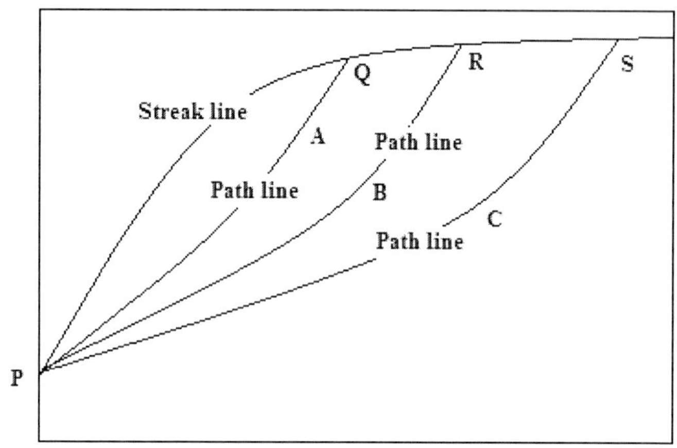

Figure 4.9. Path Lines and Streak Line at any Instant of Time.

For steady flow, streamlines, streak lines, and path lines are all identical. However, for unsteady flow, these three can be quite different. While it is easy to generate streamlines analytically, experimental techniques are more appropriate to obtain path lines and streak lines.

Time Line

It is a set of fluid particles that form a line segment at a given instant of time. It is a line drawn through adjacent particles in flow at any instant of time. For example, consider simple shear flow between parallel plates. A timeline follows the location of a line of fluid particles. A time line represents an integrated flow pattern, as it continually distorts with time (Figure 4.10). The top of the time line moves with the top plate, *i.e.* at velocity V to the right. The bottom of the timeline, however, stays at the same location at all times, because the bottom plate is not moving.

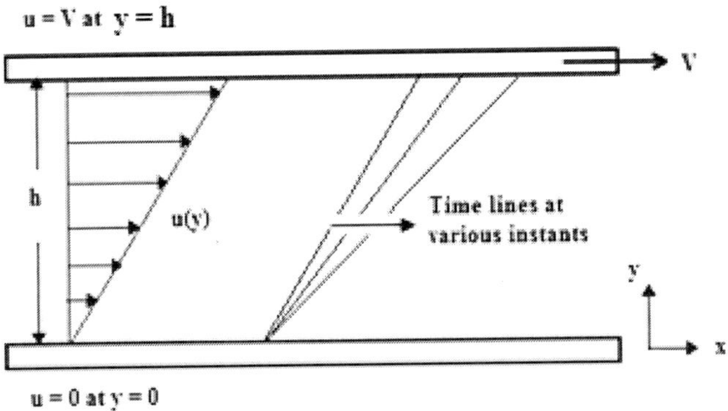

Figure 4.10. Time Lines in a Flow Field between Two Parallel Plates with Moving Upper Plate.

SURFACE FLOW AND REFRACTIVE TECHNIQUES

In experimental fluid dynamics, there are several methods to visualize the flow (Macagno, 1969). Some of the commonly used methods are discussed in the following sections.

Surface Flow Visualization

This reveals the flow streamlines in the limit of approaching a solid surface. Coloured oil applied to the surface of a wind tunnel model is one example where the oil responds to the surface shear stress to form a pattern.

Particle Tracer Methods

Particles, such as smoke or microspheres, added to a flow helps to trace the fluid motion. Illumination of the particles with a sheet of laser light helps to visualize a slice of a complicated fluid flow pattern. Assuming that the particles faithfully follow the streamlines of the flow, we can also measure its velocity using the particle image velocimetry or particle tracking besides visualizing the flow. Particles with densities matching the fluid flow exhibit the most accurate visualization.

Optical Methods

Some flow patterns can be visualized by way of changes in their optical refractive index. In this case, optical methods such as shadowgraph, schlieren photography, and interferometry help to visualize the flows. More directly, dyes may be added to flows and its concentrations measured to reveal the flow pattern. Light attenuation or laser-induced fluorescence techniques are used to measure the concentrations.

Analytical Methods

In the analytical methods, one analyzes a given flow to derive properties like streamlines, streak lines, and path lines. The flow is either given a finite representation or treated as a smooth function.

Texture Advection Methods

Texture advection methods bend the textures or images according to the flow. As the image is always finite (the flow could be given as a smooth function), these methods visualize the real flow only approximately.

Plots of Data

Various kinds of plots are also used to visualize the flow patterns. The three commonly used plots are:

- A profile plot indicates how the value of a scalar property varies along some desired direction in the flow field. It is possible to examine the relative behaviour of all variables in a multivariate data set.
- A vector plot is an array of arrows indicating the magnitude and direction of a vector property at an instant in time. Although, streamlines indicate the direction of instantaneous velocity field, but does not directly indicate the magnitude of velocity. Thus, a useful flow pattern for both experimental and computational fluid flow is the 'vector plot' that indicates both the magnitude and direction of instantaneous vector property.
- A contour plot is a two-dimensional plot of a three-dimensional surface showing lines where the surface intersects planes of constant elevation. Thus, they are curves with constant values of scalar property (or magnitude of vector property) at an instant of time. The contours filled with different colours or shades represent the magnitude of the property.

KINEMATIC DESCRIPTION IN FLUID MECHANICS

An element may undergo four fundamental types of motion (Figure 4.11).

- Translation
- Rotation
- Linear strain
- Shear strain

The translation is the movement of the fluid particle from one point to another in space. Rotation of the fluid particle can occur in any of the orthogonal axis. In the case of linear strain, the particle's side can stretch or shrink. If the angle between the sides of the particle changes then it is shear strain. The motion and deformation during fluid motion is best described in terms of rates *i.e.* velocity: rate of translation, angular velocity: rate of rotation, linear strain: rate of linear strain and shear strain: rate of shear strain (Figure 4.11). These rates expressed in terms of velocity and derivatives of velocity provide useful information. We refrain to discuss these issues being out of scope of this book.

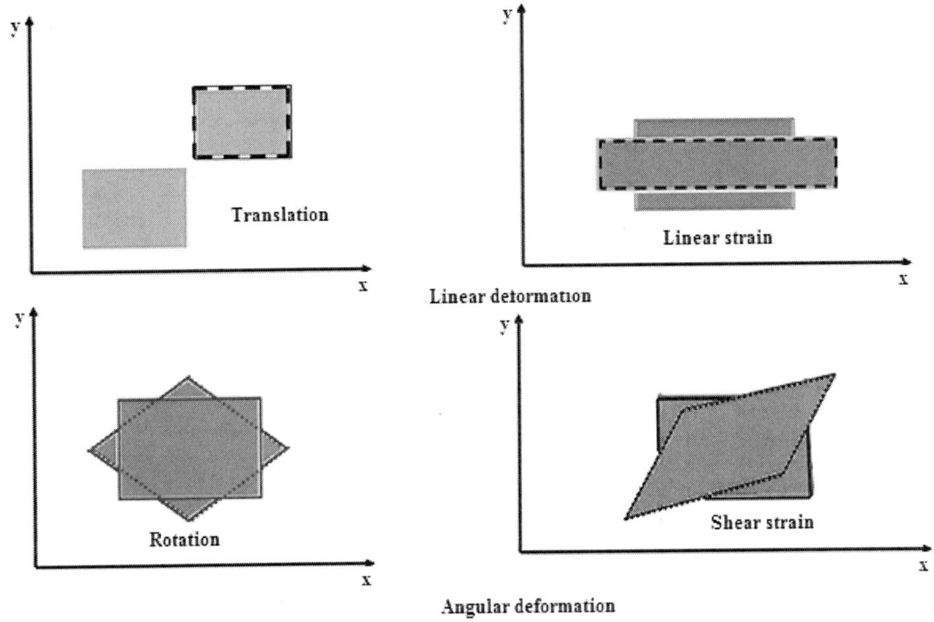

Figure 4.11. Basic Deformation Characteristics of Fluid Mass.

REYNOLDS TRANSPORT THEOREM (RTT)

A system is a quantity of matter of fixed identity. No mass can cross a system boundary. On the other hand, a control volume is a region in space chosen for the study. Mass can always cross a control surface. The fundamental conservation laws (conservation of mass, energy, and momentum) apply directly to systems. However, in most fluid mechanics problems, control volume analysis is preferred over system analysis (for the same reason that the Eulerian description is preferred over the Lagrangian description). Therefore, we need to transform the conservation laws from a system to a control volume. The Reynolds transport theorem (RTT), helps to relate system concepts to control volume concepts. It provides a relationship between the time derivative of system property and the rate of change of that property within a control volume. For example; for the one-dimensional fixed control volume, equation states that the time rate of change of any extensive property for a system is equal to the rate of change of that extensive property inside the control volume plus the net efflux of the property through the control surface. This is known as Reynolds Transport Equation, which relates the change of a property of a system to the change of that property for a control volume. This and other relationships establish a link between the system and control volume approaches. The derivation and discussion on these relations is beyond the scope of this book.

TYPES OF FLOW

In fluid mechanics, the knowledge of flow behaviour is important as the analysis and calculations depend on the flow conditions. We discuss some of the recognized flows in the following paragraphs (Figure 4.12).

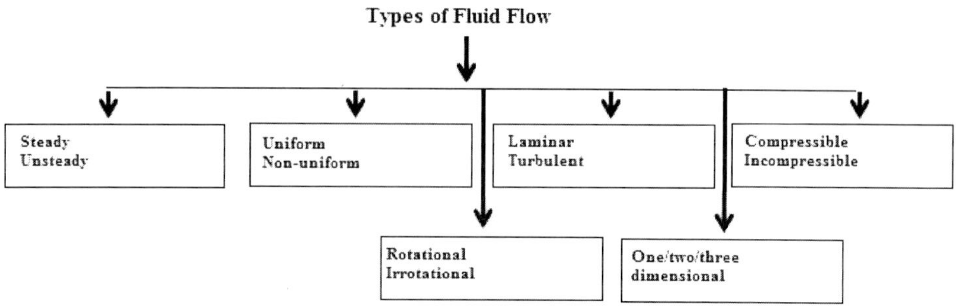

Figure 4.12. Kinds of Flow.

Steady and Unsteady Flow

A steady flow is one in which the conditions (velocity, density, pressure, acceleration and cross-section) may differ from point to point but do not change with time. Mathematically

$$\partial v/\partial t = 0;\ \partial \rho/\partial t = 0;\ \partial P/\partial t = 0 \text{ and } \partial a/\partial t = 0 \qquad (4.24)$$

Here v is the velocity, P is the pressure, ρ is the density, a is the acceleration and t is the time. In practice, there is always slight variation in velocity and pressure, but the flow is steady if the average values are constant.

In unsteady flow, fluid parameters such as velocity, density, pressure, acceleration *etc.* at a point change with time.

$$\partial v/\partial t \neq 0;\ \partial \rho/\partial t \neq 0;\ \partial P/\partial t \neq 0 \text{ and } \partial a/\partial t \neq 0 \qquad (4.25)$$

Uniform and Non-uniform Flow

If the velocity at a given instant of time is same in both magnitude and direction at all points in the flow, the flow is said to be uniform flow.

$$\partial v/\partial s = 0;\ \partial \rho/\partial s = 0;\ \partial P/\partial s = 0 \text{ and } \partial a/\partial s = 0$$

When the velocity changes from point to point in a flow at any given instant of time, the flow is termed as non-uniform flow.

$$\partial v/\partial s \neq 0;\ \partial \rho/\partial s \neq 0;\ \partial P/\partial s \neq 0 \text{ and } \partial a/\partial s \neq 0$$

Going by this definition, every fluid that flows near a solid boundary will be non-uniform - as the fluid at the boundary must take the speed of the boundary, usually zero. However, if the size and shape of the cross-section of the stream of fluid is constant then the flow may be treated as uniform.

Based on these kinds of flow, other flow types are as follows.

Steady Uniform and Steady Non-uniform Flow

In steady uniform flow, conditions do not change with position in the stream or with time. For example, the flow of water in a pipe of constant diameter at constant velocity is steady uniform flow.

In a steady non-uniform flow, conditions change from point to point in the stream but do not change with time. For example, flow in a tapering pipe with constant velocity at the inlet. Velocity will change as one move along the length of the pipe toward the exit.

Unsteady Uniform and Unsteady Non-uniform Flow

At a given instant in time the conditions at every point are the same, but will change with time. For example, a pipe of constant diameter connected to a pump pumping at a constant rate, but switched off later.

In the case of unsteady non-uniform flow, conditions of the flow may change from point to point and with time at every point. The waves in a channel are an example.

Compressible and Incompressible Flow

The flow in which density of the fluid varies during the flow is compressible fluid flow ($\rho \neq$ constant).

In incompressible fluid flow, the density of the fluid remains constant during the flow (ρ = constant). Practically, all liquids are incompressible.

Pressurized and Gravity Flow

In pressurized flows, fluid flows under pressure *e.g.* liquid flowing in pipes under pressure. On the other hand, if the fluid flows under gravity as in rivers and canals, it is known as gravity flow.

One, Two and Three Dimensional Flows

In general, all fluids flow three-dimensionally, with pressures, velocities, and other flow properties varying in all directions. In many cases, the greatest changes occur only in two or even only in one direction. In these cases, analysis is much simpler if one ignores the changes in the other directions.

One-dimensional flow

Flow is one dimensional if the flow parameters (such as velocity, pressure, depth *etc.*) at a given instant in time only vary in the direction of flow and not across the cross-section. The single space coordinate is usually the distance measured along the center line (not necessarily straight) in which the fluid is flowing. For example, the flow in a pipe is one-dimensional when variations of pressure and velocity occur along the length of the pipe. Any variation across the cross-section is negligible. The flow is unsteady, if the parameters vary in time but not across the cross-section. It is to note that in reality, flow is never one-dimensional because viscosity causes the velocity to decrease to zero at the solid boundaries. Nonetheless, valuable results may emerge even from one-dimensional analysis, since the non-uniformity of the actual flow is not too great.

Two-dimensional flow

Flow is two-dimensional if the flow parameters are functions of time and vary in the direction of flow and in one direction right angle to this direction. No variation occurs in z direction. In other words, streamline patterns are same in all planes perpendicular to z direction at any instant.

Three-dimensional flow

The hydrodynamic parameters are functions of three space coordinates and time.

Rotational and Irrotational Flows

Rotational flow: The fluid particles while flowing also rotate about their own axis.

Irrotational flow: The fluid particles while flowing do not rotate about their own axis.

Laminar Flow

In laminar flow, the motion of the fluid particles is very orderly with all particles moving along well-defined paths or streamlines. Their paths do not cross each other (Figure 4.13 left).The fluid layers move smoothly over the adjacent layers. The value of Reynolds Number (Re) determines the type of flow. The characteristics of laminar flow in pipes are as follows.

- ☆ Fluid particles move in straight lines
- ☆ If dye is added, it does not mix with water
- ☆ Low velocity
- ☆ Re < 2000 (The limits is a soft number and varies from 2000-2300 being dependent on shape of pipe, roughness, and shape of pipe inlet)
- ☆ Mathematical analysis is simple
- ☆ Rare in practice in water systems

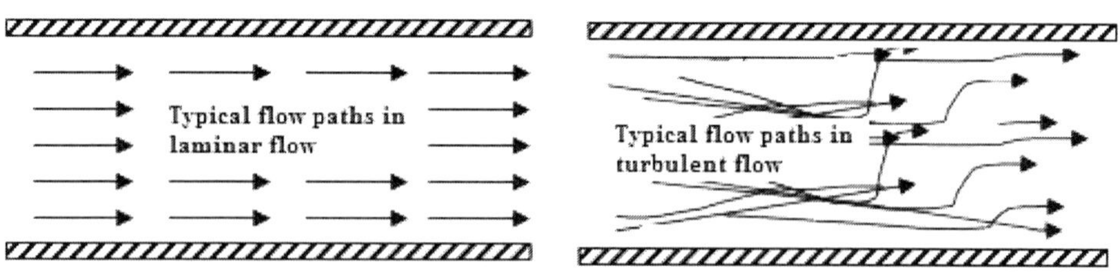

Figure 4.13. Distinctive Character of Laminar Flow (Left) and Turbulent Flow (Right).

Turbulent Flow

In turbulent flow, fluid particles move in a random and zigzag manner (Figure 4.13 right). Each particle has fluctuating velocity components superimposed on its mean velocity both transverse to and in the direction of the net flow. Turbulence is characterized by the formation of eddies. The characteristics of turbulent pipe flow are:

- ☆ Particle paths are completely irregular
- ☆ Average motion is in the direction of the flow
- ☆ Dye mixes rapidly and completely
- ☆ High velocity
- ☆ Re > 4000
- ☆ Cannot be observed by the naked eye
- ☆ Mathematical analysis is difficult

Transition Flow

In such a case, the flow is very unpredictable and often changes back and forth between laminar and turbulent states. The process where a laminar flow transits to turbulent flow is laminar–turbulent transition. In the reverse case where turbulent flow transits to laminar flow, the flow is turbulent-laminar transition. Modern experimentation has demonstrated that this type of flow may comprise short bursts of turbulence embedded in a laminar flow. The characteristics of transitional flow are:

- ☆ Dye stream wavers in water - mixes slightly
- ☆ Medium velocity
- ☆ 2000 > Re < 4000 (Lower value may vary from 2000-2300)
- ☆ Changes/fluctuations are very difficult to detect.

Reynolds Number (Re)

Since a reference to Reynolds number appears in the preceding section, this section just introduces this number here although detailed discussion will follow in a subsequent chapter. According to the definition of the Reynolds number, it is the ratio of inertia forces of the flowing fluid to the viscous forces of the fluid. In pipe flow, following equation calculates the Reynolds number.

$$Re = \rho V D/\mu \tag{4.26}$$

Here, Re is the Reynolds number, ρ is the density of the fluid, V is the velocity of fluid, D is the diameter of pipe and μ is the dynamic viscosity of fluid.

Example 4.1

A pipe of 80 mm internal diameter carries oil of density 825 kg/m^3 at a rate of 10 kg/s. The dynamic viscosity is 0.025 $N\ s/m^2$. Show that the flow is turbulent in nature.

Velocity of flow = $Q/A = (10/825)/[\pi (0.04)^2] = 2.4$ m/s

Re = $\rho V D/\mu = 825 \times 2.4 \times 0.08/0.025 = 6336$

Since Re is more than 4000, flow is turbulent

STREAM FUNCTION

Stream function, derived by the French mathematician Joseph Louis Lagrange in 1781, is a very useful tool to study fluid dynamics. The concept is directed to solve problems associated with incompressible fluid flows although a slightly modified version can also be used for steady state compressible fluid flow problems. A stream function (with regard to 2-dimensional incompressible fluid flows or in three dimensions with axi-symmetry) is an arbitrary function whose derivatives give velocity components of a particular flow situation. Mathematically, it is defined in such a way that lines of constant stream function are streamlines (hence the name stream function). The stream function defines a function $\Psi(x, y, t)$, which satisfied the two dimensional continuity equation for a incompressible steady flow.

$$\partial u/\partial x + \partial v/\partial y = 0 \qquad (4.27)$$

The idea of a stream function is to eliminate two velocity components u and v into a single variable. Besides, the stream function Ψ (x, y, t) relates to the velocity components in such a way that continuity equation (4.27) is satisfied. In that case

$$u = \partial \Psi/\partial y \text{ and } v = -\partial \Psi/\partial x \qquad (4.28)$$

Substituting these into eq. (4.27), we have

$$\partial/\partial x(\partial\Psi/\partial y) + \partial/\partial y\,(-\partial\Psi/\partial x) = 0 \qquad (4.29)$$

Since $\Psi(x,y)$ is a continuous function of x and y, we have

$$\partial^2\Psi/\partial x\partial y = \partial^2\Psi/\partial y\partial x \qquad (4.30)$$

Thus, the continuity equation is satisfied. Clearly, if we can find a stream function, Ψ (x, y, t), then the continuity equation need not be solved.

By definition, a streamline in 2-dimensional plane (x,y) is written as

$$u\,dy - v\,dx = 0 \qquad (4.31)$$

or

$$dy/dx = v/u$$

Stream function has the following two main characteristics.

- ☆ Ψ = C, a constant is a streamline
- ☆ $d\Psi$ between two streamlines is proportional to the volumetric flow

To prove the first characteristic, let us consider a line given by Ψ =C, a constant as shown in (Figure 4.14). We have

$d\Psi = 0$

Ψ = Constant

Such that

$$(\partial\Psi/\partial y)\,dy + (\partial\Psi/\partial x)\,dx = 0$$

With the definition of $\Psi(x, y)$ defined before (eq. 4.28)

$$udy - vdx = 0 \tag{4.33}$$

Going back to eq. (4.31), it is the equation of a streamline. Thus, it proves that Ψ =C, a constant is a streamline of the flow. This property helps to visualize the streamlines readily (Figure 4.14). If we can determine a stream function $\Psi(x, y)$ in the flow, we can sketch the streamlines by setting a different constant each time for Ψ and plotting the Ψ as shown in Figure (4.15). Here $\Psi1$, $\Psi2$, $\Psi3$ and $\Psi4$ are different numerical constants (*e.g.* 1, 2, 3, 4 *etc.*...). Thus, it is possible to draw infinite number of streamlines, which are always parallel to each other. A family of such streamlines is useful in visualizing the flow patterns.

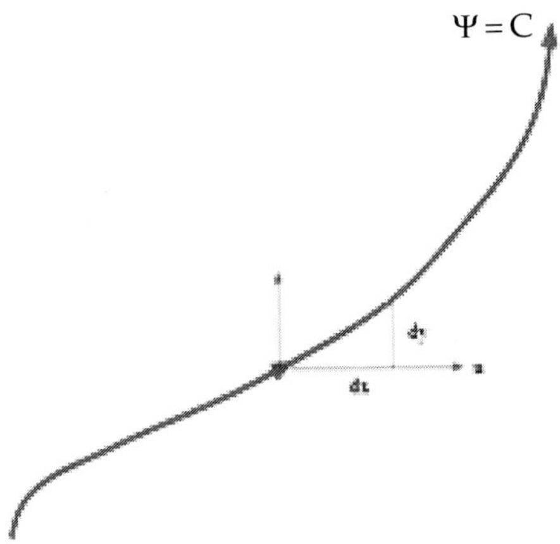

Figure 4.14. A Streamline in a Fluid Flow.

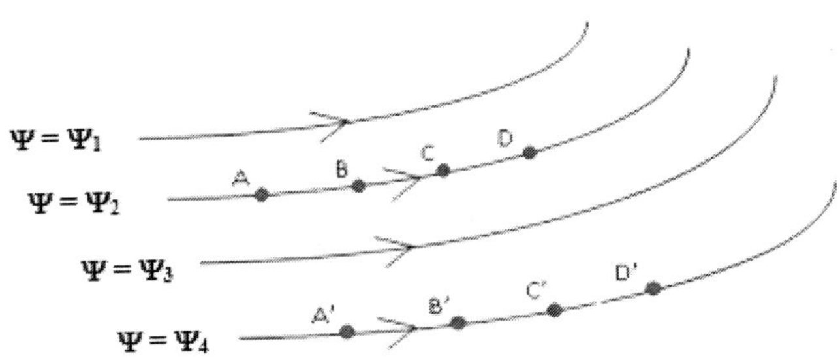

Figure 4.15. Plot of Streamlines for known Stream Function.

To prove the other useful characteristic of stream function, we determine the volumetric flow rate per unit depth between the two streamlines. Consider the volumetric flow through a small element of thickness ds placed on a streamline (Figure 4.16). The volumetric flow through the element is then

$$dQ = udy - vdx = (d\Psi/dy)\, dx - (-d\Psi/dx)\, dy = d\Psi \tag{4.34}$$

If we now integrate eq. (4.34) between two stream lines $\Psi_1 = C$ and $\Psi_2 = D$, we have,

$$Q = \int_1^2 dQ = \int_{\psi 1}^{\psi 2} d\Psi = \Psi_2 - \Psi_1 \tag{4.35}$$

It indicates that the volumetric flow rate is proportional to the difference between stream functions. If the upper streamline Ψ_2 has a value greater than the lower Ψ_1, then the volume flow rate is positive *i.e.* flow takes place from left to right.

To summarize, if one is able to formulate the function $\Psi(x,y)$ for a 2-D incompressible flow, then it is possible to visualize the flow (by sketching streamlines) as well as to determine the volumetric flow rate. Note that a streamline exists in 3-D flow but a corresponding stream function does not exist.

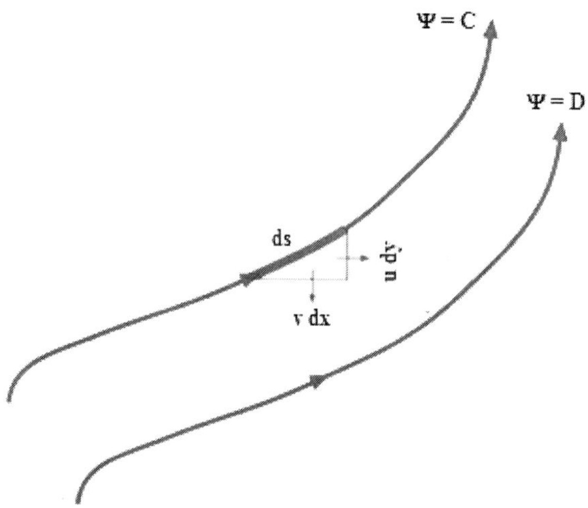

Figure 4.16. Schematic Depiction of Flow between Two Streamlines.

FLOW NETS

The Laplace equation for two-dimensional steady-state groundwater flow is written as:

$$\partial^2 h/\partial x + \partial^2 h/\partial y = 0 \tag{4.36}$$

Here h is the head and x and y are the two coordinates. To solve this equation for practical field problems, appropriate boundary conditions are essentially required. In some simple cases where boundaries are straight forward, it is possible to derive closed form solutions of the equation. For other cases, one needs to go for numerical modelling. For most geotechnical problems having complex boundaries, approximate graphical methods also provide good results. One such technique is based on a graphical representation of the Laplace equation *i.e.* the flow net technique. It is a quasi-empirical graphical method of solving the Laplace equation. Main uses of flow nets are:

- ✰ Developing a sense of how water movement takes place in porous media
- ✰ Determining the effects of seepage on engineering works and natural slopes
- ✰ Evaluating the effects of alternate seepage control measures

It is possible to make reasonable estimates of discharge using even crude flow nets. However, to estimate seepage pressures and gradients, one requires more refined flow nets as these estimates depend on the manner in which pressures are distributed and dissipated. This section mainly deals with problems in homogeneous isotropic systems, although transformation techniques are available to solve problems in non-homogeneous and anisotropic mediums (Nelson and Sundberg, 1973).

Structure of a Flow Net

A flow net consists of two sets of lines, which are always orthogonal (perpendicular to each other). One of these is flow lines, which show the direction of flow, and the other equipotential lines (lines of constant head), which show the distribution of potential energy (Figure 4.17).

Flow Line: The line along which a water particle moves (flow path) through a permeable soil medium (Figure 4.17). Flow cannot occur across flow lines. Moreover, a flow line cannot intersect another flow line.

Flow Channel: Strip between two flow lines is a flow channel (Figure 4.17). The rate of flow in a flow channel is constant. Flow quantity between all pairs of adjacent flow lines (in all flow channels) is the same.

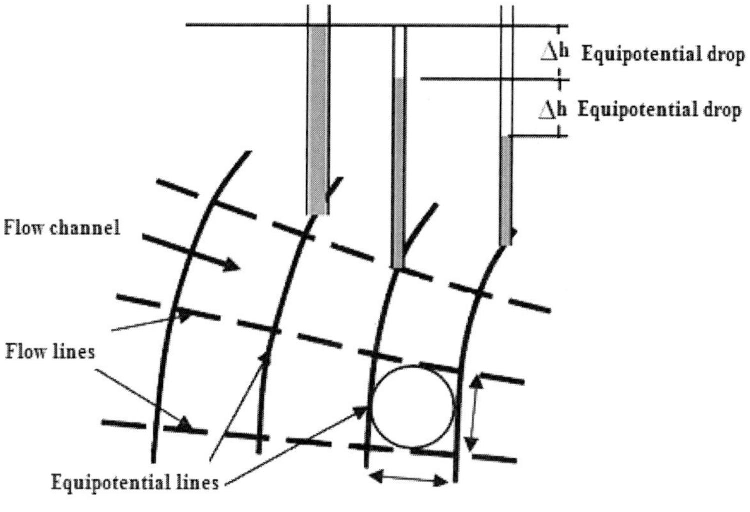

Figure 4.17. Salient Features of a Flow Net.

Equipotential Line: It is an imaginary line in a flow field such that the total head is the same for all points on the line. Therefore, the direction of flow is perpendicular to the line at all the points (Figure 4.17). An equipotential line cannot intersect another equipotential line. The difference in head between two equipotential lines is the potential head drop or head loss. Energy loss between all pairs of equipotential lines is the same.

Construction of a Flow Net

Following assumptions are made during construction of a flow net.

- ✰ Materials are homogeneous and isotropic
- ✰ Materials are saturated (all voids are filled with water)
- ✰ Material and water are incompressible (no volume change - steady state flow)
- ✰ Darcy' s law is valid

The most common approach to draw flow nets is the trial-and-error sketching. To speed up the construction of a flow net following stepwise procedure is helpful.

1. Draw cross sections of the system under consideration, water elevations, and soil deposit profiles on a convenient 2-D scale.
2. Establish boundary conditions *i.e.* identify impermeable and permeable boundaries. Indicate/label the positions of the water table, of any impermeable boundaries, of any points of known head or known pressure. For example; any surface of constant head (*e.g.* bottom of a flat-bottomed reservoir) is by definition an equipotential, and flow lines must meet it at right angles. Since flow cannot cross impermeable boundaries, the flow at such a boundary must be parallel to it, *i.e.* impermeable boundaries are flow lines, and equipotentials must meet them at right angles. As an illustration, f-g in Figure (4.18) is an impermeable boundary, a flow line. Similarly, a-c-d, the sheet pile is also an impermeable boundary. Equipotential lines must intersect these at right angles. Likewise, a-b and d-e in Figure (4.18) are equipotential lines. Flow lines must intersect them at right angles.
3. The water table, by definition, is the surface where P = 0. It is an equipotential only if it is horizontal. At any point on the water table (no matter whether it is flat or sloping) h = z, where z is the elevation of the water table above the datum. If there is no seepage percolating down to the water table, water table acts as a flow line. In a more general case (sloping water table, seepage across it), the water table is neither a flow line nor an equipotential. As such, flow lines will intersect it at an angle.
4. After defining the boundary conditions, draw one or two flow lines and equipotential lines near the boundaries as per rules described in the previous steps. Be sure that the flow lines and equipotential lines always intersect at right angles. Ensure that area between a pair of flow lines and

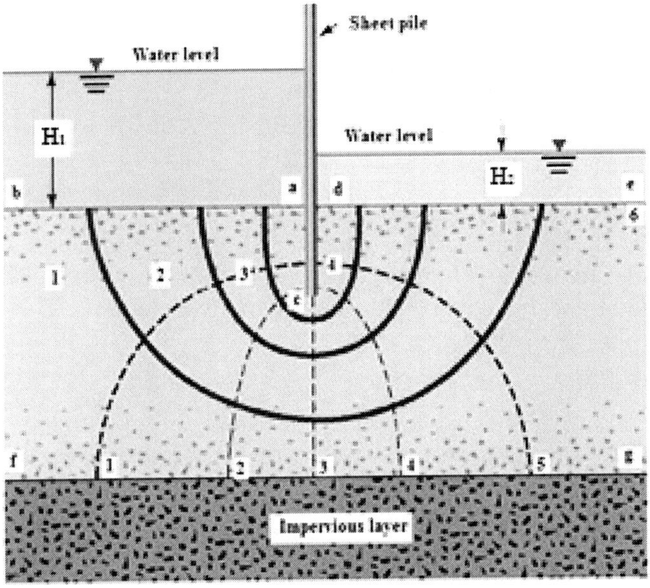

Figure 4.18. Schematic of a Flow Net for Flow around a Sheet Pile in a Homogenous Isotropic Media.

a pair of equipotential lines is approximately a curvilinear square grid (Figure 4.19, note the circle in the diagram). Where flow direction is a straight line, flow lines are equal distance apart and parallel. In addition, the flow net in confined areas between parallel boundaries usually consists of flow lines and equipotential lines that are symmetrical and elliptical in shape. Continue sketching and refining until having a good set of squares, which satisfy the boundary conditions. Try to avoid making sharp transition between straight and curved sections of flow and equipotential lines.

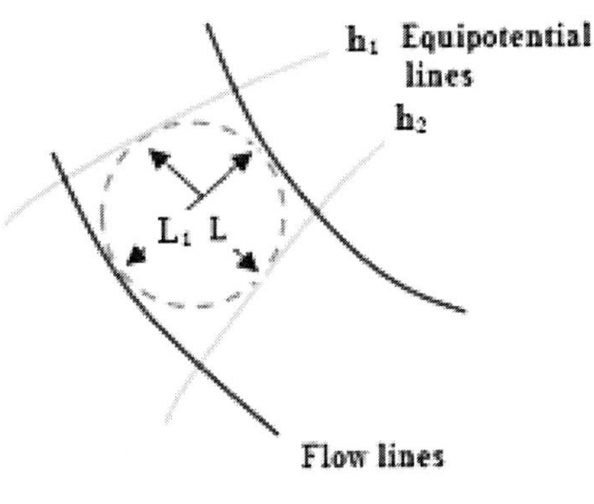

Figure 4.19. Schematic of a Curvilinear Square Grid.

5. Successive trials will result in a reasonably consistent flow net. In most cases, 3 to 8 flow lines are sufficient. Depending on the number of flow lines, the number of equipotential lines will automatically be fixed by geometry and grid layout.

Calculations

Head at any point

Determine the head at the left-most and right-most equipotentials and subtract them to get the total head difference across the net, H. Now determine N_d, the number of potential drops between these two equipotentials. The value of each potential drop, Δh is given as:

$$\Delta h = H/N_d \tag{4.37}$$

Knowing this, it is possible to label each equipotential with its correct value of h. Referring to Figure (4.18), the value of H is $(H_1 - H_2)$ and $N_d = 6$. Use the following equation to calculate the total head at any point P in the soil mass.

$$h = H - n \times \Delta h \tag{4.38}$$

Here h is the total head at any point P, H is the total head causing the flow, n is the number of equipotential drops up to point P from the first equipotential line (Depending upon the location of P, n may not be an integer as the point P may be located at some point within the square).

Seepage gradients

If one square within the flow net is considered, the loss of head due to flow through the square is equal to the change in head from one equipotential line to the next (Δh). The gradient at any point is the loss in head between equipotential lines (Δh) divided by the distance between equipotential lines (L) measured along the flow path, normally taken as midway between flow lines. Then discharge or seepage gradient (I_d) is:

$$I_d = \Delta h/L \tag{4.39}$$

The seepage gradient may change throughout the flow net because squares vary in size. The largest gradient occurs along the shortest flow path. The point of concern is usually where the gradient is the largest along the discharge face. This occurs mostly at the downstream toe of a dam. The shortest flow path discharges at this point. Known as exit gradient, it is the hydraulic gradient for the flow field on the downstream end of the hydraulic structure. To prevent piping failure, the exit gradient should be less than the critical hydraulic gradient. The quick conditions occur at a critical upward hydraulic gradient I_c, when the seepage force just balances the buoyant weight of an element of soil. The critical hydraulic gradient is typically around 1.0 for many soils.

Discharge/seepage

Again referring to (Figure 4.18), count the number of flow channels (N_f) and the number of equipotential drops (N_d). Calculate the shape factor (S_f) from the ratio of the number of flow channels to the number of equipotential drops (N_f/N_d). Then, compute rate of discharge as follows:

Apply Darcy's law to calculate the velocity in the tube as:

$$V = k\, I_d = k\, h/(N_d\, L) \tag{4.40}$$

Such that k is the hydraulic conductivity and L is the distance between the two equipotential lines. Let L_1 be the distance between the two flow lines. The flow passing per m width of soil is:

$$\Delta q = \text{Area} \times \text{velocity} = L_1.1.\ k.h/(N_d.\ L) \tag{4.41}$$

If we draw the flow net such that $L = L_1$ then

$$\Delta q = k\, h/N_d \tag{4.42}$$

The total flow will then be equal to

$$\text{Discharge (q)} = k\, h\, (N_f/N_d) = kh\, S_f \tag{4.43}$$

Eq. (4.43) is for the discharge per unit width. For a dam W m wide, multiply the value of q calculated with the help of eq. (4.43) with W to yield the total flow under the dam.

As a rule, construction of flow nets requires that all flow elements are approximately squares. However, it may not always be possible. The next best option is to construct flow nets with all the flow elements drawn as rectangles. In that case try to keep the length-to-width ratio of the flow nets as constant, such that

$$L_1/L = m$$

In that case, eq. (4.43) is written as:

$$\text{Discharge (q)} = k\, h\, (N_f/N_d)\, m = m\, k\, h\, S_f \qquad (4.44)$$

Seepage pressure

The viscous drag of water flowing through a soil imposes a seepage force on the soil in the direction of flow. The seepage force (J) acting on any square in the flow net is the product of the gradient across the square, the unit weight of water, and the volume of the square. The pore water pressures are partly supporting the weight of water $\gamma_w \times L \times L_1$ in the element per unit width and partly providing the seepage force. Then the seepage force, J is:

$$J = I_d\, \gamma_w\, L\, L_1 \tag{4.45}$$

The seepage pressure, P_s in the direction of flow is then given as:

$$P_s = I_d\, \gamma_w \tag{4.46}$$

QUESTIONS (THEORY)

1. Discuss the Langrangian and Eulerian description of fluid flow.
2. Discuss the streamlines, stream tubes, path lines, streak lines, time lines, refractive techniques and surface flow techniques in flow visualization.
3. Discuss the role of Reynolds transport theorem in studies related to fluid flow.

4. Differentiate between the following: steady and unsteady flow, uniform and non-uniform flow, laminar and turbulent flow, rotational and irrotational flow and one and three dimensional flow
5. What do you understand by a stream function? Discuss its applications.
6. What is a flow net? List the general guidelines for the construction of a flow net. Discuss its application in calculating seepage flow.

Chapter 5

Fluid Dynamics: Mass and Energy Conservation

Fluid dynamics, a sub-discipline of fluid mechanics, deals with the flow of fluids – gases and liquids. While the term used for flow of gases is aerodynamics for liquids the term is hydrodynamics. The solutions to fluid dynamics problems involve the calculation of various fluid properties such as pressure, density and flow velocity *etc.* as functions of space and time. As against this, we were only concerned with density and pressure for fluids at rest discussed in previous chapters. Foundation of fluid dynamics rests on conservation laws notably conservation of mass, conservation of momentum and conservation of energy *etc.* This chapter deals with the conservation of mass and energy given respectively by the following equations.

- ☆ Continuity equation
- ☆ Bernoulli's equation

We will look at the derivations and applications of these equations to describe steady state incompressible fluid flows.

CONTINUITY EQUATION

Continuity equation is one of the widely used equations in fluid mechanics. Derived based on principle of conservation of mass, it states that mass is neither created nor destroyed. Conservation of mass is basic to the control mass system (closed system). For a control volume, the principle of conservation of mass states that

Rate at which mass enters = Rate at which mass leaves the region +
Rate of accumulation of mass in the region (5.1)

The mathematical expression for the continuity equation is:

$\Delta m/\Delta t = 0$, where m = mass of the system

The above statements can be expressed analytically in terms of velocity and density field of a flow, the resulting expression being the continuity equation.

Let us consider a fluid flowing through a tapering pipe *i.e.* of a varying cross sectional area (Figure 5.1). Let the area on the left end of the pipe be A_1 and the area on the right end be A_2. Let us also say that the velocity of the fluid entering the pipe from the left is V_1 and the velocity of the fluid leaving the pipe from the right is V_2. For a compressible fluid, we can write

$$\text{Mass of fluid flowing per second at section XX} = \rho_1 A_1 V_1 \tag{5.2}$$

$$\text{Mass of fluid flowing per second at section YY} = \rho_2 A_2 V_2 \tag{5.3}$$

Here ρ_1 and ρ_2 are the fluid densities at section, XX and YY respectively. If there is no storage in the pipe, then applying the law of conservation of mass, (eq. 5.1), we have

$$\rho_1 A_1 V_1 = \rho_2 A_2 V_2 \tag{5.4}$$

For a incompressible fluid, $\rho_1 = \rho_2 = \rho$, we have

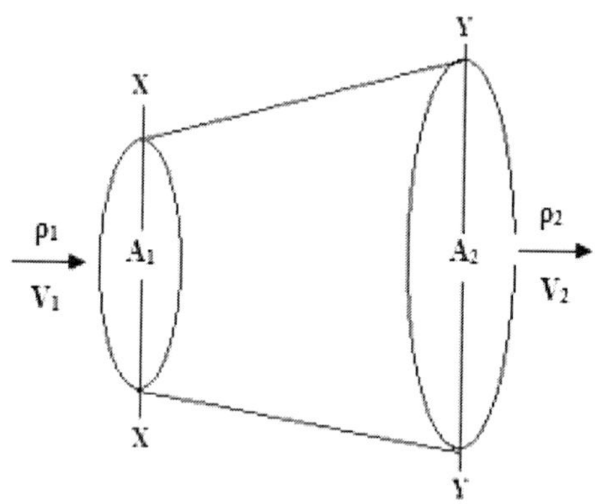

Figure 5.1. Schematics of Flow through a Tapering Pipe.

$$A_1 V_1 = A_2 V_2 \tag{5.5}$$

$$V_1/V_2 = A_2/A_1 = d_2^2/d_1^2 \tag{5.6}$$

Here d_2 and d_1 are the diameters of pipe at the inlet and the outlet.

For an incompressible fluid, A multiplied by V is the discharge. Thus, discharge, Q can be written as

$$Q = Q_1 = Q_2 = AV \tag{5.7}$$

Here A is the cross sectional area and V is the average velocity.

For a control volume with N inlets and N outlets, equation in summation form is:

$$\sum_{i=1}^{N} A_i V_i = 0 \tag{5.8}$$

In eq. (5.8), all inflows are positive and outflows negative. For compressible fluids, rate of flow is the mass of fluid flowing across a section per second.

$$\text{Mass flow rate (m)} = (\rho AV)\ \text{kg/s} \tag{5.9}$$

Since a large number of natural quantities such as mass, energy, momentum, and electrical charge follow the law of conservation under their respective appropriate conditions, continuity equation is capable of describing a variety of such physical phenomena. As such, continuity equations exist in the literature in various forms. Following section discusses the continuity equation for a rectangular Cartesian coordinate system in a detailed manner.

Continuity Equation - Differential Form

To derive the continuity equation, let us consider a point enclosed by an elementary control volume, a rectangular parallelepiped in a rectangular Cartesian frame of coordinate system (Figure 5.2). Let the dimensions of parallelepiped be dx, dy and dz. The inflow, outflow and the rate of accumulation of mass within the control volume is calculated across each surface. There are no sink or sources. Net outflow of mass along x -axis must be the excess outflow over inflow across faces normal to x -axis.

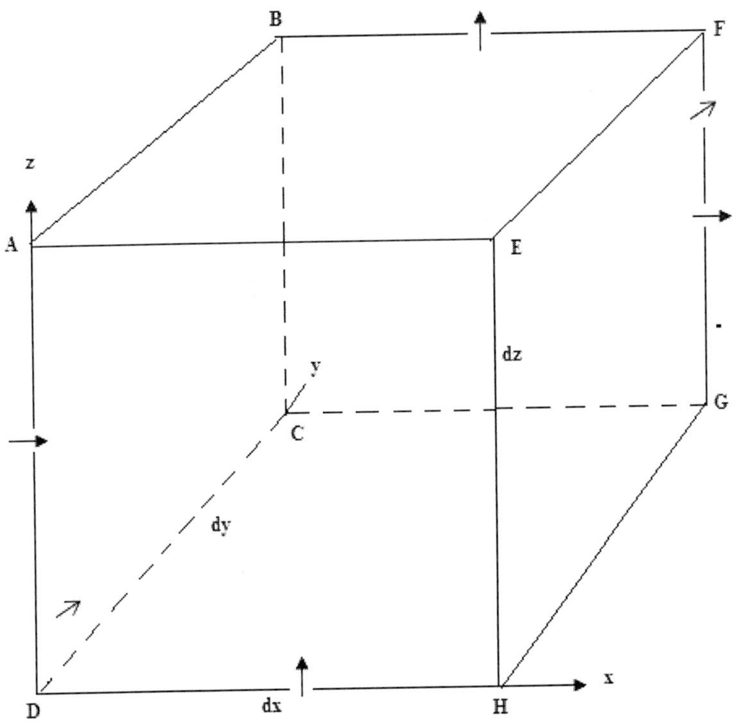

Figure 5.2. Control Volume for Deriving Continuity Equation in Rectangular Cartesian Coordinates.

Let the fluid enter across one of such faces ABCD with a velocity u and density ρ. The velocity and density with which the fluid will leave the face EFGH will be $u + (\partial u/\partial x)\,dx$ and $\rho + \partial\rho/\partial x)\,dx$ respectively.

Therefore, the rate of mass entering the control volume through face ABCD

$$\rho u\,dy\,dz \tag{5.10}$$

The rate of mass leaving the control volume through face EFGH will be $[u + (\partial u/\partial x)\,dx][\rho + (\partial\rho/\partial x)\,dx]\,dy\,dz$. After neglecting the higher order terms, it is equal to

$$[\rho u + (\partial/\partial x\,(\rho u)\,dx]\,dy\,dz \tag{5.11}$$

Similarly, influx and efflux take place in the y and z directions also. These are directly written as:

$$[\rho v + (\partial/\partial y\,(\rho v)\,dy]\,dx\,dz \tag{5.12}$$

$$[\rho w + (\partial/\partial z\,(\rho w)\,dy]\,dx\,dy \tag{5.13}$$

Here v and w are flow velocity components in y and z coordinates. Rate of accumulation for a point in a flow field is written as:

$$\partial m/\partial t = \partial/\partial t\,(\rho\,dV) = (\partial\rho/\partial t)\,dV \tag{5.14}$$

Here dV is the volume of parallelepiped. Application of principle of mass conservation leads to

$$\rho u\,dy\,dz + \rho v\,dxdz + \rho w\,dy\,dx = [\rho u + (\partial/\partial x\,(\rho u)\,dx]\,dy\,dz + [\rho v + (\partial/\partial y\,(\rho v)\,dy]\,dx\,dz + [\rho w + (\partial/\partial z\,(\rho w)\,dy]\,dx\,dy + \partial\rho/\partial t\,dV \tag{5.15}$$

Taking all terms on one side and simplifying by neglecting higher order terms, and incorporating

$dV = dx\,dy\,dz$, we get

$$[\partial\rho/\partial t + (\partial/\partial x\,(\rho u) + \partial/\partial y\,(\rho v) + \partial/\partial z\,(\rho w)]\,dV = 0$$

Since dV is not zero, we can write

$$\partial\rho/\partial t + (\partial/\partial x\,(\rho u) + \partial/\partial y\,(\rho v) + \partial/\partial z\,(\rho w) = 0 \tag{5.16}$$

It is the continuity equation for a compressible fluid in a rectangular Cartesian coordinate system.

For steady flow, $\partial\rho/\partial t = 0$ and therefore

$$\partial/\partial x\,(\rho u) + \partial/\partial y\,(\rho v) + \partial/\partial z\,(\rho w) = 0 \tag{5.17}$$

The continuity equation for an incompressible flow where ρ is constant can be written as

$$\partial u/\partial x + \partial v/\partial y + \partial w/\partial z = 0 \tag{5.18}$$

Continuity Equation: Integral Form

Let us consider a control volume V bounded by the control surface S (Figure 5.3). We can write the efflux of mass across the control surface S as:

$$\iint_s \rho\, u.dS \tag{5.19}$$

Where u is the velocity vector over the boundary of volume V (It is a vector with positive direction along the normal drawn outward from the surface). The rate of mass accumulation within the control volume becomes

$$\frac{\partial}{\partial t}\iiint_v \rho \, dV \tag{5.20}$$

Here dV is an elemental volume, and V is the total volume bounded by the control surface S. From mass balance, the continuity equation becomes

$$\frac{\partial}{\partial t}\iiint_V \rho \, dV + \iint_S \rho u.dS = 0 \tag{5.21}$$

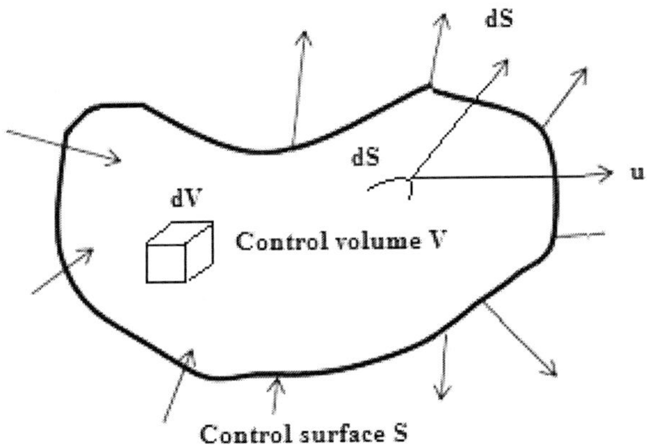

Figure 5.3. Control Volume to Derive Continuity Equation in Integral Form.

To convert the second term of eq. (5.21) into a volume integral, we use the Gauss divergence theorem such that

$$\iint_S \rho u\, dS = \iiint_V \nabla(\rho u) dV \tag{5.22}$$

Since the volume V does not change with time, the sequence of differentiation and integration in the right hand side of eq. (5.22) can be interchanged. Therefore, eq. (5.22) can be written as

$$\iiint_V \frac{\partial \rho}{\partial t} + \nabla(\rho u) dV = 0 \tag{5.23}$$

Equation (5.23) is valid for any arbitrary control volume irrespective of its shape and size. So we can write

$$\frac{\partial \rho}{\partial t} + \nabla(\rho u) = 0 \tag{5.24}$$

For a particular case of incompressible fluids, the above equation reduces to

$$\nabla (V_1) = 0 \tag{5.25}$$

This implies that the velocity field of an incompressible fluid is divergence free.

Continuity Equation: Cylindrical Polar Coordinates

The equation for a compressible fluid, unsteady flow in the 3-dimensions r, θ and z is directly written as follows (subscript with v indicate the velocity vector in r, θ and z directions).

$$\frac{\partial(\rho v_r)}{\partial r} + \frac{1}{r}\frac{\partial(\rho v_\theta)}{\partial \theta} + \partial\frac{\rho v_z}{\partial z} + \frac{\rho v_r}{r} + \frac{\partial \rho}{\partial t} = 0 \tag{5.26}$$

For steady state 2-dimensional flow field, z component and $\partial\rho/\partial t$ will vanish such that eq. (5.26) reduces to

$$\frac{\partial(v_r)}{\partial r} + \frac{1}{r}\frac{\partial(v_\theta)}{\partial \theta} + \frac{v_r}{r} = 0 \tag{5.27}$$

Example 5.1

Four pipes of equal radius carry air into a bigger pipe. The diameter of bigger pipe is 8 times that of smaller pipes. If the velocity of air in bigger diameter pipe is V then what is the velocity of air in the smaller pipes?

We assume that the diameter of smaller pipes is x cm and velocity is u cm/s.

Diameter of bigger pipes is 8 x

Using the continuity equation

$V(\pi/4)(8x)^2 = 4u(\pi/4)x^2$

$64x^2 V = 4ux^2$

$u = 16V$

Example 5.2

A compressible liquid is flowing in a diverging pipe (See Figure 5.1). The diameter of section XX and YY is 50 mm and 100 mm respectively. The specific gravity of the fluid at XX and YY are 0.998 and 0.992 kg/m³. If the velocity of liquid at section XX is 5 m/s, find the velocity at section YY.

Using eq. (5.4)

$\rho_1 A_1 V_1 = \rho_2 A_2 V_2$

Incorporating the values, we have

$0.998\times1000\times(\pi/4)(0.05)^2\times5 = 0.992\times1000\times(\pi/4)(0.1)^2\times V_2$

$V_2 = 0.998\times(0.05)^2/[0.992\times(0.1)^2] = 0.998\times0.0025/(0.992\times0.01) = 0.252$ m/s

BERNOULLI'S THEOREM

Application of the second Newton's law of motion along streamlines of fluid flow leads to a very famous equation in fluid mechanics known as Bernoulli equation

(BE), named after the Swiss scientist Daniel Bernoulli. He published this principle in his book Hydrodynamica in 1738. According to this postulation, liquids possess three types of energy namely potential energy, kinetic energy and pressure energy. Bernoulli's Theorem states that these three energies change from one form to another form, but the sum total of these energies remains constant. Clearly, the Bernoulli's theorem is a statement of the conservation of energy in fluid flows (Batchelor, 2000). Several instruments such as Pitot tube, and Venturimeter *etc.* have been designed to measure fluid flow. All these instruments use the principle of Bernoulli's theorem.

Derivation

Following assumptions are made to derive the BE.

1. The flow is inviscid or frictionless. Loss due to viscous effects is negligible compared to the magnitudes of the other terms in Bernoulli's equation. Thus, BE is valid for low viscosity fluids such as water and air
2. The flow is steady, irrotational and continuous. It means that the fully developed flow pattern does not change with time. Steady state velocity of the flow is not a function of time, but can be a function of position
3. The flow is incompressible (No change in density of the fluid), which is valid for all fluids and low speed gas of Mach 0.3 or below since the change in gas density is less than 5 per cent
4. BE can only be used between points on the same streamline. Conservation laws of physics more accurately simulate the variations in properties of fluid molecules travelling in the same path.

The BE as already stated, is a statement of the conservation of energy principle appropriate for flowing fluids. It is defined as 'in a steady ideal flow of incompressible fluids, the sum of pressure energy, kinetic energy and potential energy remains constant at every section provided no energy is added or taken out by an external source'. It also means that for an inviscid flow, an increase in the speed of the fluid occurs simultaneously with a decrease in pressure or a decrease in the fluid's potential energy. Following relation expresses the principle of conservation of energy.

$$\text{Pressure energy} + \text{Kinetic energy} + \text{Potential energy} = \text{constant} \quad (5.28)$$

$$\text{Energy per unit volume before} = \text{Energy per unit volume after} \quad (5.29)$$

Considering the flow in Figure (5.4), and applying Bernoulli theorem, we have

$$P_1 + \rho V_1^2/2 + \rho g\, Z_1 = P_2 + \rho V_2^2/2 + \rho g\, Z_2 \quad (5.30)$$

It can be stated that along streamline pressure energy, P_1 plus kinetic energy, $\rho V_1^2/2$ plus the potential energy, $\rho g\, Z_1$ (Since datum is taken at the center line, Z_1 and Z_2 are zero and are not shown in the diagram) at point 1 is equal to pressure energy plus kinetic energy plus the potential energy at point 2. Here P is the pressure, V is the velocity, ρ is the density, g is the gravitational acceleration and Z is the elevation of the point above a reference plane, with the positive Z-direction pointing upward and negative in the direction opposite to the gravitational acceleration. The term ρgZ is the hydrostatic pressure that represents the potential energy for fluid molecules at the same point and changes with elevation.

Alternate and equivalent forms of BE are:

$$P + \rho V^2/2 + \rho g\, Z = \text{Constant} \tag{5.31}$$

Or

$$P/\rho g + V^2/2g + Z = \text{Constant} \tag{5.32}$$

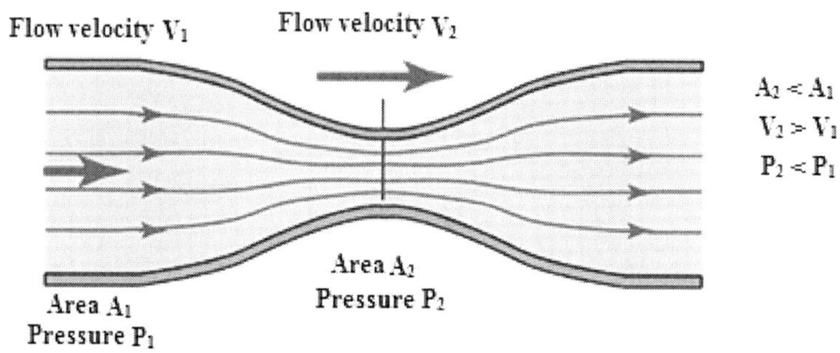

Figure 5.4. Illustrative Diagram for the Derivation of Bernoulli's Equation.

Eq. (5.32) states that the sum of the pressure head, $P/\rho g$, Velocity head, $V^2/2g$ and elevation head, Z is constant. $P/\rho g + Z$ is the piezometric head or hydraulic head denoted by h.

As already discussed under the assumptions, the above equation holds true for an ideal fluid (*i.e.* when there is no loss of head between two points). In practice, some energy is lost due to friction. As the fluid travels from point 1 to point 2, denoting the head loss due to friction as H_f, we have

$$P_1/\rho g + V_1^2/2g + Z_1 = P_2/\rho g + V_2^2/2g + Z_2 + H_f \tag{5.36}$$

In spite of the assumptions made in deriving BE, Bernoulli equation is quite useful, because it is simple to use and it can give insight into the balance between pressure, velocity and elevation.

Euler and Bernoulli Equations

Bernoulli equation for incompressible fluids can also be derived using the Euler's equations of motion following severe restrictions. The Euler's equation for steady flow of an ideal fluid along a streamline with gravity as the only body force is a relation between the velocity, pressure and density of the moving fluid.

$$V\, dV/ds = -\,1/\rho\, (dP/ds) - g\, dZ/ds \tag{5.33}$$

Here the notations have the same meaning as described earlier. Application of a force through distance ds along the streamline would physically imply work interaction. Therefore, by integrating eq. (5.33) with respect to ds, we get an equation for conservation of energy along a streamline as follows.

$$\int V\, (dV/ds)\, ds = -\int 1/\rho\, (dP/ds)\, ds - \int g\, (dZ/ds)\, ds$$

$$V^2/2 + \int dP/\rho + gZ = C \tag{5.34}$$

Here C is a constant along a streamline. For an incompressible flow, eq. (5.34) reduces to eq. (5.32) as follows:

$$V^2/2 + P/\rho + gZ = C \tag{5.35}$$

Stagnation Pressure

In fluid dynamics, a stagnation point is a point in the flow field where the local velocity of the fluid is zero. Stagnation points exist at the surface of objects in the flow field, where the object brings the fluid to rest. Consider a steady flow impinging on a perpendicular plate (Figure 5.5). One streamline divides the flow such that all the flow above this streamline goes upward of the plate and all the flow below this streamline goes downward of the plate. Along this dividing streamline, the fluid moves towards the plate. Since the flow cannot pass through the plate, the fluid must come to rest at the point where it meets the plate. The fluid along the stagnation streamline slows down and eventually comes to rest without deflection at the stagnation point. Bernoulli's equation along the stagnation streamline is:

$$P_e + \rho V_e^2/2 = P_0 + \rho V_0^2/2 \tag{5.37}$$

Here P_e and V_e are the pressure and velocity of flow, and P_0 is the stagnating pressure. Since V_0 is zero, we have

$$P_e + \rho V_e^2/2 = P_0 \tag{5.38}$$

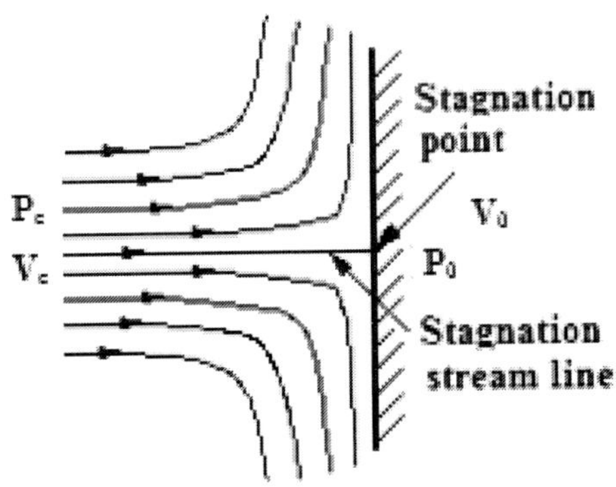

Figure 5.5. Stagnation Point in a Flow Field.

The stagnation or sum of the static and dynamic pressure (total pressure), P_0 is the pressure measured at the point where the fluid comes to rest isentropically. The word isentropically (reversible adiabatic process) implies that the entire kinetic energy of the fluid particle is utilized to increase its pressure only. It is the highest pressure found anywhere in the flow field. The dynamic pressure is not really a pressure at all: it is simply a convenient name for the quantity (half the density times the velocity squared), which represents the decrease in the pressure due to the velocity of the fluid. Rearranging eq. (5.38), we can calculate the velocity of flow.

$$V_e = [2(P_0 - P_e)/\rho]^{1/2} \tag{5.39}$$

Therefore, one can find out the velocity of flow from measurements of the static and stagnation pressure in a flowing fluid.

In practice, it is difficult to measure the stagnation pressure for a real fluid because some loss occurs due to friction. The pressure measured in the stagnation tube will always be less than P_0, since a part of the kinetic energy converts to intermolecular energy because of the fluid friction. An empirical factor C takes care of this, such that eq. (5.39) is written as:

$$V_e = C[2(P_0 - P_e)/\rho]^{1/2} \tag{5.40}$$

Hydraulic and Energy Grade Lines

Both the hydraulic and the energy grade line are graphical presentations of the Bernoulli equation. Hydraulic grade line (HGL), a measure of flow energy, is a line coinciding with the level of flowing water at any point along an open channel (Figure 5.6 left). On the other hand, HGL for the fluid flowing under pressure in a pipe system, is the level to which fluid will rise in a vertical tube open to the atmosphere at any point along the pipe. Thus, HGL is the locus of points located a distance $p/\rho g$ above the center of the pipe. In practice, HGL is the line, which gives the sum of pressure ($p/\rho g$) and datum head (Z) of a flowing fluid in a pipe with respect to some reference. It is also the line obtained by joining the top of all vertical ordinates, showing the pressure head of a flowing fluid in a pipe with center of the pipe as reference.

$$HGL = p/\rho g + Z \tag{5.41}$$

The total head (which is also total energy per unit weight), with respect to any arbitrary datum, is the sum of the elevation (potential) head, pressure head and velocity head. Fluid flowing along the pipe loses some energy because of friction and therefore total energy decreases in the direction of flow. The total energy at various points along the axis of the pipe plotted and joined by a line is the energy gradient line or energy grade line (EGL) or total energy line (TEL).

$$EGL = H = p/\gamma + V^2/2\,g + Z = \text{constant along a streamline} \tag{5.42}$$

The HGL and the EGL lines for an open channel uniform flow and a pipe flow are depicted in Figure 5.6. For ideal fluid flow [no losses due to friction (major losses) or components (minor losses)], the energy line remains at a constant level along the flow. In practice, the energy line decreases along the flow path because of friction losses. Moreover, some equipment adds energy to the system whereas other consumes energy. For example, a turbine in the flow reduces the energy line and a pump or fan in the line increases the energy line.

Drawing HGL and EGL

Engineers find it useful to employ the HGL and the EGL while working with the piping systems. These imaginary lines help the engineers find trouble spots in the system, more often the points of low pressure. Following guidelines facilitates the drawing of the HGL and the EGL.

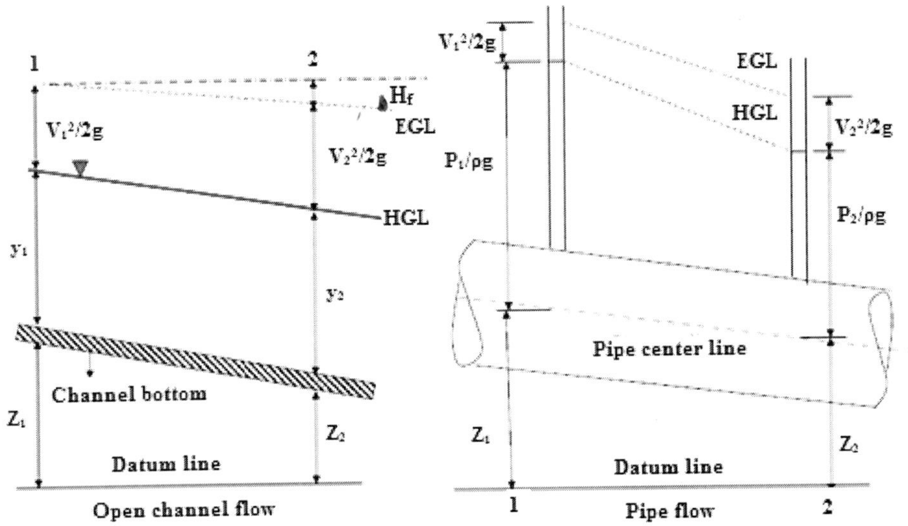

Figure 5.6. HGL and EGL in Open Channel (Left) and Pipe Flow (Right).

- ☆ HGL is always below the EGL, the EGL being at a distance $V^2/2g$ above the HGL. As the velocity approaches zero, the EGL approaches the HGL. In a reservoir, HGL and EGL are identical and lie on the water surface.
- ☆ The EGL and HGL slope downward in the direction of the flow due to the head loss in the pipe. The greater the loss per unit length, the higher is the slope. As the average velocity in the pipe increases, the loss per unit length increases.
- ☆ A sudden change occurs in the HGL and the EGL whenever a loss occurs due to a sudden geometry change as represented by a valve or a sudden enlargement, or reduction.
- ☆ A jump occurs in the HGL and the EGL whenever useful energy is added to the fluid say with a pump. A drop occurs if useful energy is extracted from the fluid, say the presence of a turbine.
- ☆ HGL coincides with pipe system when pressure = 0 *i.e.* $P/\rho g = 0$. It means the pressure is zero at points where the HGL passes through the center line of the pipe.
- ☆ A vacuum in the pipe occurs if the pipe lies above the HGL. As far as possible, never allow this condition to occur in the design of piping systems. An exception is the design of a siphon.

To illustrate the above points, let us consider a case where fluid flows through a pipe whose diameter is decreasing along the flow (Figure 5.7). Let us take four point 1, 2, 3 and 4 having an area of A_1, A_2, A_3 and A_4, the last being the area at the outlet.

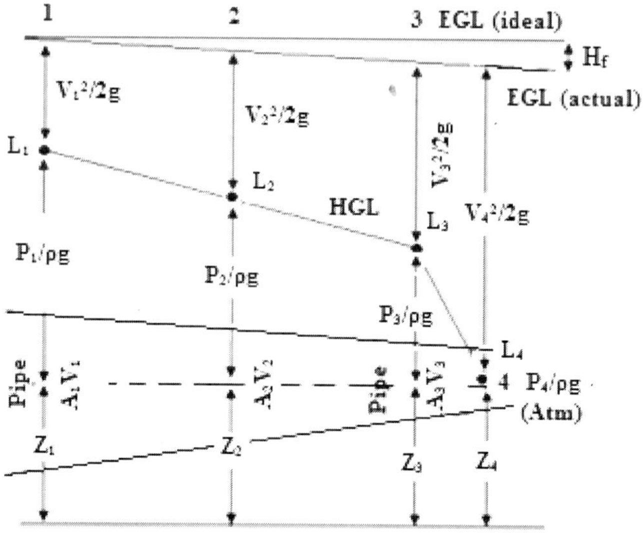

Figure 5.7. Energy Grade Line (EGL) and Hydraulic Grade Line (HGL) for a Tapering Pipe.

We go step wise to plot the HGL and EGL.

- ☆ Points 1, 2, 3, and 4 having areas A_1, A_2, A_3 and A_4 are located at a height of Z_1, Z_2, Z_3 and Z_4 above the datum and represent the potential heads.
- ☆ For flow in converging pipe, there is a decrease in the pressure head (P/ρg) and consecutively there is a rise in velocity head.
- ☆ At points 1, 2 and 3, piezometer tube measures the pressure head (P/ρg) by rise of liquid in the piezometer. Points L_1, L_2, and L_3 represent the water level in the piezometer and are located at a height of $(Z_1+P_1/\rho g)$, $(Z_2+ P_2/\rho g)$ and $(Z_3+ P_3/\rho g)$ above the datum.
- ☆ At tip of pipe (point 4) the liquid is exposed to atmospheric pressure and therefore, pressure head becomes zero. Since entire pressure energy converts into kinetic energy at this point, it results in increased flow velocity. Since pressure at point A is atmospheric, the HGL drops to zero at this point and is just above the datum given as Z_4. There is a sharp decline in HGL from L_3 to L_4. The line joining points L_1, L_2, L_3 and L_4 is the HGL and $(P/\rho g+ Z)$ is the hydraulic gradient.
- ☆ Total energy line is line joining points denoting total head represented as $P/\rho g +V^2/2g + Z$

  Note that EGL is constant in an ideal fluid although in practice it declines by an amount H_f, which represents the friction losses in the pipe system.

Example 5.3

Water is flowing in a hose with a velocity of 1.0 m/s and a pressure of 198000 N/m². Pressure decreases to atmospheric pressure (101300 N/m²) at the nozzle. If there is no change in height, use the Bernoulli equation to calculate the velocity of the water exiting the nozzle. Take density of water as 1000 kg/m^3.

Applying the Bernoulli's equation (eq. 5.30)

$0.5\,\rho V_1^{\,2} + \rho g Z_1 + P_1 = 0.5\,\rho V_2^{\,2} + \rho g Z_2 + P_2$

Since the height does not change (Z_1=Z_2), the height term cancels out from both sides, and we have

$0.5\,\rho V_1^{\,2} + P_1 = 0.5\,\rho V_2^{\,2} + P_2$

Solving for V_2, we have

$V_2 = [(0.5\,\rho V_1^{\,2} + P_1 - P_2)/0.5\rho]^{1/2} = [0.002\,(0.5\times1000\times1^2+198000 - 101300)]^{1/2}$

$V_2 = [0.002\,(0.5\times1000+198000 - 101300)]^{1/2} = [0.002(500 + 96700)]^{1/2}$

$V_2 = 13.95$ m/s

Example 5.4

Oil is flowing in a pipe at a velocity of 1 m/s and a pressure of 101300 Pa. It is proposed to use the oil at a pressure of 2 atm at a lower level. If there is no change in the velocity then how far must the pipe drop in height in order to achieve the required pressure? The density of oil is 800 kg/m^3.

Applying the Bernoulli's equation (eq. 5.30)

$0.5\,\rho V_1^{\,2} + \rho g Z_1 + P_1 = 0.5\,\rho V_2^{\,2} + \rho g Z_2 + P_2$

Since the velocity does not change (V_1=V_2), the velocity terms cancels out from both sides. Therefore

$\rho g Z_1 + P_1 = \rho g Z_2 + P_2$

Rearrange algebraically to solve for change in height $(P_1 - P_2)/\rho g = Z_2 - Z_1 = \Delta Z$

$\Delta Z = (101300 - 202600)/(800\times 9.81) = -12.9$ m

Or pipe should go 12.9 m below.

Example 5.5

Water is flowing through an inclined pipeline of diameter 20 cm and 40 cm at section A and B respectively. Sections A and B are located at height of 4 m and 4.5 m respectively from ground level. The discharge through pipe is 40 L/s. If the pressure at A is 20 kPa, find the pressure at point B.

Area at section A = $\pi\, d^2/4 = \pi\times0.20^2/4 = 0.0314$ m^2

Area at section B = $\pi\, d^2/4 = \pi\times0.40^2/4 = 0.1256$ m^2

Q= 0.04 m^3/s

$Q = A_1 V_1 = A_2 V_2$

So $V_1 = 0.04/0.0314 = 1.274$ m/s

$V_2 = 0.04/0.1256 = 0.318$ m/s

Using the Bernoulli equation

$V_1^2/2g + Z_1 + P_1/\rho g = V_2^2/2g + Z_2 + P_2/\rho g$

$(1.274)^2/(2\times9.81) + 4 + 20000/9810 = (0.318)^2/(2\times9.81) + 4.5 + P_2/9810$

$0.083 + 4 + 2.04 = 0.005 + 4.5 + P_2/9810$

$P_2/9810 = 1.618$

$P_2 = 15872.6$ Pa or 15.9 kPa

Principle of a Hydraulic Siphon

Siphon is a pipe in the form of inverse U. It is used to drain water from a container placed at a height into another container placed at some lower level (Figure 5.8). To initiate the flow, completely fill the inverted U tube with the same liquid that is being transferred. Dip one end of it into the container from which water is to be transferred. The other end is open and vertically below the free surface of the liquid in the container. Liquid flows continuously from the upper container to the lower one through the siphoning action. Let us look at the justification of flow towards the top of the tube (Figure 5.8). Bernoulli's equation explains this phenomenon. Applying the Bernoulli's equation between point A and B, we can write

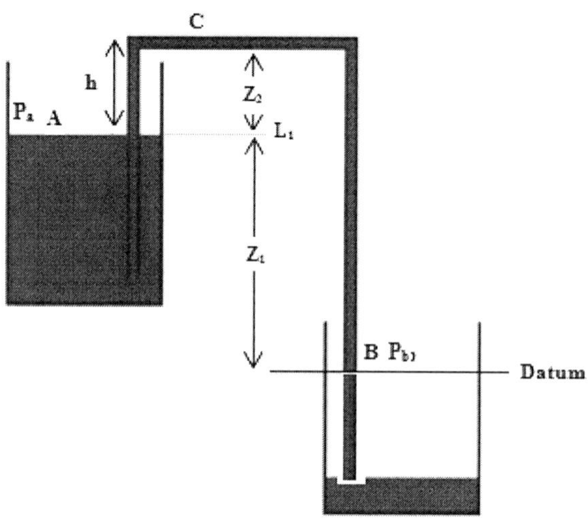

Figure 5.8. Schematic Diagram of a Siphon Illustrating its Principle of Working.

$P_a/\rho g + V_a^2/2g + Z_1 = P_b/\rho g + V_b^2/2g + 0$ (Datum)

The pressure at A and B are the same, being atmospheric. It is possible to neglect the velocity at A, being a big container compared to the area of the bore of the tube. The simplified equation yields the velocity of water at B as:

$$V_b = \sqrt{2gZ_1} \quad (5.43)$$

The above expression shows that siphoning creates a velocity head at B at the expense of the potential head difference between A and B. Applying the Bernoulli's equation between point A and C, we can write

$$P_a/\rho g + V_a^2/2g + Z_1 = P_c/\rho g + V_c^2/2g + Z_1 + Z_2$$

Since V_a is zero, V_c is equal to V_b (continuum and assuming the pipe cross-section to be uniform), we have

$$P_a/\rho g - V_b^2/2g - Z_2 = P_c/\rho g$$

$$P_a/\rho g - V_b^2/2g - h = P_c/\rho g \qquad (5.44)$$

Clearly, pressure at C is below the atmospheric pressure. Minimum pressure occurs at point C, being at the top having the maximum potential head. For siphon to function smoothly, the pressure at this point should not fall below the vapour pressure of the liquid. If that happens, flow may stop because of the creation of vapour pockets. It means the height of the tube, h cannot be increased at will but will be governed by eq. (5.44).

Example 5.6

A siphon comprising a pipe of 70 mm diameter discharges water to the atmosphere, as shown in Figure (5.8). Calculate the velocity and discharge of flow, if the values of Z_1 and Z_2 are 2.5 and 1.5 m respectively? What is the absolute pressure at the top of the siphon? Atmospheric pressure is 9.81 kN/m².

Applying eq. (5.43)

$$V_b = \sqrt{2gZ_1} = \sqrt{49.05} \sim 7.0 \text{ m/s}$$

$$Q = V_b \times A = 7.0 \times (\pi\,(0.07)^2/4) = 0.027 \text{ m}^3/\text{s}$$

Applying eq. (5.44)

$$P_a/\rho g - V_b^2/2g - Z_2 = P_c/\rho g$$

$$P_c = P_a - \rho V_b^2/2 - \rho g Z_2 = 98100 - 24500 - 14715 = 58885 \text{ N/m}^2$$

APPLICATIONS OF BERNOULLI'S EQUATION

Bernoulli's equation has been widely applied in fluid mechanics. Number of equations have been derived using the concept propounded in BE. Besides, number of flow measuring devices have been designed that incorporate the principle of BE. Some of these devices are as under.

- ☆ Pitot tube (Stagnation tube and Pitot static tube)
- ☆ Venturi meter
- ☆ Nozzles and extended nozzles

Pitot Tube

The Pitot tube, first invented by Henri Pitot in 1732, is used to measure the velocity of flowing fluids - liquids or gases. A differential pressure flow meter, Pitot tube measures two pressures namely static and stagnating pressures. The static pressure in an open channel flow is equal to the hydrostatic pressure due to its depth below the free surface. In a pipe flow, static pressure is the operating pressure in the pipe, duct, or the environment upstream of the Pitot tube. It is a measure of the amount that fluid pressure exceeds local atmospheric pressure. While measuring this pressure, velocity of the flow should remain undisturbed. Contrarily, stagnation pressure (also called total pressure) is the pressure at a stagnation point in the fluid flow. This pressure is created at a point by bringing the velocity of the flowing fluid to zero in a frictionless process. The higher value of the stagnating pressure is at the cost of kinetic or dynamic head loss. Thus, we have

$$P_{st} = P_s + P_d$$

Here P_{st} is the stagnating pressure and P_s and P_d are static and dynamic pressures respectively.

In its simplest form, Pitot tube is a glass tube bent at right angle. Such a simple Pitot tube is capable of measuring the stagnating pressure, which alone is sufficient to measure flow in open channels (Figure 5.9). It is because the static pressure in an open channel flow is equal to the hydrostatic pressure due to its depth below the free surface. The difference in level between the liquid in the glass tube and the free surface becomes the measure of dynamic pressure (Figure 5.9). For the simple Pitot tube illustrated in Figure 5.9, we use the Bernoulli's theorem at points 1 and 2.

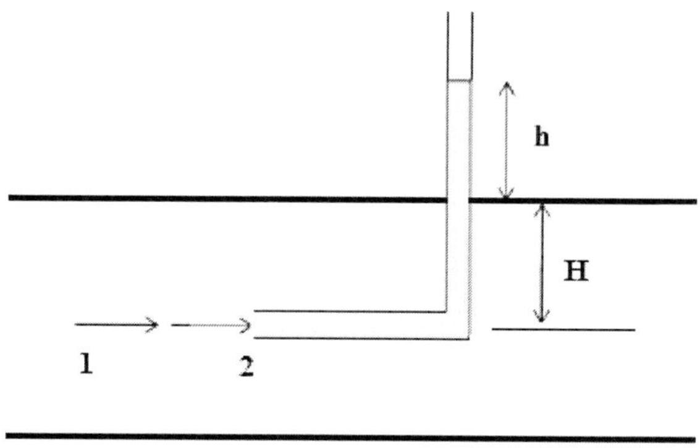

Figure 5.9. Measurement of Stagnating Pressure Using Pitot or Stagnating Tube.

$$P_1/\rho g + V_1^2/2g = P_2/\rho g + V_2^2/2g$$

At point 1: $P_1/\rho g = H$ and $V = V_1$

At point 2: $P_2/\rho g = H + h$ and $V_2 = 0$

Incorporating these values, we have

$$H + V_1^2/2g = H + h$$

$$V_1 = \sqrt{2gh} \tag{5.45}$$

Eq. (5.45) gives the theoretical velocity. To calculate the actual velocity, we multiply the right hand side by a coefficient C_p (less than 1.0) such that

$$V_1 = C_p\sqrt{2gh} \tag{5.46}$$

Here C_p is the coefficient of the Pitot tube. The coefficient takes care of the non-idealities due to friction and in converting the dynamic head into pressure head. The value of C largely depends upon the geometry of the Pitot tube. One can determine the value of C by performing calibration tests. For calculating the volumetric flow rate, we use the following relation in which A_1 is cross sectional area.

$$Q = A_1V_1 = A_1\, C_p\sqrt{2gh} \tag{5.47}$$

While using eq. (5.47), one must remember the fact that the average velocity of the open channel may be different from the velocity determined from Pitot tube at a given depth. An additional coefficient may be required to take care of this fact.

Pitot-static Tube

For pipe flows, besides the stagnating pressure one also needs to measure the static pressure. Either of the two alternatives is used. One, provide a separate piezometer upstream of the pitot tube to measure static pressure (Figure 5.10). Second, within the Pitot tube a flat opening perpendicular to the direction of fluid flow (facing into the fluid flow) is provided as shown in Figure 5.10 at point P. Right hand side of the U-tube manometer measures the stagnation pressure at point q (Figure 5.10). It is used to calculate the dynamic pressure (also called velocity pressure), a measure of the amount that the stagnation pressure exceeds the static pressure at a point in the fluid flow. The value of this pressure is equal to h (Figure 5.10). It is why; this kind of a Pitot U-tube is named as a Pitot-static tube discussed in the next paragraph.

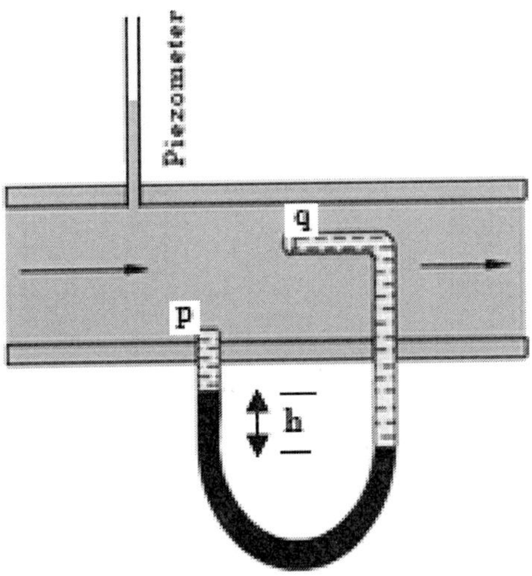

Figure 5.10. Measurement of Static, Stagnation and Dynamic Pressures.

The arrangement shown in Figure (5.10) is difficult to install. Therefore, one single instrument known as Pitot-static tube (Figure 5.11) is capable of recording the static and stagnation pressures. It comprises two concentric tubes (Figure 5.11 right). Outer body of the Pitot tube consists of ports at point M, to sense the static pressure of the fluid. A port at Q senses the stagnation pressure. The outlets M_1 and Q_1 are connected to U-tube manometer that measures the pressure difference between the points Q and M (Figure 5.11).

For pipe flow at a point along the streamline (hydrostatic pressure cancels out being equal), we have

$$P_1 + 0.5\,\rho\,V_1^2 = P_2 \tag{5.48}$$

$$P_1 - P_2 = 0.5\,\rho\,V_1^2$$

Here P_1 and P_2 are pressures at points M and Q respectively. Thus, we have

$$V_1 = \sqrt{(2\,\Delta P/\rho)} \tag{5.49}$$

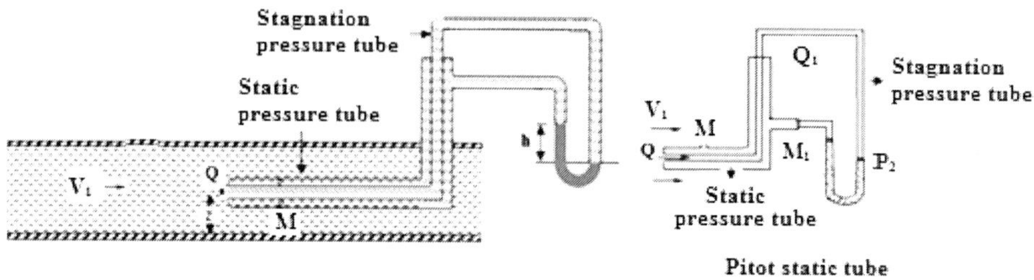

Figure 5.11. Construction of Pitot Static Tube and its Application.

If h is the difference in height of mercury column in the manometer, the pressure difference given by the manometer is:

$$\Delta P = g\,h\,(\rho_m - \rho) \qquad (\rho_m > \rho)$$

Here ρ_m and ρ are the densities of the fluid in the manometer and the pipe respectively.

Incorporating the value of ΔP in eq. (5.49), we get

$$V_1 = (2gh(\rho_m - \rho)/\rho)^{1/2} = (2gh\{(\rho_m/\rho) - 1\})^{1/2} \tag{5.50}$$

If the density of fluids is same, then $\Delta P = \rho\,g\,h$ so that the flow velocity equation is simply:

$$\text{Flow velocity, } V_1 = \sqrt{2gh} \tag{5.51}$$

As discussed previously, actual velocity is less than the theoretical velocity. Thus, it is multiplied by a constant C_p, such that

$$V_1 = C_p\sqrt{2gh} \tag{5.52}$$

Equation for volumetric flow rate is then

$$Q = A_1 V_1 = A_1\,C_p\sqrt{2gh} \tag{5.53}$$

Since the velocity of flow calculated from the Pitot-static tube is at the center of the conduit, it represents the maximum velocity as per the velocity profile in a pipe (Figure 5.12). Therefore, to calculate actual Q, we need to multiply the Q calculated with the help of eq. (5.53) by another factor to account for the average velocity in the pipe. The value of this factor is always less than 1.0. To overcome this problem, new designs of the Pitot tubes, known as Averaging Pitot-static tubes, incorporate number of ports instead of a single port. These ports distributed across the flow direction sense the average stagnation pressure obviating the need of such correction.

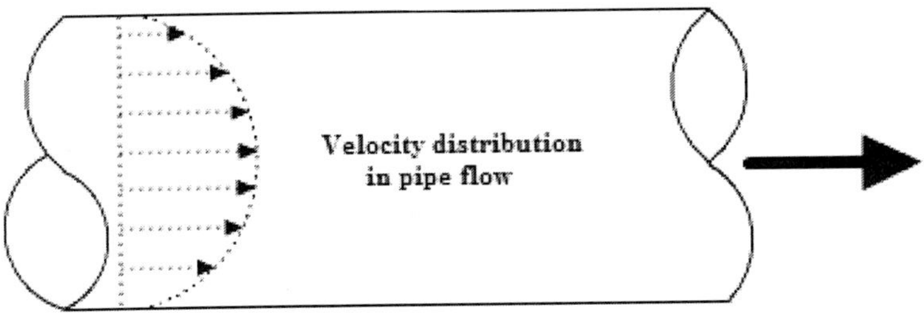

Figure 5.12. Velocity Distribution in Pipe Flow.

Advantages of pitot tubes

- It is easy and economical to install and remove, being small in size
- Minimal frictional losses as there are no moving parts
- It is less expensive
- It causes very less pressure loss
- It can work under varying environmental conditions
- It is capable of working at high temperatures.

Limitations or disadvantages

- Any foreign material in flow can change the accuracy. It is not suitable for dirty or sticky fluid like sewage
- It has low range, low sensitivity and limited accuracy, its sensitivity gets disturbed by flow direction
- It may not yield accurate results in low velocity flows
- A change in velocity profile can cause significant errors
- Too much vibration may damage the device.

Example 5.7

An aeroplane is flying at a height of about 10000 m. The Pitot tube (differential gauge) fixed in the plane measures a differential pressure of 600 N/m^2 between its static and stagnation points. What is the speed of the plane? Take density of air at the flying height as 0.43 kg/m^3.

The speed of plane is equivalent to the speed of air with respect to the plane. Pitot tube fixed in the plane measures the speed of air with respect to the plane.

Using eq. (5.49)

Speed, $V = \sqrt{(2\,\Delta P/\rho)} = \sqrt{(2\times 600/0.43)} = 52.82$ m/s

Example 5.8

A pitot static tube measures the velocity of air flowing through a duct. The manometer shows a head difference of 5 cm of water. Determine the velocity of air, if the density of air and water are 1.12 kg/m^3 and 999 kg/m^3. Assume the coefficient of the Pitot tube as 0.98.

Using eq. (5.50), we have

$V_1 = C_p\,[2gh\{(\rho_m/\rho) - 1\}]^{1/2} = 0.98\,[2\times9.81\times 0.05\times \{(999/1.12) - 1\}]^{1/2} = 28.97$ m/s

Example 5.9

An aircraft travelling against a wind velocity of 20 kmph has a Pitot static tube mounted on it. The tube registers a pressure difference of 5000 N/m^2. If the specific weight of air is 12 N/m^3, calculate the speed of the plane. Assume the Pitot tube coefficient as 0.98.

$h_{air} = 5000/12 = 416.6$ m

Using eq. (5.52)

Relative velocity, $V = C_p\,\sqrt{(2gh_{air})} = 0.98\times \sqrt{(2 \times 9.81\times416.6)} = 88.6$ m/s

Wind velocity = 20000/3600 = 5.56 m/s

Velocity of plane = Relative velocity – wind velocity = 88.6 – 5.56 = 83.04 m/s

VENTURI METER

Venturi meter is a device to measure the velocity and flow rates of fluid flowing through the pipes. A venturi tube has an entrance zone, a converging conical inlet, a cylindrical throat, and a diverging recovery cone. Small size venturi meters are made of brass or bronze and large venturi meters are made of cast iron or stainless steel. Following paragraphs describe the main sections of a venturi meter shown in Figure 5.13.

Entrance Section : It is a straight cylinder having length equal to 5 to 8 times the diameter of the pipe.

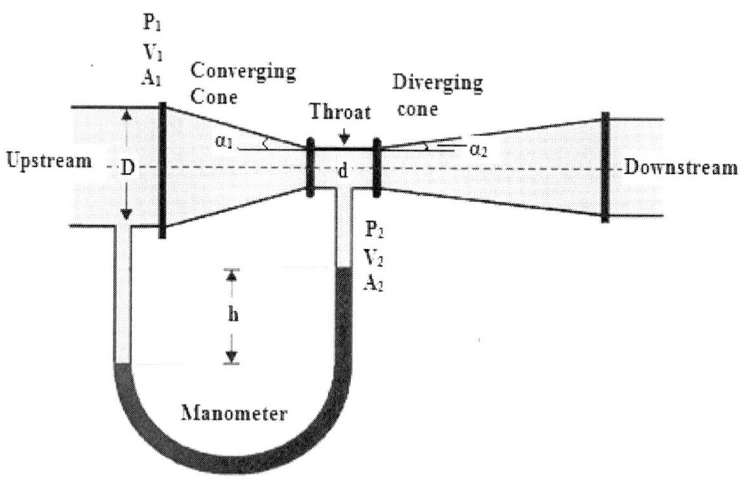

Figure 5.13. Illustrative Diagram Showing Main Sections of a Venturi Meter.

Converging Section: As the name suggest, this section is a converging section with one end attached to the entrance section, and the other with the throat. The diameter of this section decreases gradually. The angle of cone is α_1 normally 21± 2°. The flow velocity increases with a corresponding decrease in the pressure in this section.

Throat: The diameter of the venturi meter is least at the throat section. It is about 1/3rd to 1/4th of the inlet diameter. The throat length is about one throat diameter. Thus, flow experiences maximum velocity and minimum pressure in this section. The pressure difference between the inlet and the throat section is used to calculate the fluid flow rate through the pipe.

Diverging Section: The diameter of the tube gradually increases in the diverging section of the venturi meter. The cone angle α_2 ranges from 5 to 15°. Large proportion of kinetic energy converts back to pressure energy in this section.

Manometer: A U-tube manometer partially filled with mercury or some other liquid is used to measure the pressure heads. One can also use a differential pressure gauge to measure the pressure difference.

To measure the flow velocity and discharge in a pipe, attach the venturi meter in the pipe such that entrance section is on the upstream side (Figure 5.13). The measured head difference, h is used to calculate the flow velocity and discharge

Discharge Formula

Let us assume that the area of cross sections of pipe at the end where the venturi meter is connected is A_1 while the diameter at the throat is A_2. The corresponding velocity of flow is V_1 and V_2. If P_1 and P_2 are the pressure of fluid at the two places, then applying the Bernoulli's equation along a streamline from point 1 to point 2:

$$P_1/\rho g + V_1^2/2g + Z_1 = P_2/\rho g + V_2^2/2g + Z_2 \qquad (5.54)$$

According to continuity equation

$$Q = A_1 V_1 = A_2 V_2$$

Such that $V_2 = V1\ A_1/A_2$

Substituting in eq. (5.54) and rearranging gives

$$(P_1\text{-}P_2)/\rho g + Z_1 - Z_2 = [V_1^2/2g][(A_1/A_2)^2 - 1] = V_1^2[(A_1\text{-}\ A_2)^2/A_2^2\,] \qquad (5.55)$$

Rearranging for velocity V_1, we have

$$V_1 = [2g\ (P_1\text{-}P_2)/\rho g + (Z_1 - Z_2)]^{1/2}[A_2/(A_1^{2}\text{-}\ A_2^{2})^{1/2}] \qquad (5.56)$$

If the venturi meter is horizontal then Z_1 is equal to Z_2. Eq. (5.55) takes the form

$$V_1 = [2g\ (P_1\text{-}P_2)/\rho g]^{1/2}[A_2/(A_1^{2}\text{-}\ A_2^{2})^{1/2}] \qquad (5.57)$$

Since $\Delta P = g\,h\,(\rho_m - \rho)$ $\quad(\rho_m > \rho)$, we have

$$V_1 = [2gh\ \{(\rho_m/\rho) - 1\}]^{1/2}\ [A_2/(A_1^{2}\text{-}\ A_2^{2})^{1/2}] \qquad (5.58)$$

Similarly, velocity V_2 is

$$V_2 = [2gh\ \{(\rho_m/\rho) - 1\}]^{1/2}\ [A_1/(A_1^{2}\text{-}\ A_2^{2})^{1/2}] \qquad (5.59)$$

Note that the above expressions are for ideal conditions giving the theoretical velocities. Actual velocity being less than the theoretical, the velocities so calculated are multiplied by a coefficient, C_v, being less than 1.0. This issue is further discussed in the following section.

Volume Flow Rate

The volume flow rate, Q is the area of the pipe times the velocity V_1 such that:

$$Q = C_v A_1 V_1 = C_v [2gh (\rho_m/\rho)]^{1/2}[A_1 A_2/(A_1^2 - A_2^2)^{1/2}] \tag{5.60}$$

If the liquid in the manometer is lighter ($\rho_m < \rho$) than the liquid flowing through the pipe, then the term $[(P_1-P_2)/\rho g + (Z_1 - Z_2)]$ is given as

$$[(P_1-P_2)/\rho g + Z_1 - Z_2] = h (1- \rho_m/\rho)$$

$$Q = A_1 C_v V_1 = C_v [2gh (1- \rho_m/\rho)]^{1/2}[A_1 A_2/(A_1^2 - A_2^2)^{1/2}] \tag{5.61}$$

Mass flow rate can also be determined from the relation

Mass flow rate = volumetric flow rate × fluid density

Note that the discharge coefficient is not a constant but depends upon pipe geometry and Reynolds number of the flow *etc.* For a well-designed venturi, the discharge coefficient C_v is about 0.98 for pipe diameters of 5 to 20 cm and about 0.99 for larger sizes. In a properly designed venturi meter, the permanent pressure loss is about 10 per cent of the venturi differential $(P_1 - P_2)$, and 90 per cent of differential is recovered.

Venturi Meter Fitted in a Slanting Pipe

Since, the elevation term appears in the Bernoulli's equation, a question often arises whether the angle of venturi meter will affect the calculated velocity or the discharge. It can be shown that if the head difference in the venturi meter is measured using a manometer, the pipe or for that matter, the angle of the venturi meter fitted in the pipe becomes immaterial (Figure 5.14). It can be shown as follows.

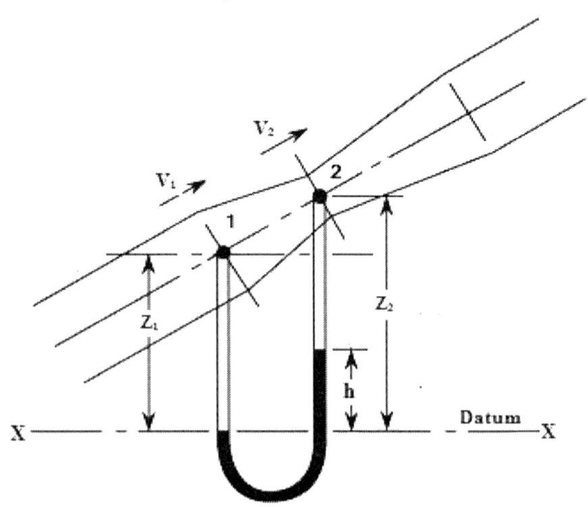

Figure 5.14. An Inclined Venturi Meter with a Manometer.

The term on the left hand side of eq. (5.55) in terms of the manometer readings for an inclined manometer can be derived from Figure 5.14 as:

$$P_1 + \rho g\, Z_1 = P_2 + \rho g\, (Z_2 - h) + \rho_m g h$$

Here ρ_m is the density of the fluid used in the manometer ($\rho_m > \rho$). Rearranging, we have

$$[(P_1 - P_2)/\rho g + (Z_1 - Z_2)] = h\, (\rho_m/\rho) - 1\}$$

Incorporating in eq. (5.56), we have

$$V_1 = [2gh\, \{(\rho_m/\rho) - 1\}]^{1/2}\, [A_2/(A_1^{2} - A_2^{2})^{1/2}] \quad (5.62)$$

$$Q = A_1\, V_1 = [2gh\, \{(\rho_m/\rho) - 1\}]^{1/2}[A_1 A_2/(A_1^{2} - A_2^{2})^{1/2}] \quad (5.63)$$

Both the expressions for V_1 and Q do not involve the elevation terms (Z_1 or Z_2). Therefore, one can use the venturi meter without bothering about its angle, provided the head difference is measured using a manometer.

Advantages of a venturi meter

- The venturi meter is suitable for all kinds of fluids such as clean, sediment laden and viscous or even for slurries
- Loss of head due to installation in the pipeline is small
- The attraction of this meter lies in its high energy recovery so that it may be used where only a small pressure head is available
- Accuracy is within 1 per cent of the full range
- Not much wear and tear is encountered
- Well established characteristics over the years

Disadvantages of a venturi meter

- Large space is required
- Expensive to install
- Its design depends upon the flow/field requirements.

Example 5.10

A gas is flowing through a cylindrical duct of 40 cm diameter with a constriction of 20 cm diameter. A venturi meter connected between two points measures the flow rate of air. Assume that the density of gas is 2.51 kg/m³ with negligible variation with position. The difference in mercury level between the two arms of the manometer is 4.5 cm. Calculate the mass rate of flow of air.

The velocity of fluid at position 2 in a venturi meter as per eq. (5.59) is

$$V_2 = [2gh\, \{(\rho_m/\rho) - 1\}]^{1/2}\, [A_1/(A_1^{2} - A_2^{2})^{1/2}]$$

Given $\rho_m = 13.6 \times 10^3$ kg/m³, $\rho = 2.51$ kg/m³ and h = 4.5 cm or 0.045 m

$A_1 = 0.1256\ m^2$

$A_2 = 0.0314\ m^2$

$A_1^2 - A_2^2 = 0.01479$

Since $\rho_m >> \rho$, we can use the simplified equation. Taking ρ inside the bracket, we can write mass flow as

$$Q = A_1A_2[2gh\rho_m/\rho)]^{1/2}/[(A_1^2 - A_2^2)^{1/2}]$$

Mass flow = $0.1256 \times 0.0314 \times 2.51 \, [(2\times 9.81\times 0.045\times 13.6\times 10^3/2.51)/0.01479]^{1/2}$ = 5.63 kg/s

Example 5.11

The difference in pressure across the venturi meter is equivalent to 8 m of the flowing liquid. The pipe diameter is 19 cm. Calculate the discharge in L/min if the throat diameter of the venturi meter is 11 cm. Assume the coefficient of discharge for the venturi meter as 0.97.

Since the fluid in the manometer and pipe are same, $\Delta P = \rho gh$. If we incorporate this in eq. (5.57) for velocity of flow, then the discharge equation is:

$$Q = C_v \, A_1 \, A_2 [2gh]^{1/2}/[(A_1^2 - A_2^2)]^{1/2}$$

$$A_1 = (\pi/4) \times 0.19^2 = 0.0284 \text{ m}^2$$

$$A_2 = (\pi/4) \times 0.11^2 = 0.0095 \text{ m}^2$$

$$Q = (0.97\times 0.0284\times 0.0095) (\sqrt{(2\times 9.81\times 8)})/\sqrt{(0.00081-0.000090)}$$

$$Q = 0.122 \text{ m}^3/\text{s}$$

$$Q = 7320 \text{ L/min}$$

Example 5.12

A venturi meter with a throat diameter of 150 mm is connected to a pipe of 250 mm diameter to measure the discharge in the pipe. The pressures at the beginning of the connection and at the throat are 140 kPa and 80 kPa respectively. Determine the flow rate. Take the coefficient of discharge as 0.98.

$$Q = C_v \, A_1 \, A_2 \, [2g \, (P_1 - P_2)/\rho g]^{1/2}/[(A_1^2 - A_2^2)]^{1/2}$$

$$A_1 = (\pi/4) \, d_1^2 = 0.787\times 0.0625 = 0.0491$$

$$A_2 = (\pi/4) \, d_2^2 = 0.0177$$

$$\rho = 1000 \text{ kg/m}^3$$

$$g = 9.81 \text{ m/s}^2$$

P_1 and P_2 are 140000 and 80000 Pa respectively. Incorporating the values

$$Q = 0.98\times 0.0491\times 0.0177[2\times 9.81 \, (140000-80000)/9810]^{1/2}/[(0.00241 - 0.000313)]^{1/2}$$

$$Q = 0.00085[2\times 9.81 \, (140000-80000)/9810]^{1/2}/0.0458$$

$$Q = 0.0186\times 10.95 = 0.204 \text{ m}^3/\text{s}$$

Another way is to calculate the head of water from the pressure. 140000 Pa is 14.276 m of water and 80000 Pa is 8.158 m of water. Then

$$Q = C_v \, A_1 \, A_2 \, [2gh]^{1/2}/[(A_1^2 - A_2^2)]^{1/2}$$

The head, h = 14.276 – 8.158 = 6.118 m. Incorporating in the following equation

$$Q = 0.0186 \, [2gh]^{1/2} = 0.0186\times 10.956 = 0.204 \text{ m}^3/\text{s}$$

FLOW THROUGH NOZZLE

Figure (5.15) shows a reservoir fitted with a nozzle on the side. Point A corresponds to free water surface and point B at the tip of nozzle opening.

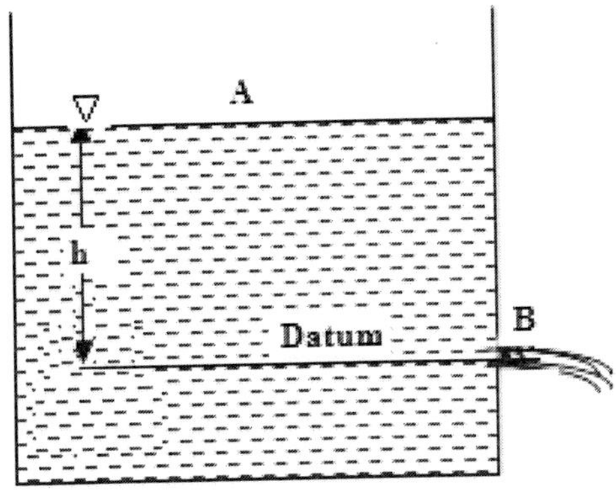

Figure 5.15. Flow through a Nozzle Fitted in a Reservoir.

At point A

Velocity $V_1 = 0$ (since the top water level is moving very slowly compared to velocity of water at B.

Pressure $(P_1) = 0$ being open to atmosphere

Height of point A from datum line = h (Datum line set at the center of the nozzle)

At point B

Velocity = V_2

Pressure $(P_2) = 0$ being open to atmosphere

Height of point B from datum line = 0

Applying Bernoulli's Theorem:

$$0 + 0 + h = 0 + V_2^2/2g + 0 \tag{5.64}$$

$$V_2^2 = 2gh$$

$$V_2 = \sqrt{2gh}$$

If we extend the nozzle further as shown in Figure 5.16, then the velocity V_3 can be shown to be:

$$V_3 = \sqrt{2g(h+h_1)} \tag{5.65}$$

Example 5.13

Fire hose used to control fire has inside diameters of 6.40 cm. The hose carries a flow of 40.0 L/s at a gauge pressure of 1.62×10^6 N/m^2. The hose goes 10.0 m up a ladder to a nozzle having an inside diameter of 3.0 cm. Neglecting the frictional resistance, show that pressure in the nozzle is atmospheric?

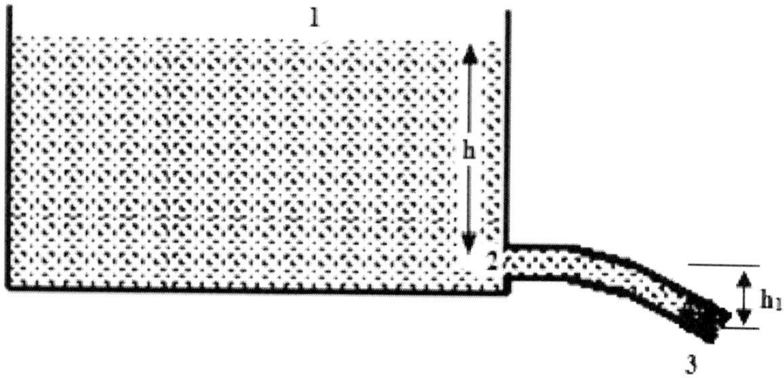

Figure 5.16. Flow Through an Extended Nozzle Fitted in a Reservoir.

As water exits into the atmosphere without change in its conditions, the nozzle pressure is equal to the atmospheric pressure. Thus, gauge pressure should be zero. To prove this, use the Bernoulli's equation:

$$\frac{1}{2}\rho V_1^2 + P_1 + \rho g\, h_1 = \frac{1}{2}\rho V_2^2 + P_2 + \rho g\, h_2$$

Taking the datum at the bottom of the ladder or ground, $h_1 = 0$ and $h_2 = 10.0$

Calculate the velocities by dividing the discharge with the areas at the two ends.

$$V_1 = 0.040/(\pi \times 0.064^2/4) = 12.4 \text{ m/s}$$

Similarly $V_2 = 0.040/(\pi \times 0.03^2/4) = 56.6$ m/s

Incorporating the values in equation

$$\frac{1}{2}\rho V_1^2 + P_1 - \rho g\, h_2 - \frac{1}{2}\rho V_2^2 = P_2$$

$$P_2 = 1.62.10^6 + 0.5 \times 1000(12.4^2 - 56.6^2) - 9.81 \times 1000 \times 10$$

$$P_2 = 1.62.10^6 + 0.5 \times 1000(12.4^2 - 56.6^2) - 98100$$

$$P_2 = 1620000 + 500\ (153.76 - 3203.56) - 98100$$

$$P_2 = 1620000 + 500\ (153.76 - 3203.56) - 98100$$

$$P_2 = -3000 \text{ (It is nearly zero)}$$

Power

The power in fluid flow is given by the equation

$$\text{Power} = Q\,(P + \frac{1}{2}(\rho V^2) + \rho gh)$$

Here the first term is power associated with pressure, the second is power associated with velocity, and the third is power associated with height. For example, PQ is the power supplied to a fluid, perhaps by a pump, to give it pressure. Similarly, $Q \times \frac{1}{2}(\rho V^2)$ is the power supplied to a fluid to give it kinetic energy and $Q \times \rho gh$ is the power going to gravitational potential energy.

Example 5.14

A 150-mm pipeline conducts water from the reservoir and discharges at a lower elevation through a nozzle, which has a discharge diameter of 50 mm (Figure 5.17). The water surface in the reservoir is at an elevation of 30 m, the pipe intake 2 and 3 at elevation 25 m and the nozzle 4 and 5 at elevation 0. The head losses are: from 1 to 2, 0; from 2 to 3, 0.6 m; from 3 to 4, 9 m; from 4 to 5, 3 m. Compute the discharge and make a table showing elevation head, pressure head, and total head at each of the five points.

Since head losses (HL) for various points are given, total head loss is the sum of the various head losses.

$$HL_{1-5} = HL_{1-2} + HL_{2-3} + HL_{3-4} + HL_{4-5} \tag{5.66}$$

$$HL_{1-5} = 0 + 0.6 + 9 + 3 = 12.6 \text{ m}$$

Energy equation between 1 and 5

$$E_1 - HL_{1-5} = E_5$$

$$V_1^2/2g + P_1/\gamma + Z_1 - HL_{1-5} = V_5^2/2g + P_5/\gamma + Z_5 \tag{5.67}$$

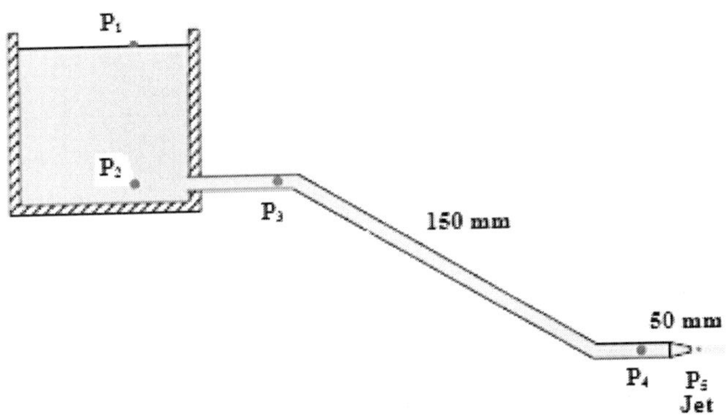

Figure 5.17. Water Flow from a Reservoir through the Pipeline.

Since $V_5 = Q/(\pi/4 \times D_5^2)$, We have

$$(0+0+30) - 12.6 = 8Q^2/(\pi^2 g D_5^4) + 0 + 0$$

$$17.4 = 8Q^2/\pi^2(9.81)(0.05^4)$$

$$Q = 0.0363 \text{ m}^3/\text{s} = 36.3 \text{ L/s}$$

Velocity heads

$$V^2/2g = 8Q^2/\pi^2 g D^4$$

$$V_3^2/2g = V_4^2/2g = 8(0.0363)^2/\pi^2(9.81)(0.15)^4 = 0.2151 \text{ m}$$

$$V_5^2/2g = 8(0.0363)^2/\pi^2(9.81)(0.050)^4 = 17.4202$$

Pressure heads

$$P_2/\gamma = E_1 - E_2 = 30 - 25$$

$P_2/\gamma = 5$ m

Energy equation between 2 and 3

$E_2 - HL_{2-3} = E_3$

$V_2^2/2g + P_2/\gamma + Z_2 - HL_{2-3} = V_3^2/2g + P_3/\gamma + Z_3$

$(0+5+25) - 0.6 = 0.2151 + P_3/\gamma + 25$

$P_3/\gamma = 4.1849$ m

Energy equation between 3 and 4

$E_3 - HL_{3-4} = E_4$

$$V_3^2/2g + P_3/\gamma + Z_3 - HL_{3-4} = V_4^2/2g + P_4/\gamma + Z_4 \quad (5.68)$$

$(0.2151 + 4.1849 + 25) - 9 = 0.2151 + P_4/\gamma + 0$

$P_4/\gamma = 20.1849$

The results in tabulated form are as follows.

Point	Elevation Head (m)	Velocity Head (m)	Pressure Head (m)	Total Head (m)
1	30	0	0	30.0
2	25	0	5	30.0
3	25	0.2151	4.185	29.4
4	0	0.2151	20.185	20.4
5	0	17.4202	0	17.4

ROTAMETER

A Rotameter is a variable area meter device. It measures the flow rate of liquids or gases in a closed tube. As the cross-sectional area through which fluid travels varies, it results in some measurable effect, measured value being related to the flow rate.

Construction

A rotameter consists of a variable area tube (tapered) with a float inside the tube. The glass-tapered tube has a scale on the surface. If not, a scale is fixed near the tube (Figure 5.18). The tube is provided with suitable inlet and outlet connections. The specific gravity of the float is higher than that of the fluid. Float made of stainless steel are commonly used, but floats made of other metals such as lead or aluminium *etc.* are also common. Floats may vary in shape as per application needs, density of fluid and the metal used. Floats with spherical shape are appropriate to measure small flows.

Principle of Working

The tapered tube is placed vertically in the flow channel in such a way that the conical shape is inside (Figure 5.18). As the fluid flows through the tube, it moves the float upward against gravity because of a combination of forces namely the buoyancy of the liquid and the velocity head of the fluid. In gases, buoyancy being negligible, the float response is mainly from the velocity head. More the flow, higher goes the float. At some point, the float will be in equilibrium *i.e.* at this point the area around float generates enough drag equal to weight - buoyancy. A change in flow rate upsets this balance of forces. The float then moves up or down, changing the annular area until it again reaches a position where the forces are in equilibrium. To satisfy the force equation, the rotameter float assumes a distinct position for every constant flow rate. The float's height in the tube is used to determine the flow rate on a calibrated reference.

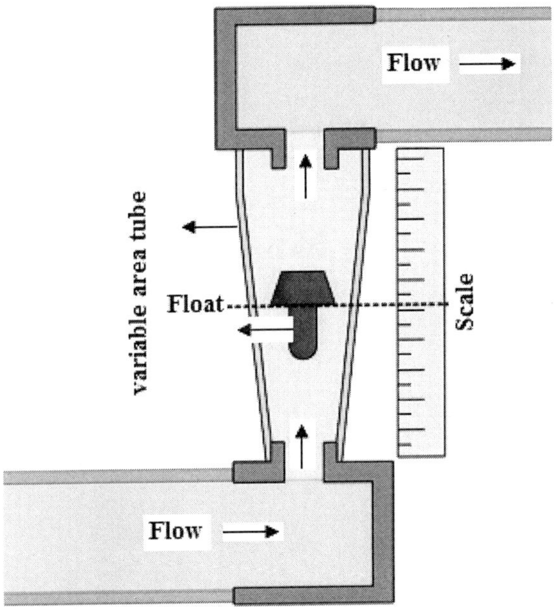

Figure 5.18. A Schematic View of a Rotameter: Note the Vertical Placement of the Meter.

As the float weight and gravity are constant, the distance to which float is displaced upward is proportional to the flow velocity of the fluid passing through the tapered tube. The discharge (Q) is given as:

$$Q = C_d\, a_{ft}\left[2g\, V_f \frac{(\rho_f - \rho)}{A_f \acute{A}}\right]^{0.5} \tag{5.69}$$

Here, Q is the discharge, C_d is the discharge coefficient usually less than 1, ρ_f is the density of fluid, ρ is the density of float, A_f is the maximum cross sectional area of float, a_{ft} is the annular area between float and tube and V_f is the volume of float. If design of the tube is such that a_{ft} varies linearly with the displacement, one has a linear relationship of the form

$$Q = K_1 + K_2\, x \tag{5.70}$$

Here x is the height of the float. Thus, the scale of the tube can be graduated linearly in terms of flow rate. The scale of the rotameter can be calibrated for direct reading of flow, or it may have a scale to read a percent of range or an arbitrary scale to be used with conversion equations or charts.

Advantages

1. Simple reliable design
2. No external power or energy required for its operation
3. Easy to handle and install
4. Relatively low cost
5. Can handle corrosive fluids
6. Scale is approximately linear
7. Low pressure drop
8. High repeatability
9. Wide range of measurements

Disadvantages

1. Because the float position is gravity dependent, rotameters must be vertically oriented and mounted
2. Difficult to handle the glass type rotameters
3. Glass tube rotameters have limitations for use in high pressure applications
4. Changes in viscosity may affect the results
5. Inability to measure opaque fluids because the float will not be visible.

Types of Rotameters

On the basis of construction, rotameters are classified in following three types.

- Glass tube rotameters
- Metal tube rotameters
- Plastic tube rotameters

The basic principle of operation of all these rotameters is the same.

Glass tube rotameters

It is the basic rotameter. The tube is precision formed of borosilicate glass, and the float is precisely machined from metal, glass or plastic. The metal float to resist corrosion is usually made of stainless steel. The float has a sharp metering edge where the reading is observed by means of a scale mounted alongside the tube. These are manufactured with end fittings of metal or plastic to take care of the chemical characteristics of the fluid being metered. Safety-shielded glass tube rotameters are in general use throughout the industry to measure flows of liquids and gases. These meters are not suited for fluids, which attack glass metering tubes, such as water with high pH and/or over 90°C (194°F) and wet steam both having softening effect on glass, caustic soda, which dissolves glass; and hydrofluoric acid, which etches glass.

Metal tube rotameter

Metal tubes are used for applications involving high pressures and temperatures beyond the practical range of glass tubes. These tubes are manufactured in aluminium, brass or stainless steel. The position of the float is observed through magnetic or mechanical followers that are located outside the metal metering tube. Similar to glass tube rotameters, fittings and materials of construction are chosen to satisfy application demands. These meters are used for services where high operating pressure or temperature, water hammer, or other forces are likely to damage glass metering tubes. These can be used for most fluids, including corrosive liquids and gases. They are particularly suited for steam, where glass tubes are unacceptable.

Plastic tube rotameters

Plastic tubes are used in some rotameters due to their low cost and high impact strength. They are typically constructed of polycarbonate, with either metal or plastic end fittings. With plastic end fittings, one must be careful during installation so as not to distort the threads. Rotameters with all plastic construction are available for applications where metal parts cannot be tolerated, such as with deionized water or corrosive chemicals.

QUESTIONS (THEORY)

1. Discuss continuity equation. Derive the continuity equation in Cartesian coordinate system.
2. Derive the Bernoulli's' equation. What does this equation states?
3. Define hydraulic grade line and energy grade line. Show by a neat sketch the two lines in the case of open channel and pipe flow.
4. Draw a neat diagram showing the hydraulic and energy grade lines for flow through a diverging pipe in the direction of flow.
5. Use the Bernoulli's equation to show that pressure at the top of the siphon is less than the atmosphere pressure making it possible for the water to flow.
6. Derive the discharge equation of a venturi meter. List the advantages and disadvantages of a venturi meter.
7. Discuss the construction and principle of working of a rotameter. Write down the discharge equation for a rotameter.
8. Differentiate between the following: The Pitot tube and the Pitot-static tube, static and stagnating pressure

Chapter 6

Flow Measurement: Weirs Notches and Orifices

INTRODUCTION

Fluid flow and level measurements are increasingly becoming important in all industries dealing with any kind of fluid. Its importance in water sector especially in irrigation is crucial to manage this critical resource. A large number of instruments/ equipments/meters are in the market to measure various kinds of fluids, some being specific to specific industrial requirement. In the irrigation sector, we use current meters, calibrated or rated channel cross sections, flumes or standardized weirs/notches and orifices to measure water flows. The measurements are generally recorded either manually or through automatic recorders. This chapter discusses some of the water measurement equipment and procedures of measurements. The discussions are limited to weirs/notches, orifices, mouthpieces and water level measuring instruments.

IRRIGATION SECTOR AND FLOW MEASUREMENT

Irrigation water is an essential input for crop production. On the other hand, mismanagement of irrigation water created many secondary problems across all economic sectors. In agriculture, irrigation efficiencies resulting from high seepage losses from the conveyance system, poor land development and mismanagement of the irrigation water resources are often debated. Over- exploitation of this natural resource has resulted in depleting groundwater table in high productivity areas. Such a scenario calls for water auditing at the farm level, which utilizes about 80 per cent of the water used in the country. Systematic water measurement at various scales, proper recording and interpretation is prerequisite of any large-scale water auditing. Besides, water measurement helps in:

- Accurate accounting of the precious resource
- Allocate equitable share of water between competitive users
- Accurate water measurement provides on-farm irrigation users to design conveyance systems, achieve the best use of the irrigation water with minimal negative environmental impacts
- Higher yields by avoiding under-irrigation or over irrigation of the crops
- First step in volumetric supply and pricing of water by government departments
- Evaluation of seepage losses in channels and determine cost benefits of proposed canal and ditch improvement programmes
- Important in other studies such as testing of wells, runoff measurement and planning of soil conservation structures according to anticipated flow rates.

Unit of Measurement

While quantity of water is usually measured in volumetric units, flows are described in terms of rate *i.e.* volume of water per unit time. Units to report irrigation water measurements vary from one nation to another, metric system being common in India. In metric units, volume is reported in terms of Liter (L), cubic meter (m^3) or cubic cm (cm^3) or hectare centimetre (ha-cm) or hectare meter (ha-m). Flow rates are measured in Liter per second (L/s) and cubic meter per second (m^3/s or commonly written as cumecs). Volume or flow rates even now are recorded in British units because dimensions of many structures are still in these units. Conversion tables are then used to convert the British units to the metric units (Annexure I).

Methods of Discharge Measurement

Commonly used devices/techniques to measure irrigation water in open channels are grouped into five categories (Miller, 1984):

1) Volumetric measurements
2) Velocity-area methods
    a) Float method
    b) Water meters
3) Tracer methods
4) Measuring structures
    a) Orifices
    b) Weirs/notches
    c) Flumes
5) Coordinate and other methods for pipe flows

This chapter deals only with the notches, weirs, orifices and mouthpieces. For other methods readers may refer to any other textbook of hydrology, irrigation water management or similar other books (Gupta, 2020).

Notch

A notch is an obstruction over which liquid flows. The depth of flow above the base of the notch being related to the discharge, it acts as a flow-measuring device. When measurements are made from a tank or reservoir, the opening is provided at the side of the tank. In such a case, the liquid surface in the tank is below the top edge of the opening. Thus, it acts as a large opening having no upper edge. It has a variable area depending upon the level of the free surface. Notch is generally of smaller size and is made of iron plate, steel, tin *etc.*

Weir

A weir is also a notch but of large size used for measuring the flow of a river, canal *etc.* It is a concrete or masonry structure of substantial breadth built across the river in the direction of flow. This allows the excess water to flow over its entire length to the downstream side. A structure, used to dam up a stream or river, over which water flows, is also a weir. Thus, a weir is similar to a small dam constructed across the river. The only difference being that in a dam the excess water flows downstream only through a small portion called spillway but in the case of weir, the excess water flows over its entire length.

Difference between a notch and a weir

A list of few differences between a notch and a weir are listed in tabular form below.

Notch	*Weir*
A notch is an opening provided in the side of tank or vessel or small rivulets or channels such that the liquid surface in tank is below the top edge of opening	A weir is any regular obstruction in open stream over which the flow takes place
Small structure	Large structure
A notch is usually made of metallic plates	A weir is made of cement concrete or masonry
Measures small flow rates	Measures large flow rates
It is more accurate	It is relatively less accurate
Primarily used to measure water flow	Primarily used as a part of bigger structure, water measurement is secondary

The conditions of flow, in the case of a weir, are practically the same as those of a notch. It is why many times a notch is called a weir and vice-versa. As long as we know, what we are studying, it should not matter whether we call it a notch or a weir. Going by the tradition of using the term weir in agricultural engineering, we use the term weir for both small and large structures. Wherever needed, some clarification is appended for the benefit of the readers.

Classification

Weirs/notches are classified in a number of ways. The most important classification is based on the shape of the opening. On this basis, notches/weirs have four groups (Figure 6.1).

- Rectangular weir
- Triangular notch/weir
- Trapezoidal weir (Cipolletti weir)
- Compound weir

Another classification based on the nature of the crest, categorizes the weirs into

- Sharp crested
- Broad crested
- Narrow crested
- Ogee shaped weir

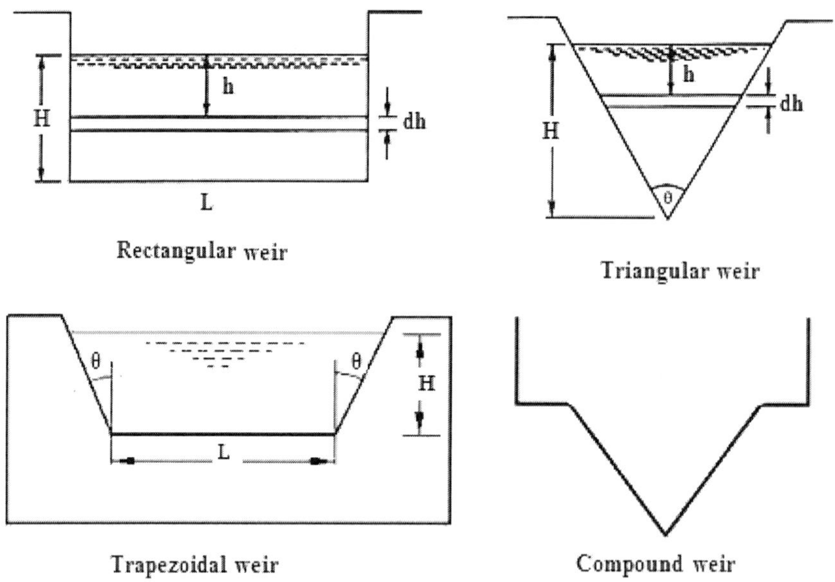

Figure 6.1. Classification of Weirs/Notches Based on the Shape of the Opening.

Sharp and broad crested weirs

Sharp crested weir: A weir that has a thin-edged crest so that the over-flowing sheet of water has minimum surface contact with the crest is a sharp crested weir. The notches are generally sharp edged (called sharp crested weir or notch) bevelled on the downstream side (Figure 6.2 left).

Broad crested weir: A broad crested weir has a substantial breadth (Figure 6.2 right) in the direction of flow. It is normally a flat-topped obstruction, which extends across the entire channel. A true broad-crested weir flow occurs when upstream head above the crest is between the limits of about 1/20 and 1/2 the crest length in the direction of flow. For example, a thick wall or a flat wooden log can act like a sharp-crested weir when the approach head is large enough that the flow springs from the upstream corner. If upstream head is small enough relative to the top profile length, the wooden log acts like a broad-crested weir.

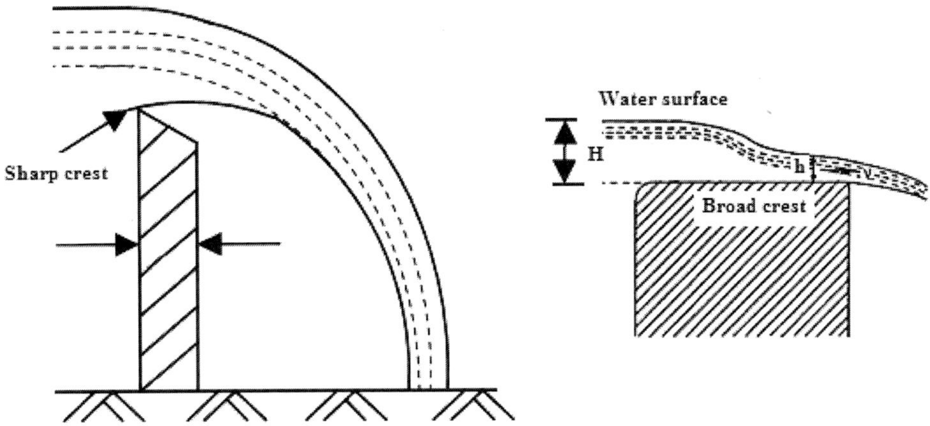

Figure 6.2. A Sharp Crested (Left) and a Broad Crested (Right) Weir.

Sharp crested weirs (or notches) are applicable to measure the discharge in small open channels where higher accuracy is required. Compared to the sharp crested weirs, a broad crested weir operates on a very different principle. Therefore, one must not confuse between the two types of weirs. A narrow crested weir is one where height of the water above the weir crest is greater than two times the width of the crest of the weir. A narrow weir is hydraulically similar to ordinary weir and therefore, formulae derived for ordinary weirs hold good in this case as well.

According to the effect of the sides on nappe (or vein), weir/notches are classified as

- ☆ Weir/notch without end contraction or suppressed weir/notch
- ☆ Weir/notch with end contraction or contracted weir/notch

Suppressed weirs

When the length of crest of the weir is same as the width of the channel, the weir is said to be a suppressed weir (Figure 6.3 left). Thus, in suppressed weir, length of weir crest is equal to the width of channel. Suppressed weirs eliminate or suppress the effects of sides of the weir/notch on flow.

Contracted weirs

End Contraction: An end contracted weir means that the ditch leading up to the weir is wider than the weir opening (Figure 6.3 right).

Bottom Contraction: When the crest of the weir is elevated above the floor of the channel, it results in bottom contraction. In this case, the flow coming along the bottom of the weir pool and up a sufficiently high bulkhead springs up in a curved, underside jet surface causing crest or bottom contraction. A weir is a fully contracted weir, if approach conditions allow full contractions at the ends and at the bottom, (Figure 6.3 right).

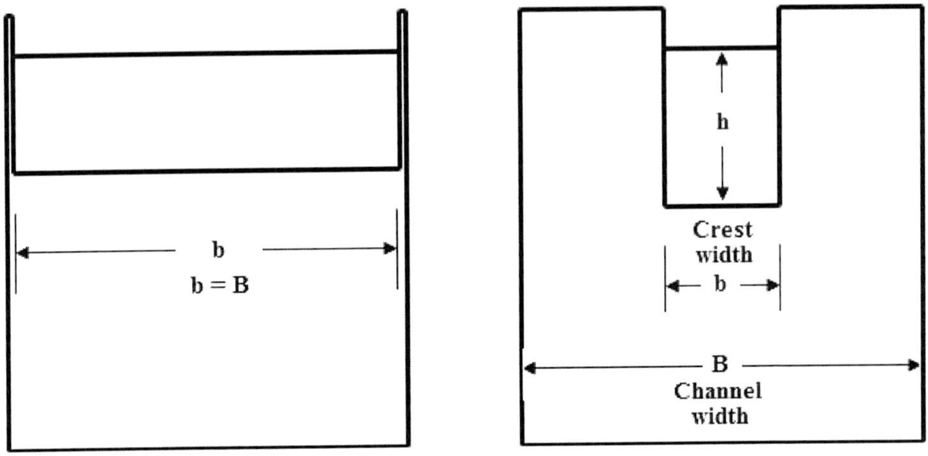

Figure 6.3. A Suppressed (Left) and a Contracted Rectangular Weir (Right).

Free and submerged flow

Based on the nature of the discharge, weirs are of two types namely ordinary or free flow weir and submerged or drowned weir.

If the downstream conditions do not restrict the discharge from the weir then it is a free flow or a modular flow weir. Free-flow exists when the downstream water level is lower than the lowest point of the weir crest elevation (Figure 6.4 left). Here the nappe freely discharges into the air. The flow is submerged flow, if the downstream water level is higher than the lowest point of the weir crest elevation (Figure 6.4 right). The discharge of water over the weir in this case is less than the calculated discharge for free flow condition. The free and submerged flow conditions are governed by submergence ratio.

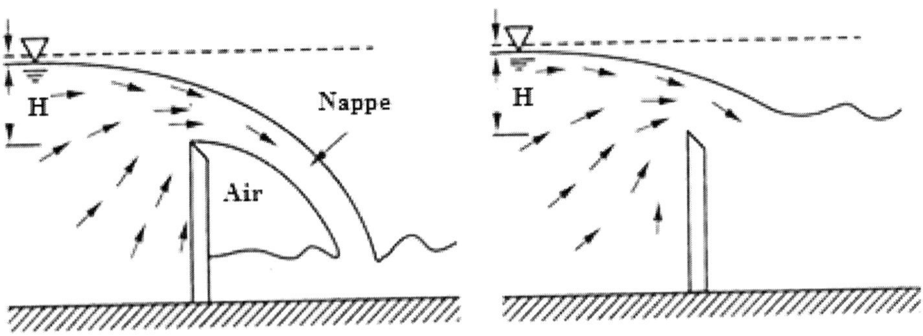

Figure 6.4. Free Flowing (Left) and Submerged Weir/Notch (Right).

Submergence ratio (SR)

Submergence ratio defined as the ratio of heads available on the downstream side of the weir, H_2 to the head available in the upstream side, H_1 plus velocity head, is written as:

$$SR = H_2/(H_1 + V_1^2/2g) \tag{6.1}$$

Both sharp and broad crested weirs are susceptible to submergence depending on the values of SR. The sharp crested weirs behave like a free weir only up to SR value of 0.66. The corresponding values for broad crested weirs are 0.85 (Sahoo, 2009).

Commonly Used Terms

This section defines some of the commonly used terms related to weir/notches/orifices (Figure 6.5).

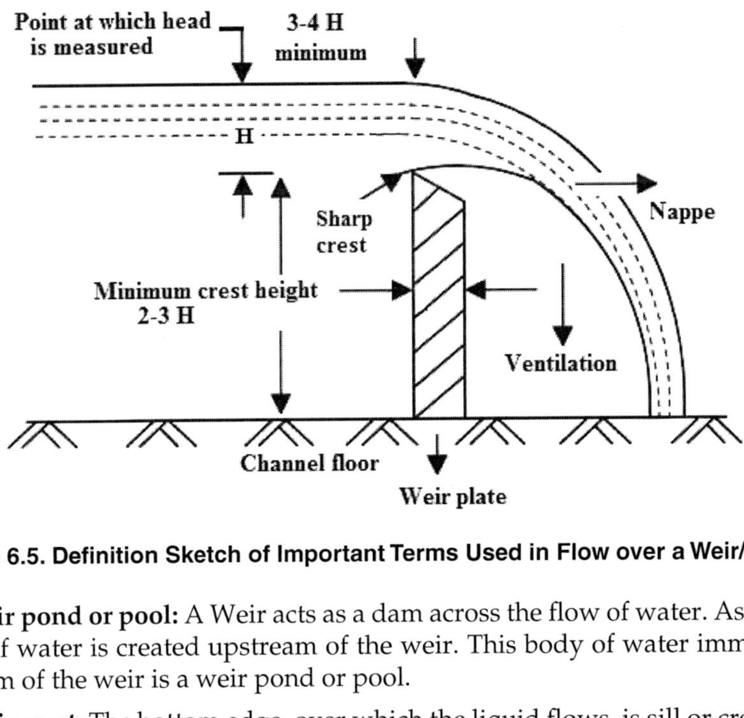

Figure 6.5. Definition Sketch of Important Terms Used in Flow over a Weir/Notch.

Weir pond or pool: A Weir acts as a dam across the flow of water. As a result, a pool of water is created upstream of the weir. This body of water immediately upstream of the weir is a weir pond or pool.

Weir crest: The bottom edge, over which the liquid flows, is sill or crest of the weir/notch.

Crest height: The height above the bottom of the tank or channel is the crest height (Figure 6.5).

Head: The depth of water flowing over the weir crest measured at some point in the weir pond.

Weir scale or gauge: The scale fastened on the side of weir or on a stake in the weir pond to measure the head on the weir crest.

Nappe: The sheet of liquid flowing over the weir (or a notch) crest, springing clear under free flow conditions is a nappe or a vein (Figure 6.5).

Difference between an orifice and a weir/notch

An orifice is an opening provided in the side or bottom of tank or vessel such that the liquid flow through the entire orifice.

A weir/notch is an opening provided in the side of tank or vessel such that the liquid surface in tank is below the top edge of opening.

DERIVATION OF DISCHARGE EQUATIONS

Most investigators derive the equation for sharp-crested rectangular weirs by mathematical integration of elemental orifice strips over the nappe (Bos, 1989). Each strip is treated as an orifice with a different head on it.

Rectangular Weir

For a weir/notch of any shape (Figure 6.6), consider a horizontal strip of width b and height dh at a depth h below the free surface. Then

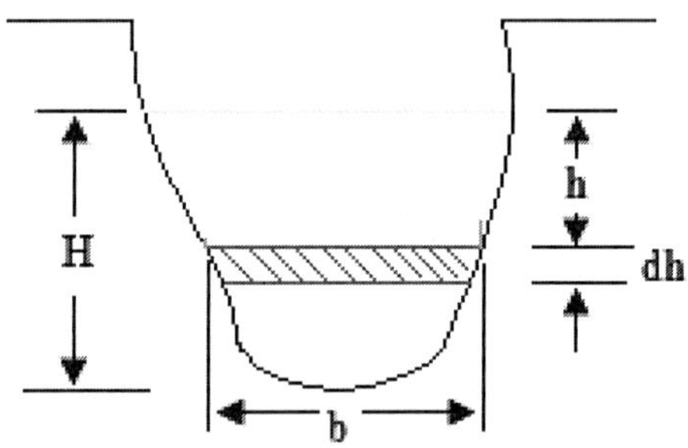

Figure 6.6. Definition Sketch to Illustrate the Derivation of Discharge Equations.

Area of strip = bdh

Velocity through strip = √(2gh)

Discharge through strip, dQ = Area × velocity = bdh√(2gh) (6.2)

To get the total theoretical discharge, Q, we integrate from h = 0 at the free surface to h = H at the bottom of the notch, such that

$$Q = \sqrt{2g}\int_0^H bh^{0.5}dh \tag{6.3}$$

Before we integrate eq. (6.3), it is necessary to express b in terms of h. Since for a rectangular notch, b is equal to L, a constant (Figure 6.1)

$$Q = \sqrt{2g}\int_0^H Lh^{0.5}dh$$

$$Q = \frac{2}{3}\sqrt{2g}LH^{1.5} \tag{6.4}$$

Number of corrections are made to eq. (6.4) to improve its measurement accuracy.

Correction for velocity of approach

The water flow approaching the notch possesses a certain velocity known as the velocity of approach (Figure 6.7). It increases the discharge over the weir. Let V_a be the velocity of approach. This condition is equivalent to a condition in which free water surface is static at a higher level, at a height $h_a = V_a^2/2g$. The head h_a is the head due to the velocity of approach. If the head of the actual flowing water level over the crest is H, the weir opening may be taken as equivalent to a large orifice in which the heads of still water above the top and bottom edges are respectively, h_a and $H + h_a$.

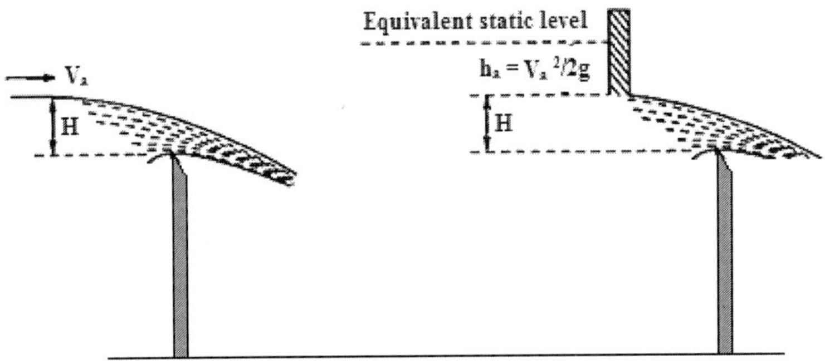

Figure 6.7. Discharge Correction due to Velocity of Approach: Schematics of Velocity Head.

Accordingly, following equation expresses the discharge.

$$Q = \frac{2}{3}\sqrt{2g}L\left[(H + h_a)^{1.5} - h_a^{1.5}\right] \tag{6.5}$$

The quantity $H + h_a$ is called still water head. The velocity of approach is determined by dividing the discharge by the area of flow of water in the stream on the upstream side of the weir. We follow an iterative approach while using the equation of discharge with velocity correction. Firstly, we calculate the discharge ignoring the velocity of approach. Then we divide this approximate value of the discharge by the area of approach to obtain an approximate value of the velocity of approach. It is used to calculate the corresponding velocity head due to velocity of approach. A more accurate value of the discharge is calculated with the help of eq. (6.5). This process is repeated until arriving at a discharge value of desired accuracy.

Corrections for end contractions

As already stated in a previous section, there is lateral contraction at each end if the length of the notch is less than the width of the stream. According to Francis, each lateral contraction is equal to 0.1 H. If the actual length of the weir is L, then the effective length of the weir will be (L – 0.2 H) for the two end contractions (Figure 6.8). In some cases, there may be intermediate obstacles like piers over the weir. In such a case if there are n lateral contractions, the effective length of the weir will be (l – 0.1 nH). Here n are the number of contractions. Hence the discharge over the weir will be

$$Q = \frac{2}{3}\sqrt{2g}(L - 0.1nH)H^{1.5} \tag{6.6}$$

Value of n is zero, if the notch covers the full width of the channel; n = 1 if the notch is narrower than the channel but is arranged with one edge coincident with the edge of the channel and n = 2 if the notch is narrower than the channel and is symmetrically located on the two sides.

If the velocity of approach is also taken into account, then eq. (6.6) is written as:

$$Q = C_d \frac{2}{3}\sqrt{2g}[L - 0.1n(H + h_a)]\left[(H + h_a)^{1.5} - h_a^{\ 1.5}\right] \tag{6.7}$$

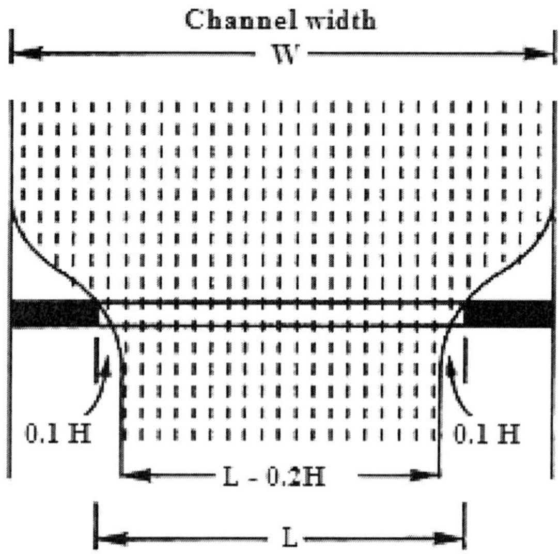

Figure 6.8. Contractions in a Two Sided Contracted Weir.

Coefficient of Discharge for Rectangular Weir

So far, we have derived the theoretical discharge equations for a rectangular notch. The actual discharge is always less than the theoretical discharge. We therefore, multiply the right hand side of each equation by coefficient of discharge, C_d to allow for energy losses and the contraction of the cross-section of the stream at the bottom and sides. For a rectangular weir, Rehbock formula helps to calculate the value of C_d (Towler, 2012).

$$C_d = 0.602 + 0.083H/P \tag{6.8}$$

Here H is the head over the crest of the weir and P is the height of the weir (Figure 6.9). The above equation is valid for P varying from 0.1 to 1.0 m and H from 0.024 to 0.6 m. Normally the value of C_d is taken as 0.623.

The most general equation of discharge after incorporating C_d is:

$$Q = \frac{2}{3}\sqrt{2g}[L - 0.1n(H + h_a)]\left[(H + h_a)^{1.5} - h_a^{\ 1.5}\right] \tag{6.9}$$

Francis Formula

Francis formula commonly used to measure the discharge in suppressed or contracted weirs without velocity correction is

$$Q = C_d \frac{2}{3}\sqrt{2g}[L - 0.1nH]H^{1.5} \tag{6.10}$$

Figure 6.9. Illustration of Parameters Used in the Rehbock Equation.

According to Francis, weir coefficient, a lumped parameter includes coefficient of discharge, acceleration due to gravity and some geometrical constant. Combining them together, we have

$$Q = 1.839[L - 0.1nH]H^{1.5} \sim 1.84[L - 0.1nH]H^{1.5} \tag{6.11}$$

Here Q is the water flow rate, m^3/s, L is the length of the weir, m, H is the head over the weir, m, and n is the number of contractions.

Submerged Weir

When the water on the downstream side of a weir/notch rises above the level of the crest, the weir is said to be a submerged weir (Figure 6.10). It is possible to divide the flow over the submerged weir into two portions:

Flow over the upper part of the line AB is weir flow designated as Q_1.

$$Q_1 = C_{d1} \frac{2}{3}\sqrt{2g}L(H_1 - H_2)^{1.5} \tag{6.12}$$

Flow through the remaining depth H_2 is the discharge through a submerged orifice designated as Q_2.

$$Q_2 = Cd_2 \ LH_2\sqrt{2g(H_1 - H_2)} \tag{6.13}$$

Here C_{d1} and C_{d2} are the coefficients of discharge for weir and orifice respectively.

Total flow rate $Q = Q_1 + Q_2$

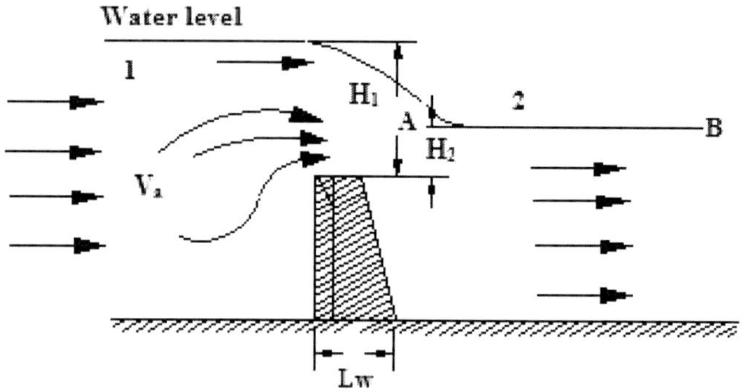

Figure 6.10. Submerged Flow Over a Weir/Notch.

Time of emptying a reservoir by a rectangular weir

Let us consider a reservoir of area, A provided with a rectangular weir of length L. We wish to find the time taken for the head of water over the weir to fall from an initial value H_1 to a value H_2. Let the head of water at any instant over the weir be h. Let the fall in water level in a small interval of time dt be dh. Then water discharged in time dt is given as

$$dQ = Adh = (2/3)\, C_d\, L\, \sqrt{2g}\, h^{3/2}\, dt \tag{6.14}$$

$$dt = \frac{Ah^{-1.5}dh}{\frac{2}{3}C_d\, L\sqrt{2g}}$$

If we integrate this equation from time 0 to time t_1 such that head at time t_1 changes from H_1 to H_2, we have

$$t_1 = \frac{A\int_{H1}^{H2} h^{-1.5}dh}{\frac{2}{3}C_d\, L\sqrt{2g}} \tag{6.15}$$

$$t_1 = \frac{2A\left(\frac{1}{\sqrt{H_2}} - \frac{1}{\sqrt{H_1}}\right)}{\frac{2}{3}C_d\, L\sqrt{2g}} \tag{6.16}$$

Example 6.1

A rectangular weir 0.5 m wide covers the whole width of the stream and discharges water at a constant head of 400 mm. Calculate the discharge over the weir in liters per second, if the coefficient of discharge is 0.62.

b = 0.5 m, H = 400 mm = 0.4 m, C_d = 0.62

$$Q_1 = \frac{2}{3} C_d\, L\sqrt{2g}(H)^{1.5}$$

Q = 2/3 [0.62× 0.5×√(2×9.81) × (0.4)$^{1.5}$]

Q = 0.667[0.62× 0.5×4.43 ×0.253] = 0.232 m³/s = 232 L/s

Example 6.2

A rectangular notch 0.8 m high stands clear across a rectangular channel 3 m wide. The depth of water in the approach channel is 1.25 m and at 10 m downstream it is 1.0 m. If C_d value of the submerged weir is 0.62, calculate the discharge.

$$Q_1 = C_{d1}\frac{2}{3}\sqrt{2g}L(H_1 - H_2)^{1.5} = 0.62\times0.667\times4.43\times3\times(0.45\text{-}0.2)^{3/2} = 5.496\times0.125$$

=0.687 m³/s

$$Q_2 = C_{d2}\; LH_2\sqrt{2g(H_1 - H_2)} = 0.62\times3\times0.2\times\sqrt{2\times9.81\times0.25} = 0.372\times2.215 =$$

0.824 m³/s

Total discharge = 0.687 +0.824 = 1.511 m³/s or 1511 L/s

V-notch

For a V-notch shown in Figure 6.1 with an included angle θ, let us begin with eq. (6.3) derived earlier

$$Q = \sqrt{2g}\int_0^H bh_1^{\,0.5}dh \tag{6.17}$$

We incorporate b as a function of h given as

b = 2(H-h) tan (θ/2), such that

$$Q = 2\sqrt{2g}\,\tan\left(\frac{\theta}{2}\right)\int_0^H (H-h)h^{0.5}dh$$

$$Q = 2\sqrt{2g}\,\tan\left(\frac{\theta}{2}\right)\left[\frac{2}{3}Hh^{1.5} - \frac{2}{5}h^{2.5}\right] \tag{6.18}$$

Incorporating the limits for terms in the square bracket from 0 to H and simplifying, we get

$$Q = \frac{8}{15}\sqrt{2g}\,\tan\left(\frac{\theta}{2}\right)H^{2.5} \tag{6.19}$$

This equation is valid for angles in the range of 25° to 100°.

As in rectangular notch, we multiply the theoretical discharge by C_d to get the actual discharge. C_d takes care of energy losses and the contraction of the cross-section of the stream at the bottom and sides.

$$Q = C_d \frac{8}{15}\sqrt{2g}\tan\left(\frac{\theta}{2}\right)H^{2.5} \tag{6.20}$$

The coefficient of discharge, C_d besides other factor depends on the vertex angle of the notch (Table 6.1). The value is more at lower heads and smaller angles of the notch/weir, ascribed to lesser contraction of the nappe. Besides, a head correction is also required in eq. (6.20). The effective head, H_e being $H_e = H + K_h$, where K_h is an empirical head correction factor. It is a function of the notch angle θ (Table 6.1).

Table 6.1. Coefficients of Discharge for a Sharp-Crested V-notch (Kindsvater and Carter, 1957)

Notch angle θ (degree)	20	40	60	80	100
C_d	0.595	0.581	0.577	0.577	0.580
K_h (mm)	2.8	1.8	1.2	0.85	0.80

For a commonly used notch having a vertex angle of 90°, we have tan 45° =1 and therefore

$$Q = C_d \frac{8}{15}\sqrt{2g}H^{2.5} \tag{6.21}$$

If we take C_d as 0.6, then

$$Q = 1.417H^{2.5} \tag{6.22}$$

This equation for a V-notch is valid for a fully contracted V notch. Following relations must hold true for a V-notch to be fully contracted (Figure 6.11).

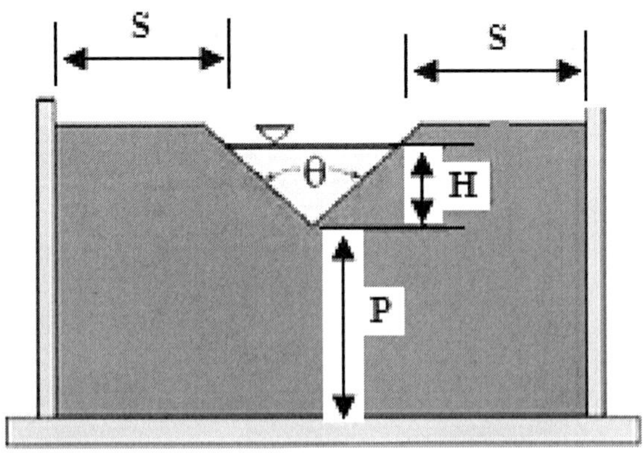

Figure 6.11. Definition Sketch to Describe the Contracted Nature of V-notch.

$P > 2H_{max}, S > 2H_{max}$

In the forgoing theory, two main assumption have been made. One the velocity of the liquid approaching the notch is very small so that it is possible to neglect the kinetic energy. The other assumption is that the velocity through any horizontal element across the notch depends only on its depth below the free surface. These are satisfactory assumptions for flow over a notch or weir in the side of a large reservoir. On the other hand, if the notch or weir is at the end of a narrow channel, the velocity of approach may be substantial. The increase in head due to the approaching velocity of flow then is given as:

$$H_a = H + V_1^2/(2g)$$

Here H_a is the corrected head, H is the head without correction and V_1 is the mean velocity of flow in the approach channel. Note that the value of V_1 is the discharge divided by the full cross sectional area of the channel and not that of the notch.

Time of emptying a reservoir of area A by a triangular notch can be derived in the same manner as for a rectangular weir. The final equation is:

$$t = \frac{2}{3}A\left(\frac{1}{H_2^{1.5}} - \frac{1}{H_1^{1.5}}\right) / \left[\frac{8}{15}C_d\sqrt{2g}\tan\left(\frac{\theta}{2}\right)\right] \tag{6.23}$$

Advantages of a triangular over rectangular notch

A triangular notch as a gauging device in a hydraulic laboratory has certain advantages over the rectangular notch/weir.

- ☆ The coefficient of discharge of a triangular notch is practically independent of the head, because the ratio of the head to the wetted length or crest is constant for all heads. However, in a rectangular weir the ratio of the head to the wetted crest length is not constant. Therefore, coefficient of discharge of a rectangular weir is not actually a constant but a function of the head over the weir.
- ☆ For small discharges, a triangular notch provides greater head than a rectangular weir. Hence, head measurement is more accurate in a triangular than in a rectangular notch/weir.
- ☆ For small discharges, a clinging nappe may be formed in a rectangular weir. For the same discharge over the triangular notch, no clinging nappe is formed because of the greater head.
- ☆ No special arrangement is needed to ventilate the nappe in a triangular weir.

Trapezoidal Weir

Consider a trapezoidal weir whose crest length is L and the sides form an angle θ with the vertical (Figure 6.1). Let H be the head of water over the crest. This weir effectively consists of a rectangular weir of length L and a triangular weir subtending an angle 2θ. The equation for flow through a trapezoidal weir is obtained by adding the discharge equations for rectangular and V-notch/weirs.

$$Q=\frac{2}{3}C_d\sqrt{2g}LH^{1.5}+\frac{8}{15}C_d\sqrt{2g}\tan(\theta)H^{2.5} \quad (6.24)$$

Cipolletti weir (Figure 6.12) is a special case of a trapezoidal weir. In this case, sides of the weir slope is one horizontal to 4 vertical, such that tan θ = ¼. After incorporating this value in eq. (6.24) and simplifying, we get

$$Q=\frac{2}{3}C_d\sqrt{2g}H^{1.5}\left(L+\frac{4}{5}\times\frac{1}{4}H\right) \quad (6.25)$$

$$Q=\frac{2}{3}C_d\sqrt{2g}H^{1.5}(L+0.2H) \quad (6.26)$$

In other words, the discharge from a Cipolletti weir is higher by an amount equal to the reduced amount in a two sides contracted rectangular weir (eq. 6.11). Thus, discharge of a Cipolletti weir can be calculated by the simple equation for a suppressed rectangular weir (eqs. 6.4 or 6.5) although Cipolletti weir is a contracted weir. There is no need to apply contraction correction in the Cipolletti weir.

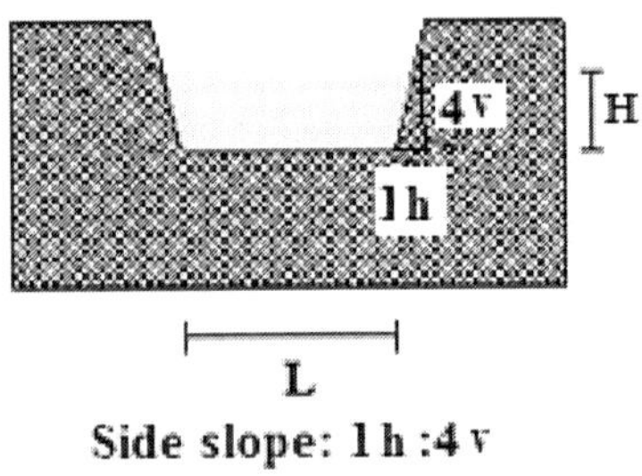

Figure 6.12. Definition Sketch of a Cipolletti Weir.

Compound Weirs

The most commonly used compound or combination weir is the V-notch at the bottom with a rectangular weir at the top (Figure 6.1). It is useful where flow fluctuation is quite wide may be small at one and large at other time. The equation of discharge is written by suitably adding the discharges at various heads as will be shown in the case of a stepped notch.

Stepped-up weir

Another common combination is a stepped weir. A stepped weir is a combination of rectangular weirs as shown in Figure 6.13. It is obvious that the discharge over such a weir will be the sum of the discharges over the different rectangular weirs.

For a two-step weir (Figure 6.13), the discharge equation is:

$$Q = \frac{2}{3} C_d L_1 \sqrt{2g} H_1^{1.5} + \frac{2}{3} C_d L_2 \sqrt{2g} (H_2 - H_1)^{1.5} \quad (6.27)$$

Example 6.3 illustrates the application of this principle for a three-step weir.

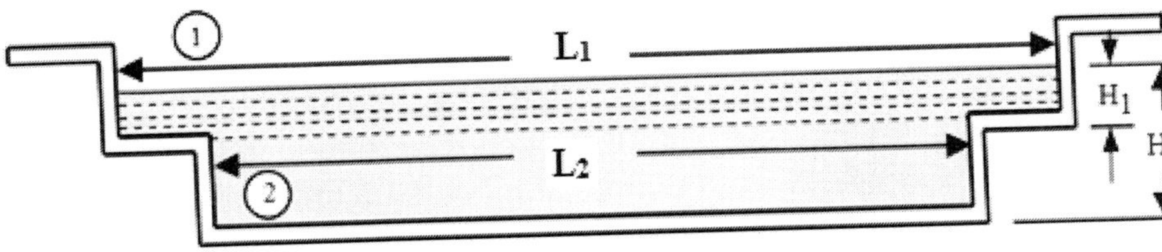

Figure 6.13. Construction Details of a Stepped-up Weir.

Example 6.3

Find out the discharge in L/s for a stepped up weir shown in Figure 6. 14. Take the value of C_d as 0.6

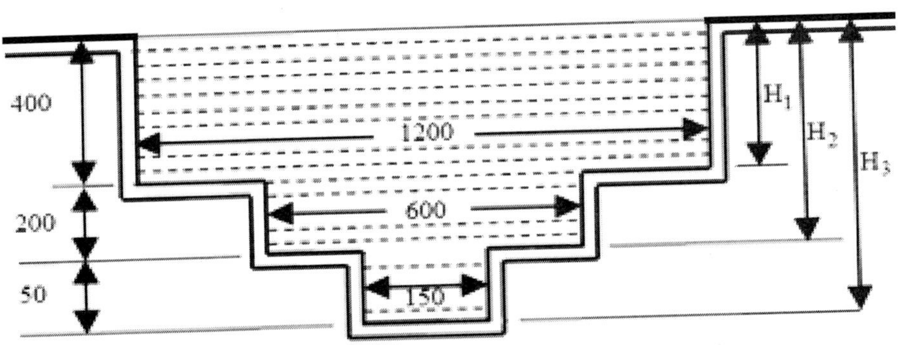

Figure 6.14. Dimensions of a Stepped-up Weir for Example 6.3.

$$Q = Q_1 + Q_2 + Q_3$$

$$Q = \frac{2}{3} C_d L_1 \sqrt{2g} H_1^{1.5} + \frac{2}{3} C_d L_2 \sqrt{2g} (H_2 - H_1)^{1.5} + \frac{2}{3} C_d L_3 \sqrt{2g} (H_3 - H_2)^{1.5}$$

Given C_d =0.6, L_1, L_2 and L_3 = 1200, 600 and 400 and H_1, H_2 and H_3 = 400, 600 and 650

Incorporating all the values in the above equation, we get

$Q = 0.538 + 0.225 + 0.016 = 0.779 \text{ m}^3/\text{s}$

Broad-crested Weir

Working of a broad-crested weir is quite different than a thin-plate and a narrow-crested weir because the flow pattern in these weirs is different. Although, broad-crested weirs can fit in channels of any cross-sectional shape, but discussion in this section is limited to rectangular channels. A broad crested weir has a very broad sill (it may be sharp edged or a smooth curve on the inflow (see Figure 6.15) as well as on the outflow side), so that the water flow over the sill is comparable to the flow of water in a channel (Tracy, 1957). Consider the broad-crested weir shown in Figure 6.15. Let y_1 be the head of water over the weir and L the length of the weir. As the water flows over the weir, it reaches a uniform depth of flow y_2, over the crest. Let V be the uniform velocity of flow over the weir. If we apply the Bernoulli's equation at sections 1-1 and 2-2 taking the crest of the weir as datum, we have

$$y_1 + V_1^2/2g = y_2 + V^2/2g \tag{6.28}$$

If the upstream velocity head is negligible, then

$$y_1 - y_2 = V^2/2g$$

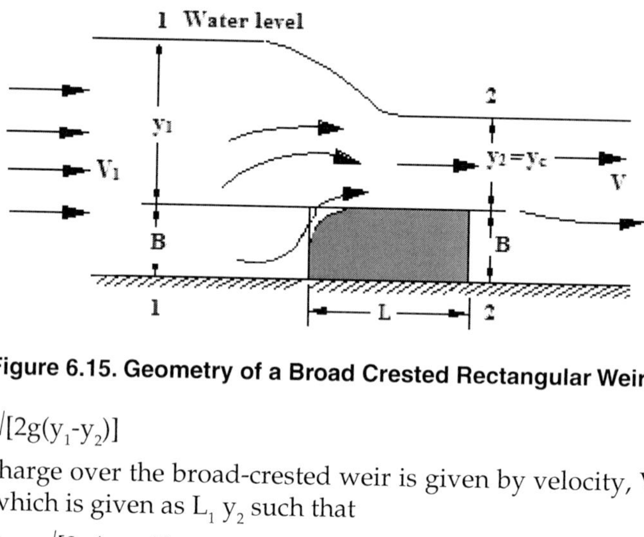

Figure 6.15. Geometry of a Broad Crested Rectangular Weir.

$$V = \sqrt{[2g(y_1-y_2)]}$$

The discharge over the broad-crested weir is given by velocity, V multiplied by the area, which is given as $L_1\ y_2$ such that

$$Q = L_1\ y_2\sqrt{[2g(y_1-y_2)]} \tag{6.29}$$

$$Q = L_1\ \sqrt{[2g(y_1y_2^2 - y_2^3)]} \tag{6.30}$$

Here L_1 is the width of the river/stream. In order to measure the discharge over the broad-crested weir, one needs to measure two heads, y_1 and y_2. Experiments, on the other hand, have shown that the flow adjusts itself to have maximum discharge for the available head y_1. For maximum Q, the term $(y_1y_2^2 - y_2^3)$ must be minimum. Differentiating with respect to y_2 and equating it to zero, we have

$$\frac{d}{dy_2}(y_1 y_2^{\,2} - y_2^{\,3}) = 0$$

$$2y_1y_2 - 3y_2^{\,2} = 0$$

$$y_2 = (2/3)\, y_1$$

This downstream head over the weir is known as critical depth. In other words, the maximum discharge over the broad-crested weir will occur when the critical depth of flow occurs over the surface of the weir crest. Incorporating the value in the discharge equation

$$Q_{max} = L_1\left(\frac{2y_1}{3}\right)\sqrt{2g}\left(y_1 - \frac{2y_1}{3}\right) = 2.3\frac{L_1\sqrt{2g}}{3}y_1^{\,1.5} \tag{6.31}$$

$$Q_{max} = 1.705\, L_1 y_1^{\,1.5} \tag{6.32}$$

To account for the velocity of approach, V_1, modified discharge equation is:

$$Q_{max} = 1.705\, L_1\left[\left(y_1 + \frac{V_1^{\,2}}{2g}\right)^{1.5} - \left(\frac{V_1^{\,2}}{2g}\right)^{1.5}\right] \tag{6.33}$$

Since $V_1^{\,2}/2g$ is normally small, it is either neglected as in eq. (6.32) or multiplied by a coefficient C_v, which can take care of the velocity head. Besides, the discharge calculated with eq. (6.33) is the theoretical discharge. Actual discharge is less than the theoretical discharge because of possible flow curvature, friction losses, and non-uniform velocity distribution. Thus, the actual discharge is theoretical discharge times the coefficient C_d. Any of the following equations gives the actual discharge.

$$Q_{max} = 1.705C_d\, L_1\left[\left(y_1 + \frac{V_1^{\,2}}{2g}\right)^{1.5} - \left(\frac{V_1^{\,2}}{2g}\right)^{1.5}\right] \tag{6.34}$$

$$Q_{max} = 1.705\, C_d\, C_v\, L_1\, y_1^{\,1.5} \tag{6.35}$$

Experimental investigations have shown that if the ratio $y_1/L > 1.6$ then the jet of water touches only the upstream edge and flows clear of the downstream. Weirs falling under this class are thin-plate weirs. On the other hand, if $0.5 \le y_1/L \le 1.6$, the jet of water remains in contact with the entire crest. These kind of weirs are narrow-crested weirs. In both cases, the flow pattern is similar to that of a rectangular weir. Broad-crested weirs are normally operated in the range, $0.08 < y_1/L < 0.5$. It ensures that nearly uniform critical flow is established in the short reach above the weir block. It is not possible to neglect head losses across the long weir blocks *i.e.* $y_1/L < 0.08$. On the other hand, for short weir blocks $y_1/L > 0.5$, the streamlines of the flow over the weir block will not be horizontal.

For a given discharge, width of the channel and depth of the flow before the weir, minimum height of the broad crested weir required to ensure critical flow over the weir crest is calculated by the following equations. Applying the energy equation at the two sections (Figure 6.15), we have

$$(y_1 + B) + V_1^2/2g = (y_2 + B) + V_c^2/2g$$
$$B = (y_1 + B) + V_1^2/2g - y_2 - V_c^2/2g$$

Here (y_1+B) is the depth of flow or channel depth before the weir. Sum of the two values is known although ultimate aim is calculate the value of B. V_c is the critical flow velocity.

$$y_2 = y_c = (Q^2/gL_1^2)^{1/3}$$
$$V = V_c = \sqrt{g y_c} = Q/Ly_c \qquad (6.36)$$

Example 6.4

Determine the discharge over a broad crested weir 50 m long. The depth of flow over the weir is 0.58 m. Take coefficient of discharge as 0.6. If the cross sectional area above the channel is 45 m^2, show that the error in the calculated discharge will be less than 5 per cent compared to the case when the velocity of approach is taken into account.

Using eq. (6.35) with $C_v = 1.0$

$$Q_{max} = C_d \times 1.705\, L_1 y_1^{1.5}$$

$$Q_{max} = 0.6 \times 1.705 \times 50 \times 0.58^{3/2} = 22.60 \text{ m}^3/\text{s}$$
$$V_1 = Q/A = 22.6/45 = 0.5 \text{ and } V_1^2/2g = 0.013$$

Using eq. (6.34)

$$Q_{max} = C_d \times 1.705\, L_1 \left[\left(y_1 + \frac{V_1^2}{2g} \right)^{1.5} - \left(\frac{V_1^2}{2g} \right)^{1.5} \right]$$

$$Q_{max} = 0.6 \times 1.705 \times 50 \times [(0.58 + 0.013)^{3/2} - (0.013)^{3.2}] = 51.15\,(0.457 - 0.0015)$$
$$= 23.3 \text{ m}^3/\text{s}$$

Error (per cent) = 100 (23.3 - 22.6)/23.3 = 3 per cent.

Example 6.5

The discharge in a rectangular channel is 9 m^3/s with a water depth of 1.4 m. If the stream is 5 m wide, what should be the minimum height of the broad crested weir so that it functions with critical depth on the weir crest?

$$V_1 = 9/(5 \times 1.4) = 1.29 \text{ m/s}$$
$$y_c = (Q^2/gL^{12})^{1/3} = (9^2/(9.81 \times 5^2))^{1/3} = 0.691 \text{ m}$$
$$V_c = \sqrt{g y_c} = \sqrt{(9.81 \times 0.691)} = 2.603 \text{ m/s}$$

V_c can also be determined as (e.g. 6.36)

$$V_c = 9/(5 \times 0.691) = 2.604 \text{ m/s}$$
$$B = (y_1 + B) + V_1^2/2g - y_2 - V^2/2g$$
$$B = 1.4 + 1.29^2/(2 \times 9.81) - 0.691 - 2.603^2/(2 \times 9.81) = 1.4 + 0.085 - 0.691 - 0.345$$
$$= 0.449 \text{ m}$$

Ogee Weir

In a sharp crested weir, the nappe leaving the crest springs or rises slightly at the lower surface. It experiences a maximum rise of 0.115 H_1 above the crest and then falls (H_1 is the head over sharp crest). Suppose the space below the bottom surface of the nappe is filled with masonry or concrete, the consequent weir formed is an ogee weir (Figure 6.16). In other words, the solid boundary of an ogee weir exactly coincides with the bottom surface of the nappe of a sharp crested weir under the designed head. The lower part of it is tangential to the upper curve to support the falling sheet of water. Ogee weirs are commonly used as spillways in storage dams.

If H_1 is the head above the sharp crest then the discharge equation is simply given by the Francis formula

$$Q = 1.84\, L_1 H_1^{3/2} \tag{6.37}$$

Alternatively, if the head H above the spillway is considered then the discharge equation is

$$Q = 2.20\, L_1 H^{3/2} \tag{6.38}$$

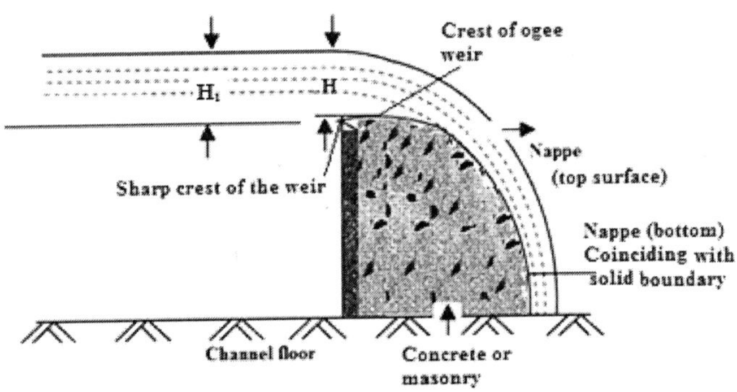

Figure 6.16. A Schematic View of an Ogee Weir.

Example 6.6

Determine the discharge over an ogee spillway of 150 m length under a head of 1.5 m above the spillway.

Using eq. (6.38), we have

$$Q = 2.20 \times 150 \times 1.5^{3/2} = 330 \times 1.5^{3/2} = 606.25 \text{ m}^3/\text{s}$$

ORIFICE

An orifice, made of steel, aluminium or any other metal plate, is an opening provided on the side or bottom of a tank or across a channel such that the liquid flows through the entire opening. Note that water flows over a weir while water passes through an orifice. While the stream of water discharged by an orifice is a jet, the sheet of water discharged by a weir or notch is a nappe or vein. A standard orifice has the following characteristics.

- It is placed perpendicular to the flow
- It has sharp edges (no more than 1-2 mm thick)
- It is contracted from the channel boundary and water surface is at least twice the smallest opening dimension, and
- The cross sectional area of the orifice is small in relation to the stream cross section. It has a cross-sectional flow area upstream of the orifice, A_1, at least 8 times the orifice cross-sectional area, A.

Orifices are classified in a number of ways. Depending upon the position in a tank, it may be side orifice or bottom orifice. Based on the size, orifices are small and large. If the head of the liquid from the center of the orifice is more than five times the depth of orifice, that it is a small orifice. If the head of the liquid from the center of the orifice is less than five times the depth of orifice, then the orifice is a large orifice. Based on shape, orifices are circular, square, rectangular and triangular, although most common shapes are circular and rectangular. The orifices may be sharp edged, square edged or bell mouthed. Based on discharge conditions, orifices are free flowing, partly submerged or fully submerged. Under free flow conditions, the flow from the orifice discharges entirely into air. These orifices find applications in measuring comparatively small streams used in irrigation methods such as border strips, furrows or check basins.

Hydraulic Coefficients

Orifices and nozzles are typically used to deliberately reduce pressure, restrict flow or to measure flow rate. The hydraulic coefficients are dimensionless numbers used to characterize the flow and pressure loss behaviour of nozzles and orifices in fluid systems. The following hydraulic coefficients or orifice coefficients are important in water measurement using orifices.

Coefficient of contraction

Coefficient of contraction, C_c is the ratio of the area at vena contracta (Figure 6.17) to the area of the orifice. It is given as follows.

$$C_c = \text{Area of the jet at vena contracta/Area of the orifice} \qquad (6.39)$$

The value of coefficient of contraction varies slightly with the available head of the liquid, size and shape of the orifice. The average value of C_c is 0.64.

Coefficient of velocity

Coefficient of velocity, C_v is the ratio of actual velocity of the jet at vena-contracta to the theoretical velocity, the theoretical velocity of jet at vena-contracta being $\sqrt{2gh}$. Here h is the head of water at vena-contracta. Mathematically it is expressed as:

$$C_v = \text{Actual velocity of the jet/Theoretical velocity of the jet} \qquad (6.40)$$

The value of this coefficient varies slightly with shapes of the edges of the orifice. The value is quite high (0.95 to 0.99) for sharp-edged orifices. For a sharp edged orifice, the value increases with the head of water.

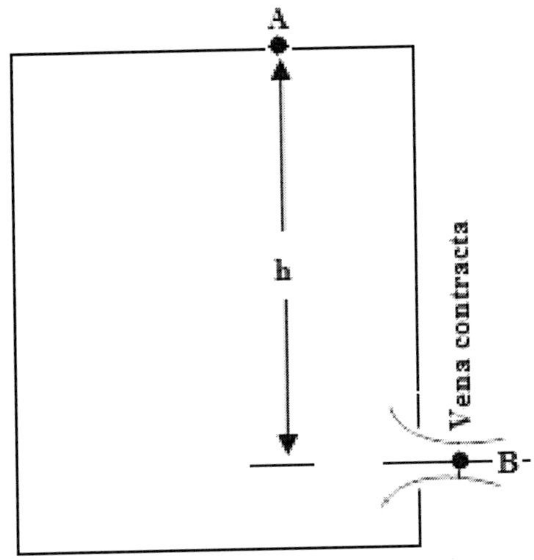

Figure 6.17. Formation of Vena Contracta and Definition Sketch to Determine Exit Velocity.

Coefficient of discharge

Coefficient of discharge, C_d is the ratio of the actual discharge through an orifice to the theoretical discharge. Mathematically coefficient of discharge is

$$C_d = \text{Actual discharge/Theoretical discharge} \tag{6.41}$$

$$C_d = \text{Actual velocity} \times \text{Actual area/Theoretical vlocity} \times \text{Theoretical area}$$

$$C_d = C_v \times C_c \tag{6.42}$$

An average value of coefficient of discharge varies from 0.60 to 0.64.

Coefficient of resistance

Coefficient of resistance, C_r is the ratio of loss of head in the orifice to the head of water available at the exit of the orifice.

$$C_r = \text{Loss of head in the orifice/head of water} \tag{6.43}$$

The loss of head in the orifice takes place, because the walls of the orifice resist the flow as it comes out of the orifice. The coefficient of resistance is close to 1.0 and therefore, has little significance while solving numerical problems.

Vena Contracta

Let us consider a circular sharp-edged orifice provided on the side of a tank that discharges directly into the atmosphere (Figure 6.17). Water jet from the orifice converges to minimum cross-section on the vertical axis passing through point B and then diverges again. The section in the fluid stream at which cross-sectional area of the jet is minimum and less than that of orifice is known as Vena contracta. The maximum contraction occurs at a section downstream of the orifice, where jet is more or less horizontal. The maximum flow velocity is at this point.

Discharge Equation

If we apply the Bernoulli's equation at points A and B (Figure 6.17), then

$$P_a/\rho g + V_a^2/2g + Z_a = P_b/\rho g + V_b^2/2g + Z_b \qquad (6.44)$$

Here P, V and Z have the same meaning as defined earlier. The subscripts a and b indicate these values at points A and B respectively. Referring to Figure (6.17), pressure at points A and B, P_a and P_b are zero (Atmospheric), Z_a =h and Z_b =0 (taking datum at point B), velocity at point A, V_a can be neglected (taken as zero) in comparison to V_b, being a big reservoir. Accordingly, simplifying eq. (6.44) we have

$$V_b^2/2g = h$$

$$V_b = \sqrt{2gh} \qquad (6.45)$$

Here h is the depth of water from surface to the center of the orifice. This derivation is known as Torricelli's theorem and the equation as Torricelli's equation.

We know discharge, Q = A V such that A is the area of the orifice and V the velocity as determined with the help of eq. (6.45). Thus, for a free flowing orifice, discharge is

$$Q = A\sqrt{2gh}$$

Area, A for a circular orifice is given as $\pi d^2/4$ such that d is the diameter of the orifice. For a rectangular orifice it is W times d such that W is the width of the orifice and d the depth. To determine the actual discharge, we use a discharge coefficient that takes care of energy loss due to fluid viscosity and formation of vena contracta resulting in reduced area of flow. Incorporating C_d, we get

$$Q = C_d\, A\sqrt{2gh} \qquad (6.46)$$

The discharge coefficient for a standard orifice is 61×10^{-2}. Therefore,

$$Q = 0.61\, A\sqrt{2gh} \qquad (6.47)$$

Note that the above equation is for a fully contracted orifice plate. Orifices installed in open channels are fully contracted, *i.e.*, the perimeter of the opening is far enough from the walls of the approach channel and water surface, and the jet fully contracts when passing through the opening (Figure 6.18 left).

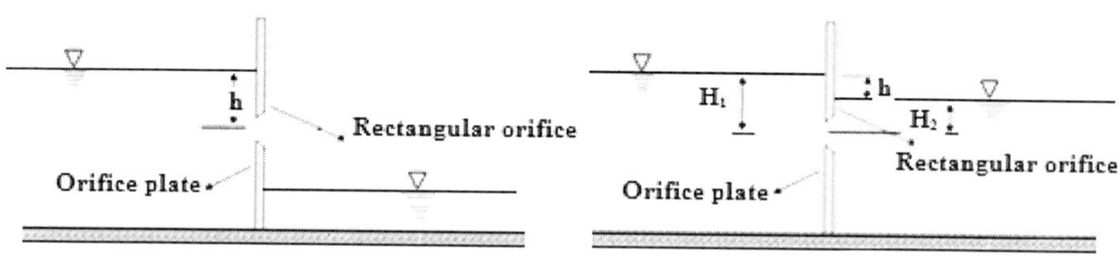

Figure 6.18. A Rectangular Free Flowing (Left) and Submerged Orifice (Right).

Discharge corrections

Discharge calculated with the help of eq. (6.47) needs to be corrected on the following two accounts.

- ☆ Head correction due to velocity of approach
- ☆ Coefficient of discharge correction due to partial suppression of the orifice

For an orifice installed in a channel if the upstream channel cross-sectional area, A_1, is less than 8 times the orifice opening area, A, then one must take into account the contribution of approach velocity. The upstream velocity head, h_v is:

$$h_v = V_1^2/2g = Q^2/(2g\ A_1^2) \quad (6.49)$$

Add h_v so calculated to the hydraulic head so that eq. (6.47) modifies to

$$Q = 0.61A\sqrt{2g}(h + h_v) \quad (6.50)$$

If a portion of the orifice is suppressed (an edge of the orifice does not protrude from, but is even with the upstream channel floor or walls), the discharge coefficient, C_d, must be increased by 1.5 per cent for every 10 per cent suppression of the orifice perimeter.

$$C_d = 0.61\ (1 + 0.15\ r) \quad (6.48)$$

Here r is the ratio of the suppressed to the contracted portion of the orifice perimeter. A common example of a suppressed orifice is a sluice gate in which the bottom and sides are suppressed and the top is contracted.

Besides what is stated in this section, if the orifice's edges are not sufficiently sharp to create full contraction of the jet, the discharge coefficient will be higher than 0.61 and may be as high as 0.80. Water passing through such orifices clings to the edges of the opening rather than springing free. An orifice, neither fully contracted nor suppressed along any part of its perimeter may also show variations in its discharge coefficient. It is necessary to calibrate the orifices if any of the two or both the conditions are encountered.

In submerged flow orifices, the downstream water level is above the top of the opening and the flow discharges into water. Thus, the total head loss or drop in the water surface is less than that for a free flowing jet (Figure 6.18 right). For a submerged orifice (Figure 6.18 right), the discharge is given as

$$Q = C_d\ A\sqrt{2gh} \quad (6.51)$$

Here, h is difference in the depth of water on the upstream and downstream faces of the orifice plate, $(H_1\text{-}H_2)$, m.

Partially Submerged Orifice

A partially submerged orifice is one that has its outlet side partially submerged under liquid (Figure 6.19). Following equation calculates the discharge through a partially submerged rectangular orifice of width, W.

$$Q = C_{d1}\ (W(H_2 - H_1))\sqrt{2gh} + \frac{2}{3}Cd_2\ W\sqrt{2g}\left[h^{1.5} - H_1^{1.5}\right] \quad (6.52)$$

The first term on RHS in eq. (6.52) represents flow through drowned (submerged portion of orifice same as eq. (6.47), whereas the second term represents discharge through free portion acting as a large rectangular weir. Note that two different coefficients of discharge one for the orifice and the other for the weir are used.

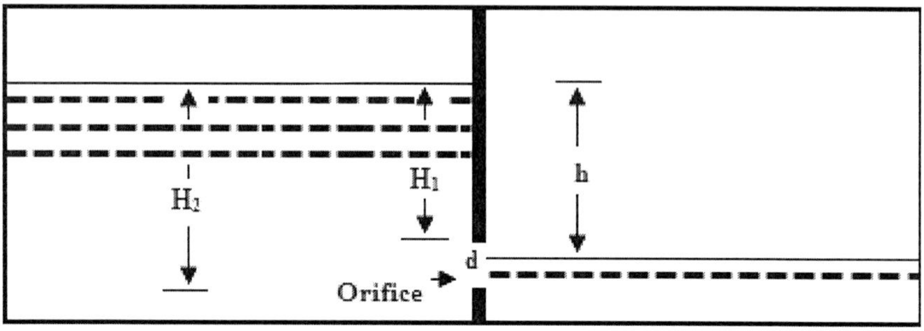

Figure 6.19. A Partially Submerged Rectangular Orifice.

Time of Emptying a Circular Horizontal Tank

Consider a circular horizontal tank of radius, R having the length, L (Figure 6.20). It has an orifice of area, a m^2 at its bottom. If H_1 is the initial height of the liquid and H_2 the final height, we wish to find out the time T for the liquid level to fall from H_1 to H_2.

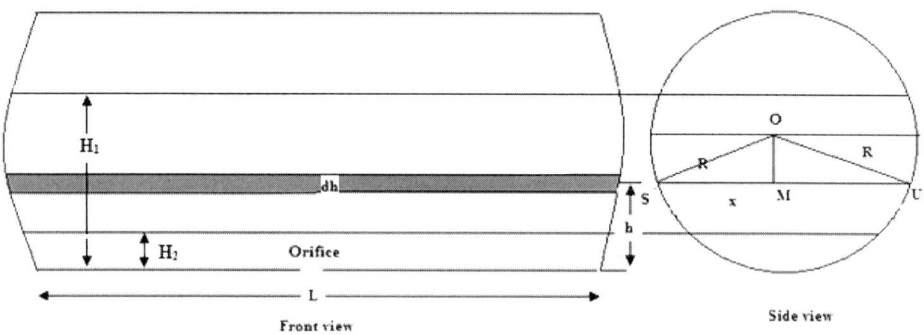

Figure 6.20. Emptying a Horizontal Circular Tank through an Orifice at its Bottom.

Suppose at an instant, height of the liquid in tank is h. It is lowered by dh in time dt. If x is the radius of liquid surface at distance h from the orifice, then volume of liquid leaving the tank in time dt.

A dh = SU × L × dh (SU is given in side view of Figure 6.20)

Since SU = 2x, we have

$$A\,dh = 2x \times L \times dh \tag{6.53}$$

Flow velocity through the orifice = $\sqrt{2gh}$

Discharge through orifice in time dt,

$$= C_d \times \text{area of orifice} \times \text{flow velocity} \times dt = C_d a\sqrt{2gh}\,dt \tag{6.54}$$

Since volume of liquid leaving the tank = volume of liquid flowing through the orifice, we have

$$-2x\ Ldh = C_d a\sqrt{2gh}\ dt \tag{6.55}$$

Note the negative sign in above equation because head on the orifice decrease with increasing time.

$$dt = \frac{-2x\ Ldh}{c_d a\sqrt{2gh}} \tag{6.56}$$

Now, we incorporate x in terms of h. From the geometry of a horizontal circular tank (Figure 6.20 right),

OS = R and OM = (R-h)

$x = SM = \sqrt{[(OS^2) - (OT)^2]} = \sqrt{R^2 - (R-h)^2} = \sqrt{(2Rh - h^2)}$

Substituting this value of x in eq. (6.56)

$$dt = \frac{-2L\sqrt{(2Rh - h^2)}dh}{c_d a\sqrt{2gh}} \tag{6.57}$$

By integrating the above expression between H_1 and H_2, total time T needed to empty the tank from height H_1 to H_2 is

$$\int_0^T dt = \frac{-2L\int_{H1}^{H2} h^{-0.5}\sqrt{(2Rh - h^2)}\ dh}{C_d a\sqrt{2g}}$$

$$T = \frac{-2L}{C_d a\sqrt{2g}}\int_{H1}^{H2}\sqrt{2R - h}\ dh \tag{6.58}$$

$$T = \frac{4L[(2R - h)^{1.5}]}{3C_d a\sqrt{2g}}$$

Incorporating the limits H_1 to H_2 in above equation, we can write as:

$$T = \frac{4L\left[(2R - H_2)^{1.5} - (2R - H_1)^{1.5}\right]}{3C_d a\sqrt{2g}} \tag{6.59}$$

Case 1: Time for emptying the tank completely ($H_2 = 0$)

$$T = \frac{4L\left[(2R)^{1.5} - (2R - H_1)^{1.5}\right]}{3C_d a\sqrt{2g}} \tag{6.60}$$

Case 2: Tank is half-full at the beginning and is being emptied

$H_1 = R$ and $H_2 = 0$

$$T = \frac{4L\left[(2R)^{1.5} - (R)^{1.5}\right]}{3C_d a\sqrt{2g}} \quad (6.61)$$

$$T = \frac{4L\,R^{1.5}[2\sqrt{2} - 1]}{3C_d a\sqrt{2g}} \quad (6.62)$$

$$T = \frac{0.5L\,R^{1.5}}{C_d a} \quad (6.63)$$

Students may derive the equation to lower the water depth from H_1 to H_2 in a circular vertically placed tank under similar conditions. The final equation for such a case is:

$$T = \frac{2A\left[(H_1)^{0.5} - (H_2)^{0.5}\right]}{C_d a\sqrt{2g}} \quad (6.64)$$

Here A is the area of the circular tank. The above equation is valid whether the orifice is at the bottom or on the side of the tank.

Example 6.7

Calculate the discharge of water in L/s at the opening of an orifice of 4 cm diameter if the orifice is located 4 m below the water surface in a tank. Assume any value that may be required. What would be the discharge if the water level on the downstream side rises to 1.5 m above the center of the orifice?

$V = \sqrt{2gh} = \sqrt{2\times9.81\times4} = 8.85$ m/s

Area $=(\pi/4)\,(0.04)^2 = 0.00126$ m^2

Assuming C_d as 0.6

$Q = C_d\,AV = 0.6\times0.00126\times8.85 = 0.0067$ m^3/s = 6.7 L/s

In the second case

$h = 4\text{-}1.5 = 2.5$ m

$V = \sqrt{2\times9.81\times2.5} = 7.00$ m/s

$Q = 0.6\times0.00126\times7.00 = 0.0053$ m^3/s = 5.30 L/s

MOUTHPIECES

A mouthpiece, also used to measure discharge, is an extended form of orifice in which a tube or pipe is attached to the orifice (Figure 6.20). Based on their position, discharge conditions, and shape, mouthpieces are classified as follows.

- ☆ Based on the position of the mouthpieces these are external and internal. In the external mouthpiece, pipe is external projecting outside the tank walls. On the contrary, in the internal, pipe is fixed inside the tank/vessel. Internal mouthpieces are also known as Borda's or re-entrant mouthpieces.

- Based on flow pattern, mouthpieces are running free and running full. In the former, water jet after contraction in mouthpiece does not touch the internal walls of the mouthpiece. In the latter, water jet after contraction in mouthpiece touches the pipe's (mouthpiece) internal walls and the tube runs full. If the length of the tube of mouthpiece is less than 2-3 times the diameter of the pipe, then it is a free flow mouthpiece. If it is greater than 2-3 times the diameter, then it is a full flow mouthpiece. A free flow external mouthpiece behaves as an orifice. If the length of the mouthpiece is more than 3 to 3.5 times the diameter, then it behaves like a pipe. The value of discharge coefficient will reduce because of the friction in the pipe.
- Based on shape, mouthpieces are cylindrical, converging, diverging, and converging-diverging *etc.*

A mouthpiece, an extended form of orifice, helps to increase the discharge compared to an orifice. The discharge from the tank through a mouthpiece compared to an orifice can be increased by running the mouthpiece full so that the jet of liquid emerging from the mouthpiece is of same diameter as that of mouthpiece or by maintaining a sufficient pressure-head in the tank to allow the mouthpiece to run full. Although, mouthpieces help to increase discharge, the use of mouthpieces is not a common practice. It is because a mouthpiece can create –ve pressure at the vena contracta. As soon as the pressure at vena contracta reaches the vapour pressure of the fluid, cavitation may occur, and mouthpiece will not function properly. This imposes a limitation on the use of mouthpieces under high head conditions.

External Cylindrical Mouthpiece

Consider a cylindrical mouthpiece of cross-sectional area A, which is attached externally to the tank (Figure 6.21). The tank contains a liquid of specific weight, γ. Let H be head of the liquid over the mouthpiece. As the liquid enters the mouthpiece, the area of flow reaches a minimum value A_c at section CC. Beyond this section, a sudden enlargement occurs and the liquid fills the mouthpiece. The area of the jet at outlet is A_o, which is obviously equal to A. Referring to Figure 6.21, if H_a and H_c represents the absolute pressure heads at section B-B (atmospheric) and at vena-contracta, C-C, H_l is the head loss through the mouthpiece, and V_c is the velocity of the jet at vena-contracta, then

$$A_c/A = C_c \text{ (coefficient of contraction)}$$

$$A_c V_c = A\,V \qquad (6.65)$$

$$V_c = V\,A/A_c$$

If A_c/A is taken as 0.62 then (Note that equations derived for this case are based on this assumption. If the value of C_c is different, then the final equation will also change accordingly).

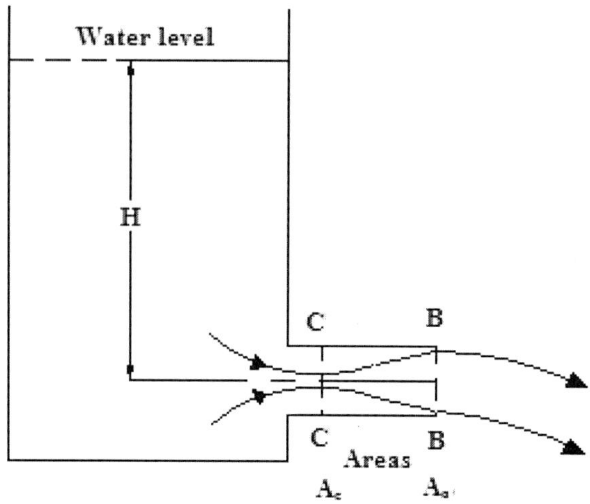

Figure 6.21. Flow through an External Cylindrical Mouthpiece.

$$V_c = V/0.62 = 1.613\ V \tag{6.66}$$

Head loss from vena contracta to the outlet, H_1 is

$$H_1 = (V_c - V)^2/2g = (1.613\ V - V)^2/2g = (1.613 - 1)^2\ V^2/2g$$
$$= 0.375\ V^2/2g \tag{6.67}$$

Applying Bernoulli's equation between the free surface of the liquid in the tank and section B-B at the outlet, we get,

$$H = V^2/2g + H_1 \tag{6.68}$$

$$H = V^2/2g + 0.375\ V^2/2g = 1.375\ V^2/2g \tag{6.69}$$

$$V = \sqrt{(2gH/1.375)} = 0.855\sqrt{(2gH)} \tag{6.70}$$

Theoretical velocity, $v = \sqrt{2gH}$

$$\text{Coefficient of velocity} = V/v = \sqrt{(2gH/1.375)}/\sqrt{2gh}$$
$$= 1/\sqrt{1.375} = 0.855 \tag{6.71}$$

Since the area of the jet at the outlet is same as that of the mouthpiece, coefficient of contraction at B-B is 1.0

$$\text{Therefore, } C_d = C_c \times C_v = 1\times 0.855 = 0.855 \tag{6.72}$$

$$Q = 0.855\ A\ \sqrt{2gh} \tag{6.73}$$

Pressure head at vena contracts, section C-C

Applying Bernoulli's equation between the free surface of the liquid in the tank and section C-C at the Vena contracta, we get,

$$H_a + H = H_c + V_c^2/2g \tag{6.74}$$

$$V_c = V/0.62$$

$$V_c^2/2g = (1/0.62)^2\ V^2/2g = 2.60\ V^2/2g \text{ and } H = 1.375\ V^2/2g \tag{6.75}$$

Incorporating these values, we have

$$H_c = H_a - 2.60\ V^2/2g + 1.375\ V^2/2g = H_a - 1.225\ V^2/2g$$
$$= H_a - 0.89\ H \qquad (6.76)$$

Thus, the head at vena contracts is less than the atmospheric pressure and it decreases with increase in head, H.

Eq. (6.76) is used to find out the maximum head that can be used under such a situation. The head at the vena contracts must not go below the separation head of the fluid, for water it being 2.5 m absolute. Taking H_a as 10.3 m, we have

$$H_{sep} = 10.3 - 0.89\ H_{max}$$
$$H_{max} = (10.3 - 2.5)/0.89 = 8.76\ \text{m}$$

Internal Cylindrical Mouthpiece

Borda's mouthpiece (Figure 6.22) are mainly of two types *i.e.* mouthpiece running free and mouthpiece running full.

Borda's mouthpiece running free

If the liquid jet after contracting does not touch the sides of the tube (Figure 6.22 left), it is a mouthpiece running free. As per momentum equation, the expression for force on the mouthpiece is:

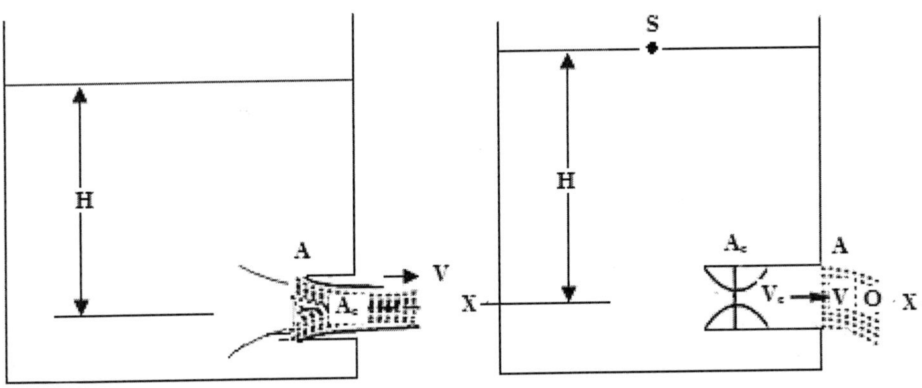

Figure 6.22. Borda's Mouthpiece Running Free (Left) and Running Full (Right).

Pressure force on mouthpiece = pressure × area of mouthpiece

$$= \gamma HA \qquad (6.77)$$

Here, γ is weight density of liquid (ρg), H is the height of liquid above the mouthpiece, A is the area of mouthpiece, A_c is the area at vena contracta and V_c is the flow velocity at vena contracta.

Mass of liquid flowing per second = $\rho\ A_c\ V_c$ (6.78)

Since momentum = mass × velocity

Rate of change of momentum = mass flowing per second × change of velocity

$$= (\rho\ A_c\ V_c) \times (\text{final velocity–initial velocity}) \qquad (6.79)$$

If initial and final velocities are 0 and V_c, then

Rate of change of momentum = $(\rho\, A_c\, V_c) \times (V_c - 0)$

$$= \rho\, A_c\, V_c^2 = \gamma A_c V_c^2/g \quad (6.80)$$

Since rate of change of momentum = Pressure force

Equating eq. (6.77) and eq. (6.80), we have

$$\gamma HA = \gamma A_c V_c^2/g$$

$$HA = A_c V_c^2/g \quad (6.81)$$

Applying Bernoulli's theorem to the free surface and jet

$$H = V_c^2/2g \quad (6.82)$$

$$AV_c^2/2g = A_c V_c^2/g \quad (6.83)$$

Thus, $A_c/A = 0.5$ or the coefficient of contraction, $C_c = 0.5$

Since there is no loss of head $C_v = 1.0$

$$\text{and } C_d = C_c \times C_v = 0.5 \times 1 = 0.5 \quad (6.84)$$

$$\text{Thus, discharge } Q = C_d\, A\sqrt{2gH} \quad (6.85)$$

$$Q = 0.5\, A\sqrt{2gH} \quad (6.86)$$

Borda's mouthpiece running full

When the mouthpiece tube is about 3 times its diameter, the liquid jet after contraction in the tube touches the internal walls of the tube. Such a condition is known as mouthpiece running full (Figure 6. 22 right)

From equation of continuity

$$V_c A_c = V\, A \quad (6.87)$$

Since $A_c/A = 0.5$ (As in the previous case)

$$A/A_c = 1/0.5 = 2.0$$

Considering XX as the datum line and using Bernoulli's equation at points S and O, the outlet of the mouthpiece (Figure 6.22 right)

$$H + V_s^2/2g = 0 + V_o^2/2g + H_1 \quad (6.88)$$

(Pressure is atmospheric at both the points)

Since $V_s = 0$ as $V_s << V$

$$\text{Head loss, } H_1 = (V_c - V_o)^2/2g \quad (6.89)$$

Therefore, eq. (6.88) can be written as

$$H = V_o^2/2g + (V_c - V_o)^2/2g$$

$$H = V_o^2/2g + V_o^2\,(V_c/V_o - 1)^2/2g \quad (6.90)$$

Since $A/A_c = 2.2$ therefore, $V_c/V_o = 2$

$$H = 2V_o^2/2g \quad (6.91)$$

$$V_o = \sqrt{gH} \quad (6.92)$$

Theoretcal velocity, $V = \sqrt{2gH}$ (6.93)

$C_v = V_o/V = 1/\sqrt{2} = 0.707$ (6.94)

Since the area of jet at exit equals the area of the mouthpiece, $C_c = 1$

$C_d = C_c C_v = 1 \times 0.707 = 0.707$ (6.95)

Discharge through a Borda's mouthpiece running full is

$Q = 0.707\ A\sqrt{2gH}$ (6.96)

Pressure head at vena contracta

Applying Bernoulli's equation at Vena contracta and the outlet

$H_c + V_c^2/2g = H_a + V_o^2/2g + H_l$ (6.97)

Since V_c is equal to $2V_o$ and $H_l = V_o^2/2g$

$H_c + 4V_o^2/2g = H_a + V_o^2/2g + V_o^2/2g$

$H_c = H_a - 2V_o^2/2g = H_a - gH/g$ (since $V_o^2 = gh$) (6.98)

$H_c = H_a - H$ (6.99)

Since the head at the vena contracts should not be less than the separation head of the fluid, the maximum head that can be used under such a situation is determined using eq. (6.99). Separation head for water is 2.5 m absolute and taking H_a as 10.3 m, we have

$H_{sep} = 10.3 - H_{max}$

$H_{max} = 10.3 - H_{sep}$

$H_{max} = 10.3 - 2.5 = 7.86$ m

Example 6.8

Find the discharge from a 100 mm diameter external mouthpiece, fitted to one side of a large vessel. If the head over the mouthpiece is 4.5 m, determine the absolute pressure head of water at the vena contracta. Take atmospheric pressure as 10.3 m of water.

d = 100 mm = 0.1 m, so that its area = 0.0079 m^2

H = 4 m

$Q = 0.855\ a\sqrt{2gH} = 0.855 \times 0.0079 \times \sqrt{2 \times 9.81 \times 4.5} = 0.0635\ m^3/s$ or 63.5 L/s

Absolute pressure head of water at vena contracta of an external mouthpiece is

$H_c = H_a - 0.89\ H = 10.3 - (0.89 \times 4.5) = 6.23$ m

Example 6.9

An internal mouthpiece having an area of cross section as 0.10 m^2 is provided on the side of a tank. The distance between the center of mouthpiece and free water surface is 2.8 m. Find out the difference in discharge if the mouthpiece is running free and running full.

Use eq. (6.86) to find out the discharge of an internal mouthpieces running free

$$Q = 0.5\ a\ \sqrt{2gH} = 0.5 \times 0.10 \times \sqrt{2 \times 9.81 \times 2.8} = 0.371\ m^3/s \text{ or } 371\ L/s$$

Use eq. (6.96) to find out the discharge of an internal mouthpieces running full

$$Q = 0.707\ a\ \sqrt{2gH} = 0.707 \times 0.10 \times \sqrt{2 \times 9.81 \times 2.8} = 0.524\ m^3/s \text{ or } 524\ L/s$$

Difference in the discharge = 524 – 371 = 153 L/s

Example 6.10

Determine the discharge through a 100 mm diameter internal mouthpiece under a head of 3 m when it flows free and in full. What will be the pressure head at vena-contracta if it runs full? Take atmospheric pressure as 10.33 m of water.

Use eq. (6.86) to find out the discharge of an internal mouthpieces running free

$$Q = 0.5\ a\ \sqrt{2gH} = 0.5 \times 0.00785 \times \sqrt{2 \times 9.81 \times 3} = 0.00393 \times 7.67$$
$$= 0.0301\ m^3/s \text{ or } 30.1\ L/s$$

To find out the discharge of an internal mouthpieces running free, use eq. (6.96)

$$Q = 0.707\ a\ \sqrt{2gH} = 0.707 \times 0.00785 \times \sqrt{2 \times 9.81 \times 2} = 0.00555 \times 7.67$$
$$= 0.0426\ m^3/s \text{ or } 42.6\ L/s$$

From eq. (6.99), we have

$$H_c = H_a - H = 10.33 - 3.0 = 7.33\ m$$

Convergent Mouthpiece

A jet of liquid always contracts at the entrance of a mouthpiece. This contraction results in some head loss. To counteract the loss of head due to contraction of the jet up to vena contract, a mouthpiece is given the same slope as the jet of liquid (Figure 6.23 left). A convergent mouthpiece is like a frustum of cone, the larger end attached to the tank wall. The jet as usual contracts but exits before it expands in the mouthpiece. Thus, a mouthpiece that terminates at the vena contracta of the jet is a convergent mouthpiece. Therefore, there is no head loss due to expansion in such a mouthpiece.

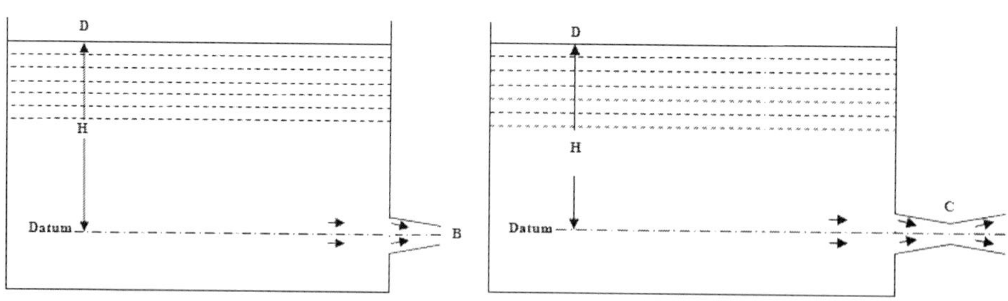

Figure 6.23. A Convergent (Left) and a Convergent-Divergent Mouthpiece (Right).

To derive the equation of discharge through a convergent mouthpiece, let us apply the Bernoulli's theorem at points D and outlet B. If the velocity at D is assumed negligible in comparison to the velocity at B, we have

$$0 + H = V_b^2/2g + 0$$

$$V_b = (2gH)^{1/2}$$

The discharge is then obtained by multiplying the velocity by the area and a discharge coefficient, C_d so that

$$Q = C_d A (2gH)^{1/2} \quad (6.100)$$

Here A is the area of the mouthpiece at B. The coefficient of discharge, C_d varies with the angle of convergence θ. For a convergent mouthpiece, the value of coefficient of velocity is 1.0 when the angle θ is about 13.5°. At this angle, the inner profile of the mouthpiece almost matches with that of the jet. Since the area of the jet at outlet is the same as that of the mouthpiece at the outlet, value of C_c is 1.0. Thus, the value of C_d for a convergent mouthpiece is 1.0 for an angle of convergence of 13.5°. In practice, a small loss of head always occurs due to friction. Therefore, coefficient of discharge for a convergent mouthpiece is in the range of 0.94 to 0.98.

Convergent-divergent Mouthpiece

In this mouthpiece, the mouthpiece is first made convergent up to the vena contracta of the jet and beyond that it is made divergent (Figure 6.23 right). For a steady flow through the outlet of the mouthpiece, the shape of the divergent portion is made according to the profile of the expanding jet. The discharge equation for a convergent-divergent mouthpiece is same as for a convergent mouthpiece, as there are no head losses. The coefficient of discharge C_d in the case of convergent-divergent mouthpiece is also 1.0. The diameter of the mouthpiece, for the purpose of calculating the discharge, is taken at the vena-contracta *i.e.*, at *C* where the convergent and divergent pieces meet.

Pressure in a convergent-divergent mouthpiece

Consider a vessel open to atmosphere at its top (Figure 6.23 right), having an orifice fitted with a convergent-divergent mouthpiece. The theoretical absolute pressure head at vena contracta is the atmospheric pressure. The jet expands from vena contracta (*C*) to outlet of the tube (*B*). However, if the divergence is made too large, the jet will not touch the walls of the mouthpiece. Applying the Bernoulli's equation at D, C, and B, we have

$$H_{at} + H = H_c + V_c^2/2g = H_{at} + V_o^2/2g$$

Here H_{at} is the atmospheric head and V_c and V_o are the velocities at vena contracta and outlet respectively.

$$V_c = \sqrt{2g}\,(H_{at} + H - H_c)$$

Theoretical velocity $V = \sqrt{(2gH)}$

Therefore,

$$V_c/V = A/A_c = \sqrt{2g}\,(H_{at} + H - H_c)/\sqrt{(2gH)} = \sqrt{[1+(H_{at}-H_c)/H]} \quad (6.101)$$

Here A is the area at the outlet. With increasing area at the outlet, V_c increases and therefore H_c decreases. On the other hand, we cannot decrease H_c at will as the pressure at the vena contracts should not go below the separation pressure of the fluid. Incorporating H_{sep} for H_c, we have

$$A/A_c = \sqrt{[1+(H_{at} - H_{sep})/H]} \quad (6.102)$$

We know that separation pressure for water is 2.5 m. Therefore

$$A/A_c = \sqrt{[1+(10.3- 2.5)/H]} = \sqrt{(1 + 7.8/H)} \quad (6.103)$$

The above expression gives the ratio of areas of divergence to the convergence of the mouthpiece. If D and D_c are the diameters of the mouthpiece at outlet and vena contracta, then it is possible to express eq. (6.102) in terms of diameters as follows:

$$D^2/D_c^2 = \sqrt{[1+(H_{at} - H_{sep})/H]} \quad (6.104)$$

Or

$$D^2/D_c^2 = \sqrt{(1 + 7.8/H)}$$

Example 6.11

Water flows through a convergent-divergent mouthpiece. The diameters at convergence and divergence are 5 and 6 cm respectively. If the mouthpiece is discharging water under a head of 5 m, show that there is no fear of separation at the vena contracta. Assume the atmospheric and separation pressures of water as 10.3 and 2.5 m respectively. Also determine the maximum diameter of the divergent portion beyond which separation is likely to occur.

Using eq. (6.104), we have

$$D^2/D_c^2 = \sqrt{[1+(H_{at} - H_c)/H]}$$

$$36/25 = \sqrt{[1 + (10.3 - H_c)/5]} = 1.44$$

$$1 + (10.3 - H_c)/5 = 2.02$$

$$10.3 - H_c = (2.02 - 1.0) \times 5 = 5.1$$

$$H_c = 10.3 - 5.1 = 5.2 \text{ m}$$

Since H_c is more than 2.5 m, there is no fear of separation.

To determine the maximum diameter beyond which separation is expected to occur, we again use eq. (6.104)

$$D^2/D_c^2 = \sqrt{[1+(H_{at} - H_{sep})/H]}$$

$$D^2/25 = \sqrt{[1+ (10.3-2.5)/5]} = 1.6$$

$$D = \sqrt{(1.6\times25)} = 6.32 \text{ cm}$$

WATER LEVEL GAUGES

One of the sources of errors with any water measuring device arise from measured head (or stage). For example with stage measured to only ± 1 mm, proportional errors at low stages and with small structures can be quite high. The USGS accuracy requirement is ± 3 mm up to an effective stage of 2 m, and above this, 0.2 per cent of effective stage (2 m × 0.002 = 4 mm). In surface water resources especially rivers, accuracy diminishes as velocity and surge increase. Therefore selection of appropriate gauges to measure water level is paramount to measure flow rates. The gauges used for water level measurements are of two types, namely non-self-registering gauges, and self-registering gauges. Non-self-registering gauges need an observer to note the levels of water surface at the time of reading. The self-registering gauges automatically record the readings and produce a continuous record of water levels. The gauge effectiveness stems from its accuracy to measure the water levels (National Environmental Monitoring Standards, 2013).

Non-Self-Registering Gauges

Some of the commonly used non-self-registering gauges are discussed in the following sections.

The staff gauge

The staff gauge consists of a graduated board about 150 mm wide firmly fixed in a vertical position. The gauge must be of sufficient length to cover the highest and the lowest water levels. The graduations should be bold enough to be visible from a distance. Piers and abutments provide excellent supports for fixing the staff gauge. A convenient local datum should be established on site using appropriate benchmarks (BM's). While setting a staff gauge, the zero division of the gauge is marked in such a way that it represents a pre-decided level as per the local datum. Sloping staff gauges are used to improve the accuracy.

The weight gauge

A staff gauge is inconvenient at inaccessibility sites posing problems while taking the readings. In such situations, a weight gauge is used so that the reading site can be set at a more approachable site. It consists of a weight attached to a graduated chain, or tape or wire. Before taking any observation, weight is lowered so that it touches the water surface. Then the observation is recorded against a fixed mark of known elevation. Alternatively, the gauge may take the form shown in Figure 6.24. In this set-up, adjust the weight at the water surface before taking the observation at the graduated board (http://www.engineeringenotes.com/fluids/water-level/measurements-of-water-level-methods-velocity-discharge-and-instruments-fluid-mechanics/47925. Opened on 01.03.2020).

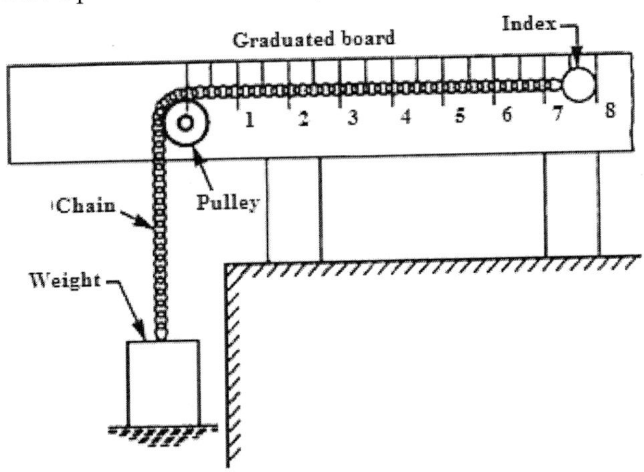

Figure 6.24. The Principle of Working of a Weight Gauge.

The Float Gauges

Ball float

This gauge provides a direct reading of the liquid level. The most practical design for the float is a hollow metal ball or sphere, which could be of any size, shape, or material provided it floats on the liquid. The ball float attached to a rod connected to a rotating shaft indicates liquid level on a calibrated scale (Figure 6.25). The operation of the equipment is simple. The ball floats on top of the liquid in the tank. With any change in the liquid level, the pointer reflects this change on the scale via the rotating shaft.

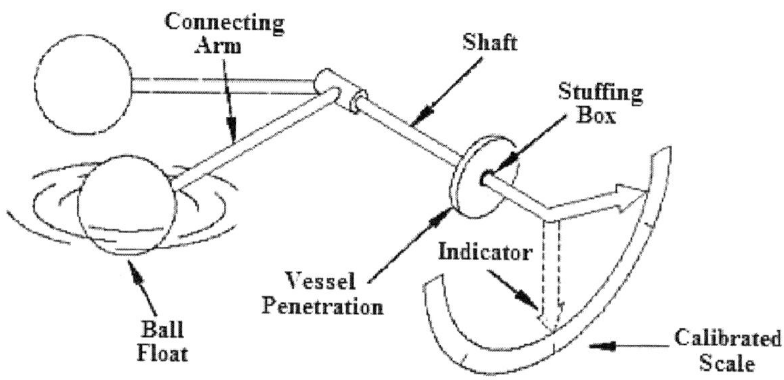

Figure 6.25. Ball Float Gauge.

Chain float

This type of float gauge has a float ranging in size up to 30 cm diameter. It is used where level changes are larger going beyond the range of ball floats. The operation of the chain float is similar to the ball float except for the positioning of the pointer and its connection to the position indicator. The float is connected to a rotating element by a chain with weight attached on the other end. The weight keeps the chain taut during any change in liquid level (Minz, 2018, Figure 6.26 left).

Float gauge with pulley

This gauge consists of a float connected to a wire that passes over a pulley. A pointer attached to the other end of the wire moves on a graduated vertical scale, which indicates the liquid level (Figure 6.26 right). These gauges can be made self-recording by providing an electro-mechanical float that will convert pulley rotation to millimeters of rise and fall into an electrical pulse. This pulse can be recorded on a data logger.

Box gauge

This gauge is enclosed in a long wooden box about 300 mm square (Figure 6.27). The bottom of the box has few holes. The float carries a graduated scale. The water level is read against an index fixed in position. In another arrangement, the rod carries a pointer, which moves against a fixed scale. In the former arrangement, the readings increase downwards; while in the latter, the readings decrease downwards.

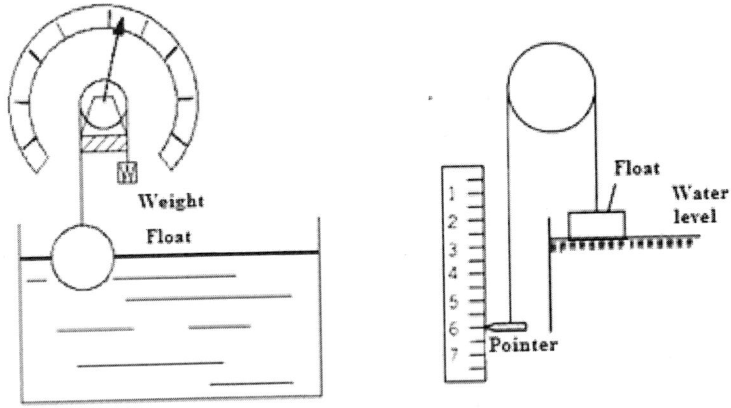

Figure 6.26. Chain Float (Left) and a Float Gauge with Pulley (Right).

Tubular glass gauge

A standard tubular glass gauge features a glass tube, end block, seals, and guard rods designed to shield the glass from any damage. This gauge is normally placed parallel to a vessel at the appropriate elevation, is mounted with fittings that seals the ends of the tube and maintains pressure. This gauge is not suited for observing the levels of hazardous fluids or fluids operating under temperatures above 212° Fahrenheit (F) and pressures above 1 bar. Fluid can easily escape into the environment if the glass tube gets fractured or there is a leak at the seals. Improved glass gauge designs feature reinforcements, protection against breakage, and improvements on the capability of these gauges. The protective measures come in the form of an outer tube that collects leaked fluid in case of a fracture, sheet metal guards, or a wire glass layer that surrounds the gauge.

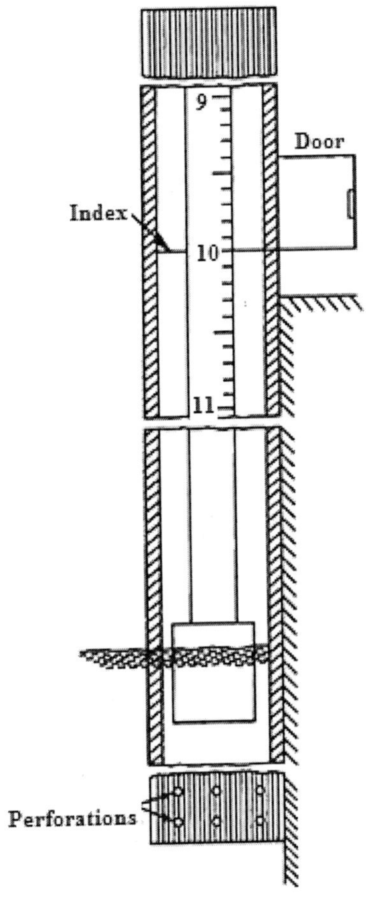

Figure 6.27. A Schematic View of Box Gauge.

Transparent Circular Gauges

Circular liquid level gauges are positioned alongside a vessel at an elevation that may cover an important level range. These gauges have a limited indication scale and are most effective in monitoring relatively small variations. A typical circular gauge features a body with a cover, nut, and stud attached to a glass segment with a gasket and an optional shielding specification

(Figure 6.28, Minz 2018). During the measuring process, the liquid may wet the body, gasket, and glass. A screen protects the glass from wetting by the process fluids. Because of the relatively short measurement range, multiple circular gauges are used on a single vessel to monitor the high, the low or the intermediate levels. These devices are capable of indicating moving fluid and fluid colour or its contamination.

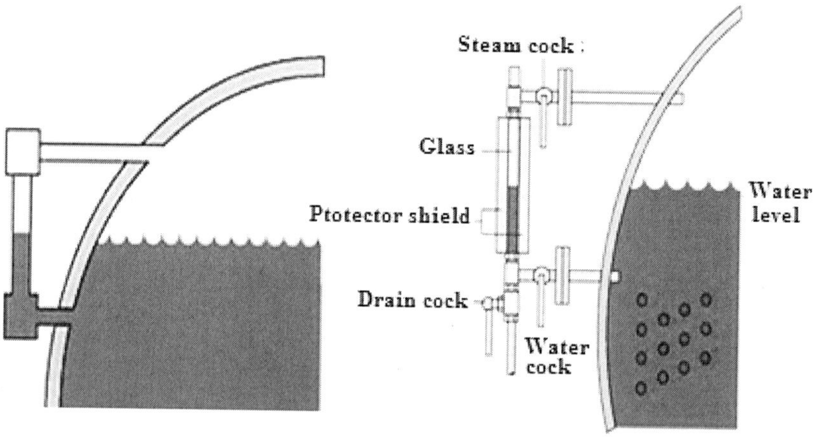

Figure 6.28. Gauge Glasses for Measuring Liquid Levels in Horizontal Tanks.

Long Transparent Gauges

A long-form flat glass transparent gauge is capable of monitoring fluid levels over a wider range compared to a circular gauge. This type of gauge consists of a glass segment with covers attached to it, a gasket, chamber housing, and optional tie bars or shields (Figure 6.29). Measuring chamber mounted with glass and covers have machined vision slots for observations. The optional tie bars augment the gauge's strength. As in circular gauges, shields protect the glass from process fluids and take care of the environmental issues. The gauge provides transparent glass panels on the front and back of the measuring chamber. Level indications illuminated from behind with either lighting panels or ambient light, allow easier visual inspections regarding liquid colour and the presence of any floating contamination. These transparent gauges are also effective in monitoring the interface between two liquids because of their visibility character. On the contrary, some unreadable areas due to the presence of tie bars or the top and bottom edges of the gauge covers may prove to be an irritant.

The Hook and Point Gauges

A typical hook and point gauge shown in Figure 6.29, consist of a graduated bar attached to a hook or pointer. The bar can slide up or down a fixed mounting. While taking a reading, the sharp point of the hook or pointer is brought very close to the water surface. The sliding bar is now clamped. By using a slow motion screw, the hook or the pointer is moved until the sharp point of the hook or pointer just touches the water surface. Verniers scale is provided for enhanced accuracy. These gauges are generally used in hydraulic laboratories for precise measurements of heads of water over notches, weirs *etc.*

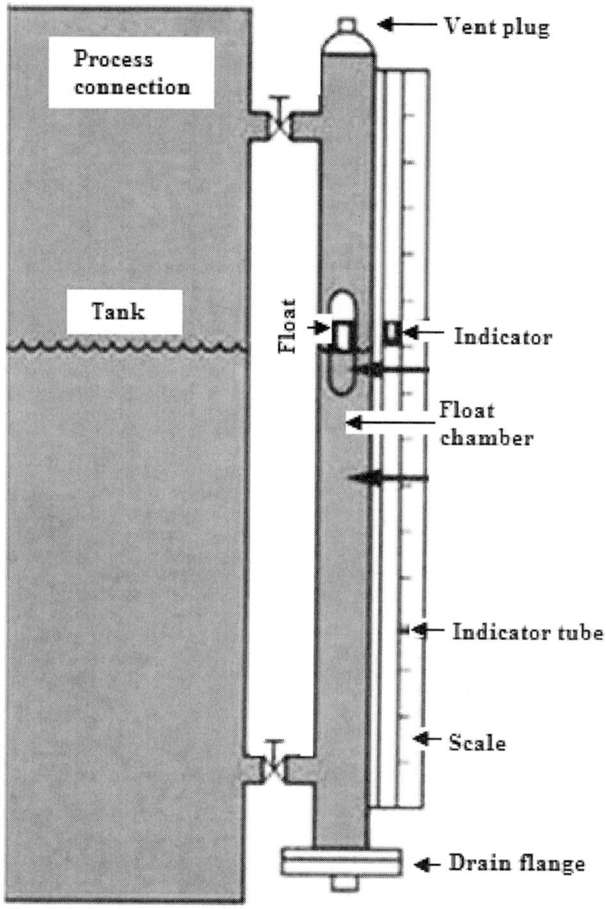

Figure 6.29. Long Gauge Glass Indicating Liquid Level in Vertical Tank.

Monitoring Groundwater Levels

There are many ways to measure the water elevations in observation wells and piezometers. Electric water level indicators are now available for smaller diameter wells. Float recorders, pressure loggers or electronic water level loggers are used for accurate and continuous measurements. Amongst the non-recording types, we have electric plumb bob, chain and bell sounding device and blowpipe.

The electric plumb bob consists of a plumb bob connected to a graduated steel tape, wound on a brass reel (Figure 6.31 left). This forms part of a low voltage electrical circuit. The ciruit is completed as soon as the plumb bob touches, the water surface. The contact is registered on a milli-ammeter, or buzzer/light, prompting the reader to take the observation.

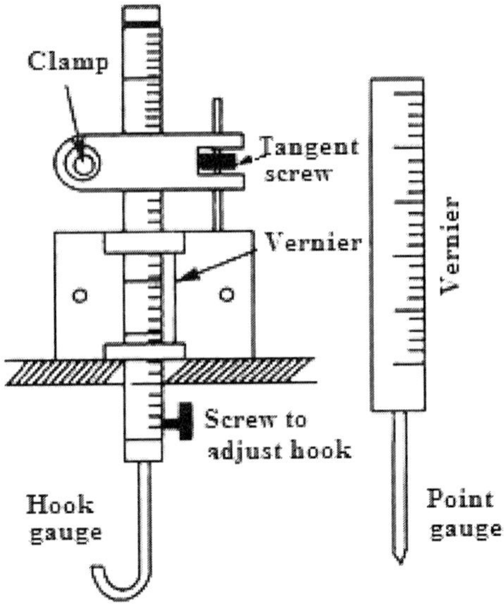

Figure 6.30. A Hook Gauge (Left) and a Point Gauge (Right).

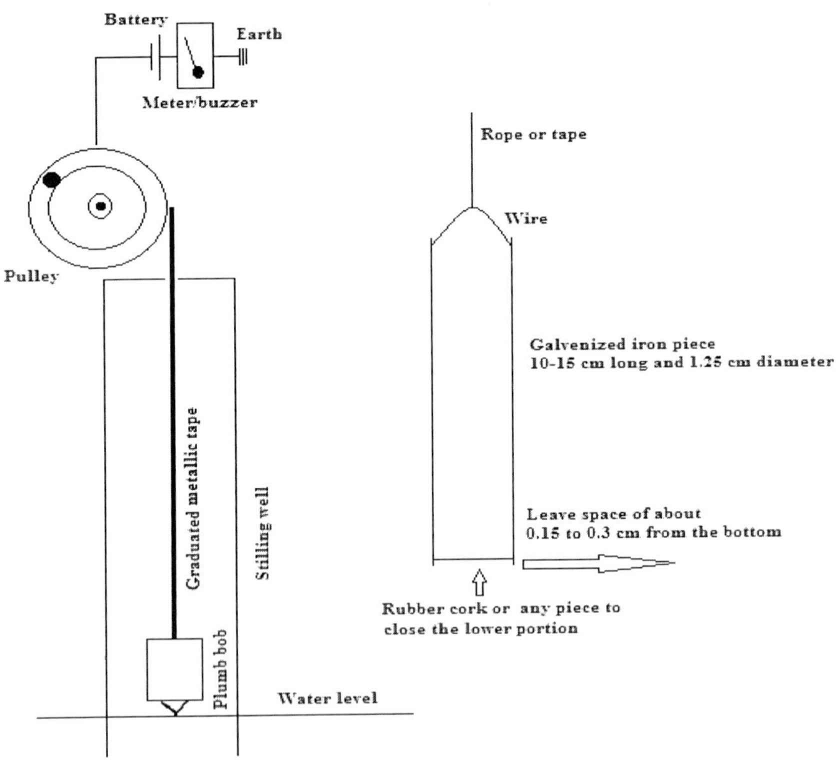

Figure 6.31. The Electric Plumb Bob (Left) and a Chain and a Bell Sounding Device (Right) for Monitoring Groundwater Levels.

The most commonly time tested method is the chain and bell sounding device fitted at the bottom end of a thin steel or metallic tape. The device is a 1-1.5 cm galvanized iron pipe of about 15-20 cm length. The bottom end is sealed with a rubber cork leaving about 1 mm gap (Figure 6.31 right). An arrangement is made to attach the tape at the upper end. When lowered into the well, it makes a clearly audible "plop" sound as it contacts the water surface. The observer records the reading upon hearing the plop sound.

Another device is the blowpipe that uses a hollow aluminium pipe of 10 mm diameter inserted in a plastic tube marked in mm scale. It is gradually lowered in the hole while blowing air through the tube at short intervals. As soon as the pipe reaches the water level, a particular sound indicates its position. The depth of water table with reference to this point is recorded.

Self-Registering Gauges

Many kinds of self-registering gauges are in use. The essential components of all these gauges are similar. Normally a device consists of a float protected from wind or wave action, which is attached to a wire wound around a wheel. The wire has a counter weight or spring on the other end that keeps the wire straight. The vertical displacement of the float transferred through the wheel is reduced in scale by a gearing mechanism. This displacement is finally transmitted to a pencil, which traces a curve on a uniformly moving sheet of paper. In most gauges, the paper mounted on a drum, revolves once in 24 hours by clockwork. Self-registering gauges are more useful than the non-self-registering gauges. The readings of these gauges are free from personal errors. These gauges are particularly useful in obtaining gauge readings during floods or other natural calamities. Options that are more sophisticated include: the shaft encoder, pressure transducer, gas purge sensor, acoustic transceiver, and Radar. Acoustic instruments may be with sound path in air or with sound path in water.

QUESTIONS (THEORY)

1. Name various classification criteria of sharp crested weirs and discuss various kinds of weir under the specified criteria.
2. Using the basic principle, derive the equation of discharge for a rectangular weir. Using the equation, derive the Francis formula.
3. What is the impact of one, two and n ends contractions on discharge equation of a rectangular weir?
4. Write down the discharge equation for a submerged rectangular weir.
5. Derive the equation for time of emptying a reservoir by a rectangular weir.
6. Using the basic principle, derive the equation of discharge for a V-notch weir.

7. Derive the equation of discharge for a trapezoidal weir. Use this equation to derive the discharge equation of a Cipolletti weir.
8. How a broad crested weir differs from sharp crested weir? Derive the equation for maximum discharge of a broad crested weir.
9. Draw neat diagrams for free flow, submerged flow and partially submerged flow through orifices. Write down the relevant equations for these cases.
10. Derive the equation for time of emptying a vertical cylindrical tank by a circular orifice.
11. What do you understand by a mouthpiece? Name and describe the various kinds of mouthpieces.
12. Derive the equation of flow through an external cylindrical mouthpiece.
13. Derive the flow equation through a Borda's mouthpiece (1) for mouthpiece running free and (2) mouthpiece running full.
14. Discuss the construction of a converging mouthpiece. Derive its discharge equation.
15. Discuss the construction of a converging-diverging mouthpiece. Explain through appropriate derivation, the limitation on the outlet area of the diverging portion of the mouthpiece.
16. Discuss the various kinds of float gauges.
17. Differentiate between the following: notches and weirs, narrow crested and broad crested weirs, suppressed and contracted weirs, free and submerged flow orifices, trapezoidal and Cipolletti weirs, and hook gauge and point gauge.

Chapter 7

Open Channel Flow

Flow in open channels or conduit (pipes) is of interest in science, engineering, and everyday life. Open-channel flows are those whose boundaries are not entirely a solid and rigid material; a part of the boundary of such flows may be fluid or nothing at all. Open-channel flow is a flow of liquid in a conduit with free surface *i.e.* a surface with local atmospheric pressure. Since the flow occurs by gravity, it is also named as free surface flow or gravity flow. Flows in closed conduits like pipes or air ducts, on the other hand, are entirely in contact with rigid boundaries. The pipe flow does not have a free water surface. Most closed conduits in engineering applications are either circular or rectangular in cross section. Table 7.1 reports major differences between the open channel and pipe flows.

Table 7.1. Comparative Characteristics of Open Channel and Pipe Flows

Open Channel Flow (OCF)	*Pipe Flow (PF)*
OCF must have a free surface	PF has no free surface
A free surface is subject to atmospheric pressure	No direct atmospheric pressure, hydraulic pressure only
The driving force is mainly the component of gravity along the flow direction	The driving force is mainly the pressure force along the flow direction
Hydraulic grade line (HGL) is coincident with the free surface	HGL is (usually) above the conduit
Flow area is determined by the geometry of the channel plus the level of free surface, which is likely to change along the flow direction as well as with time	Flow area is fixed by the pipe dimensions
The cross section may be of any form circular to irregular as in the case of natural streams, which may change along the flow direction as well as with time	The cross section of a pipe is usually circular
Relative roughness changes with the level of free surface	The relative roughness is a fixed quantity for a particular kind of material
The depth of flow, discharge and the slopes of channel bottom and of the free surface are interdependent	No such dependence

Important examples of open-channel flows are the flows in rivers, irrigation canal network including irrigation channels, even sheets of water running across the ground surface after a rain and tidal currents. Even a partially flowing pipe will be a case of open channel flow (Figure 7.1). The basic equations that govern the water flow in open channels are continuity equation, Bernoulli equation and Darcy-Weisbach equation (www.nptel.iitm.ac.in/courses/105107059/module1/./lecture1.pdf Opened on 13.02.2018). This chapter discusses some important issues related to open channel hydraulics.

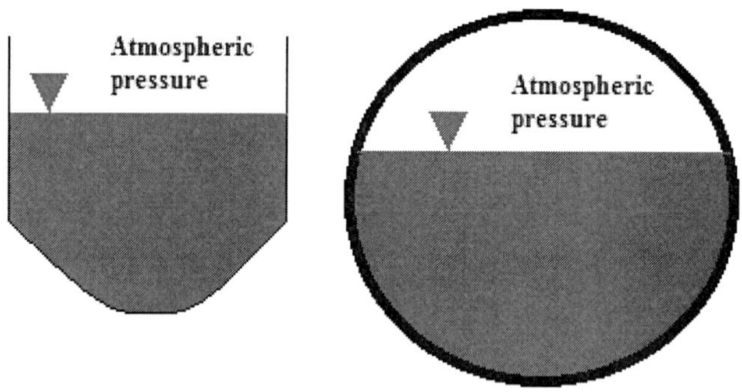

Figure 7.1. Free Surface in Open Channel Flows: Open Channel (Left) and Partially Flowing Pipe (Right).

TYPES OF OPEN CHANNELS

Prismatic and Non-Prismatic Channels

A channel in which cross sectional shape, size and the bottom slope are constant over long stretches is termed as prismatic channel. Most man-made artificial channels such as the rectangular (Figure 7.2), trapezoidal, triangular and half-circular in shape are prismatic channels. All natural channels generally have varying cross section and consequently are non-prismatic.

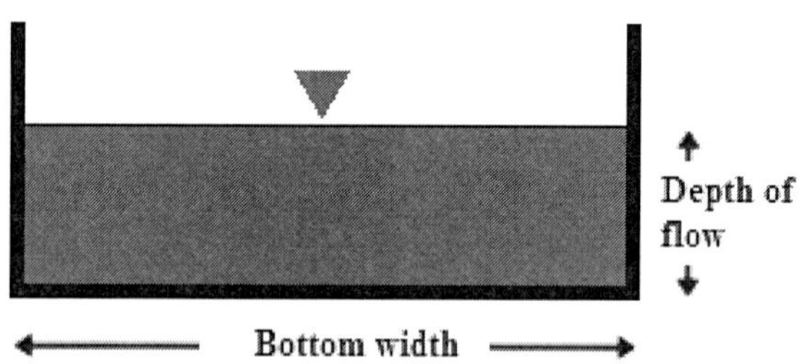

Figure 7.2. A Schematic View of a Rectangular Prismatic Channel.

Rigid and Mobile Boundary Channels

Rigid channels are those in which channel boundary or cross section is not deformable. The shape and roughness magnitudes are not functions of flow parameters. The lined channels and non-erodible unlined channels are rigid channels. Channels are mobile boundary channels, if the boundary of the channel is mobile and flow carries considerable amounts of sediment through suspension and is in contact with the bed. In the mobile channel, depth of flow, bed width, and longitudinal slope of channel undergo changes with space and time depending on the kind of flow.

Unlined and Lined Channels

Depending upon the kind of channel surface, channels are unlined and lined channels. Unlined channels are earthen channels without any lining. Lined channels on the other hand are lined with stones, masonry or even plastic sheet. While the lined channels are rigid channels, unlined channels are mobile boundary channels.

Open Channel Geometry

Depth of Flow (y): It is the vertical depth or distance from the bottom to free surface.

Depth of Flow (d): Also known as depth of flow section, it is the flow depth normal to the direction of flow or water surface. If θ is the slope of the channel bottom then

$$y = d\cos\theta \tag{7.1}$$

If θ is small then y is nearly equal to d.

Top Width (B): The width of the channel at the free surface.

Area of Cross Section (A): The flow area, A is the cross-sectional area of flow normal to the direction of flow. Area of cross section is the depth of flow times the bottom width in the case of rectangular channels (Figure 7.2). If b is the bottom width of the channel and y is the depth of water flow, the area of wetted section of channel, $A = by$.

Wetted Perimeter (P): It is the length of line of intersection of channel-wetted surface with a cross-sectional plane normal to the flow direction. It is the sum of the lengths of that part of the channel sides and bottom, which are in contact with water. The wetted perimeter for a rectangular channel (Figure 7.2) with dimensions as in the area of cross-section is given as: $P = b+2y$.

Hydraulic Radius (R): It is the ratio of area of wetted cross section to wetted perimeter. The hydraulic radius for a rectangular cross-section is:

$$R = A/P = by/(b+2y) = y/(1+2y/b) \tag{7.2}$$

Average depth (D): Also termed as hydraulic average depth, it is the area divided by top width or $D = A/B$.

Hydraulic Slope (S): The slope of the hydraulic grade line (HGL) is the hydraulic gradient or hydraulic slope. In free-surface flow, the hydraulic grade line (HGL) usually, but not always, coincides with the free surface. If the velocity head, $V^2/(2g)$, is added to the top of the hydraulic grade line, the line joining the resulting points is the energy-grade line (EGL). Here V is the mean flow velocity for the channel cross section, and g is the acceleration due to gravity. This line represents the total head at different sections of a channel. The slopes of HGL and EGL are usually denoted as S_f and S_e. The channel bottom slope is denoted as S_o. For steady uniform flows, the three slopes are equal. Contrarily, in non-uniform gradually varied flows, three slopes are not equal. For uniform flows $S_f = S_o$. Thus, in uniform and steady uniform flows, channel bottom slope is the hydraulic slope. It is the ratio of vertical drop in longitudinal channel section (h) to the channel length (l).

$$S_o = h/l$$

Hydraulic Parameters

Rectangular channel

$$B = b$$

$$A = b \times y \tag{7.3}$$

$$P = b + 2 \times y \tag{7.4}$$

$$R = b \times y/(b + 2 \times y) = y/(1 + 2y/b) \tag{7.5}$$

Here b is the bottom width of the channel and y the depth of flow.

Trapezoidal

The cross sectional area of flow in the trapezoid area, A is given as

$$A = y(b + B)/2 \tag{7.6}$$

$A = (y/2)(b + b + 2zy)$, because $B = b + 2zy$ (Figure 7.3 left).

Simplifying, we have: $A = by + zy^2$ (7.7)

Here z is the side slope, H:V = z:1

The wetted perimeter is: $P = b + 2l$ such that l is the length of the side. By the Pythagoras theorem: $l^2 = y^2 + (yz)^2$, or $l = [y^2 + (yz)^2]^{1/2}$, so the wetted perimeter is:

$$P = b + 2y(1 + z^2)^{1/2} \tag{7.8}$$

From eq. (7.7) and eq (7.8), hydraulic radius is:

$$R = (by + zy^2)/[b + 2y(1 + z^2)^{1/2}] \tag{7.9}$$

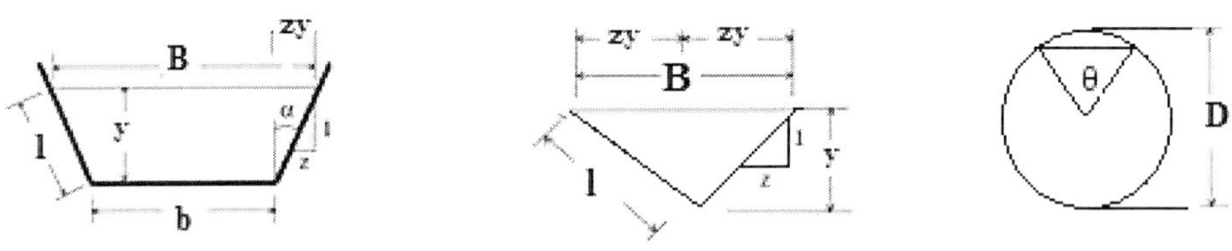

Trapezoidal Cross Section — Triangular Channel — Circular channel

Figure 7.3. Trapezoidal (Left), Triangular (Middle) and Circular Cross-sections of Open Channels.

Triangular

Figure 7.3 (middle) represents a triangular cross section with the two sides having the same slope. Compared to the trapezoidal cross-section, fewer parameters describe the triangular cross-section. The parameters, as shown in the diagram are: B, the top width of the liquid; l, the sloped length of the triangle side; y, the liquid depth measured from the vertex of the triangle; and the side slope specification, H:V; z : 1.

The triangular area, A is By/2. Since B = 2yz (Figure 7.3 middle), the area of the triangular section becomes:

$$A = y^2z = l \sin\theta \qquad (7.10)$$

The wetted perimeter, P is 2l. With $l^2 = y^2 + (yz)^2$, it becomes

$$P = 2[y^2(1 + z^2)]^{1/2}$$

The hydraulic radius is

$$R = A/P = y^2z/\{2[y^2(1 + z^2)]^{1/2}\} \qquad (7.11)$$

Circular

The hydraulic radius for a circular pipe flowing full is easy to calculate. The cross-sectional area is $A = \pi D^2/4$ and the wetted perimeter is $P = \pi D$, giving $R = D/4$. Here D is the diameter of the pipe.

If the pipe is not flowing full and the arc makes an angle θ at the center (Figure 7.3 right), then for θ radians, the area and the perimeter of the sector are calculated as

$$\text{Top width} = D \sin \theta/2 \qquad (7.12)$$

$$\text{Area, } A = (1/8)(\theta - \sin \theta)D^2 \qquad (7.13)$$

$$\text{Wetted perimeter, } P = \theta\, D/2 = r\, \theta \qquad (7.14)$$

$$\text{Hydraulic radius, } R = (D/4)(1 - \sin \theta/\theta) \qquad (7.15)$$

Parabolic

For a parabolic cross section

$$A = (2/3)\, B\, d$$

$$\text{Wetted perimeter, } P = B + 8\, d^2/(3B) \qquad (7.16)$$

$$R = 2\, By/(1.5\, B^2 + 4y^2) \qquad (7.17)$$

Here B is the top width. The approximate relation y =1.5 R is commonly used in the design of parabolic channels.

ECONOMICAL OR BEST HYDRAULIC SECTIONS

A channel section is most economical, if for the same flow, the construction cost is minimum. Since the cost of construction of a channel depends on the depth of excavation and construction for lining, minimum construction cost will be if the channel passes maximum discharge for a given cross sectional area. It is also evident from the continuity equation and uniform flow formulae that the maximum velocity of flow will be only for the maximum hydraulic radius or minimum wetted perimeter (eq. 7.18). Thus, to carry a given discharge at minimum construction cost of the channel, channel should have minimum wetted perimeter. To illustrate the point, let us look at the Manning's equation.

$$V = (1/n)\,(R^{2/3})\,(S_o^{1/2}) \tag{7.18}$$

Here V is the flow velocity, n is the Manning's roughness coefficient and other parameters are as defined earlier. This equation forms the basis to determine the dimensions of economical sections of different kinds of channels (Michael, 2008). Other terms for most economical section are the best section or hydraulic efficient section as the discharge passing through the most economical section for a given cross-sectional area (A), slope of the bed (S) and a roughness coefficient (n), is maximum.

Rectangular

For a rectangular channel, wetted perimeter given by eq. (7.4) is rewritten as:

$$P = b+2\times y = (A/y) + 2y$$

$$dP/dy = -\,A/y^2 + 2 = 0$$

$$y = \sqrt{A/2} \text{ or } b = 2\,y \tag{7.19}$$

Incorporating this in the hydraulic radius equation

$$R = A/P = By/(b+2y) = 2y2/4y = y/2 \tag{7.20}$$

Therefore, the best hydraulic rectangular section is when the bottom width is twice the depth of flow (eq. 7.19) or the hydraulic radius is half the depth of flow (eq. 7.20).

Trapezoidal

Rewriting the wetted perimeter equation (eq. 7.8) for this section, we have

$$P = b+2\times y\times \sqrt{(1+z^2)}$$

Since $b = (A/y) - z\times y$, then

$$P = (A/y)+[2\times \sqrt{(1+z^2)} - z)]\,y$$

For a minimum value of P, $dP/dy = 0$

$$dP/dy = -\,(A/y^2)+[2\times \sqrt{(1+z^2)} - z)] = 0$$

$$-\,A+[2y^2 \sqrt{(1+z^2)} - zy^2] = 0$$

$$dP/dz = 0 + 2z\,y\,(1+z^2)^{-0.5} - y$$

$$2z\,y/(1+z^2)^{0.5} - y = 0$$

$$(1+z^2)^{0.5} = 2z$$

$$4z^2 = 1 + z^2$$

$$z = 1/\sqrt{3} \tag{7.21}$$

$$\tan\alpha = 1/\sqrt{3} \text{ or } \alpha = 30^\circ$$

Here α is the internal angle that the side makes with the vertical (Figure 7.3 left). It can be shown that

$$b = 2y \tan \alpha = 2y/\sqrt{3} \tag{7.22}$$

$$R = y/2 \tag{7.23}$$

Thus, an optimum trapezoid always envelops a semi-circle whose center is the water surface and radius equal to the depth. The sides of the section are tangent to this circle. It also means that the trapezoidal section is the most economical if b and y follow the relation:

$$b = 2y \tan \theta/2 \tag{7.24}$$

Such that θ is the external angle trapezoidal side makes with the horizontal.

Circular

As already discussed before,

$$R = A/P = D/4 = r/2 \tag{7.25}$$

Here r is the radius of the circle. Therefore, a semi-circular cross-section is the best hydraulic cross-section.

N-sided Channel

The derivation for the rectangular section reveals that reflection of the rectangular optimum section about the water surface forms a square of side b. The derivation of trapezoidal cross-section reveals that the optimum section about the water surface will form a regular hexagon of side b. Thus, for an N-sided channel, the optimum hydraulic section should be in the form of half a 2N-side regular polygon, the side angle being

$$\theta = (N-1)/N \tag{7.26}$$

Triangular

$$\text{Area} = y\, l \sin \theta$$

Here θ is half angle between the two sides.

$$R = A/P = (l\, y \sin \theta)/2\, l = (y \sin \theta)/2 = (l \cos \theta \sin \theta)/2 = (l \sin 2\theta)/4$$

(Since $y = l \cos \theta$ and $\cos \theta \sin \theta = \sin 2\theta/2$)

For R to be maximum, $dR/d\theta = 0$

$$(2l \cos 2\theta)/4 = 0$$

$$\text{Cos } 2\theta = 0 \text{ or } \theta = 45^\circ$$

A triangular channel section is the most economical if each of its sloping side makes an angle of 45° with vertical or is half square described on a diagonal and having equal sloping sides.

We can also use the analogy given by eq. (7.26) for N-sided channel. For a triangular section, we have N=2 such that

$$\theta = (N-1)/N = 1/2 \text{ or } 45^\circ.$$

It may not always be possible to design a channel that meets the requirements of the most efficient cross section. As a compromise between hydraulic efficiency and other considerations, the ratios between bed width (b) and water depth (d) for trapezoidal channels are approximated (Table 7.2; Smedema and Rycroft, 1983).

Table 7.2. Recommended Bed Width, b and Depth of Flow, d Ratios for Trapezoidal Channels

Drain depth	bed width/depth of flow (b/d) ratios
Small channel, d=0.75 m	1 for clay to 2 for sand
Medium channel, d=0.75 to 1.5 m	2 to 3
Large channel, d=1.5 m	3 to 4

Example 7.1

Calculate the best hydraulic rectangular cross-section to convey $Q=10\ m^3/s$ discharge with Manning's n= 0.02 and slope = 0.0009.

For the best rectangular hydraulic cross section,

$$R = y/2,\ A = 2\ y^2$$

Using the Manning's equation (eq. 7.18) for discharge, we have

$$Q = A\ (1/n)\ (R^{2/3})\ (S_o^{1/2})$$

Here Q is the discharge in m^3/s.

$$10 = 2y^2\ (1/0.02)\ (y/2)^{2/3}\ (0.0009)^{1/2}$$

$$y^{8/3} = 5.29 \text{ or } y = 1.87^m$$

$$b = 2y = 3.74^m$$

STATE OF THE FLOW

Effect of Viscosity

Reynolds number, Re briefly described in Chapter 4, characterizes the fluid viscosity effect on pipe or open channel flows. The Re for the sake of brevity is rewritten as:

$$Re = V\ R/\nu \tag{7.27}$$

Here ν is the kinematic viscosity. Other parameters have the same meaning as previously described in this chapter. Note that in open channels, we use R the hydraulic radius rather than D the diameter of the pipe. Using the same corollary, the hydraulic radius of a circular pipe of diameter, D, will be D/4. Accordingly, the values to classify flow are one fourth of the values in pipe flow. Thus, for laminar, transitional and turbulent flow in open channels following guidelines apply (Bansal, 2009).

Laminar	$Re < 500$
Transitional	$500 < Re < 1000$
Turbulent	$Re > 1000$

Effect of Gravity

The gravity is the driving force for flow in an open channel. Therefore, it has a great bearing on the state of flow. Froude number, the square root of the ratio of the inertia to gravitational force, determines the effect of gravity in an open channel flow.

$$Fr = (\text{Inertia force}/\text{gravitational force})^{1/2}$$

$$Fr = (V^2/gD)^{1/2} \text{ or } Fr = V/\sqrt{gD} \tag{7.28}$$

Here D is the hydraulic mean depth defined previously. If Fr is equal to 1, the flow is critical. It means the flow velocity being equal to the velocity of wave propagation, downstream waves or pressure disturbances will not be able to travel upstream. Flow is supercritical if $Fr > 1$ and sub-critical if $Fr < 1$. A detailed discussion on this issue appears in a later section of this chapter.

Example 7.2

Bottom width, side slopes, and bottom slope of a trapezoidal channel are 6 m, 1:1, and 0.0002. If the depth of uniform flow is 1.5 m, comment upon the state of flow for a discharge of 12 m^3/s.

$$Fr = V/\sqrt{g D}$$

Bottom width = 6 m, Top width = 9 m, Area = 1.5 (9+6)/2 = 11.25 m^2

$$D = A/T = 11.25/9 = 1.25 \text{ m}$$

$$Fr = (12/11.25)/\sqrt{(9.81\times1.25)} = 0.307$$

Since Fr is less than 1, flow is sub-critical.

CHEZY'S EQUATION

Figure 7.4 is a schematic representation of an open channel flow. For a uniform flow, the gravity forces should exactly balance the friction forces at the boundary. The gravity force acting downward is ρgAL, which in the direction of the flow becomes $\rho gAL\sin\theta$. Here ρ is the density of the fluid, L the length and A the area of cross-section of the channel respectively. Besides, g is acceleration due to gravity, and θ angle made by the bottom of the channel with the horizontal. If τ is the mean boundary shear stress offered by the boundary of the channel then total shear stress is given as: $PL\tau$. Here P is the wetted perimeter of the channel. Equating the two, we have

$$PL\,\tau = \rho gAL\sin\theta \tag{7.29}$$

For small slopes, $\sin\theta \approx \tan\theta \approx S_o$, such that

$$\tau = \rho gAS_o/P = \rho gRS_o \tag{7.30}$$

Since τ for a turbulent flow is proportional to the square of the velocity, we have

$$\tau \propto V^2 \text{ or } \tau = KV^2$$

Here K is the proportionality constant. Incorporating τ in eq. (7.30), we have

$$V^2 = \rho g\, RS_o/K \tag{7.31}$$

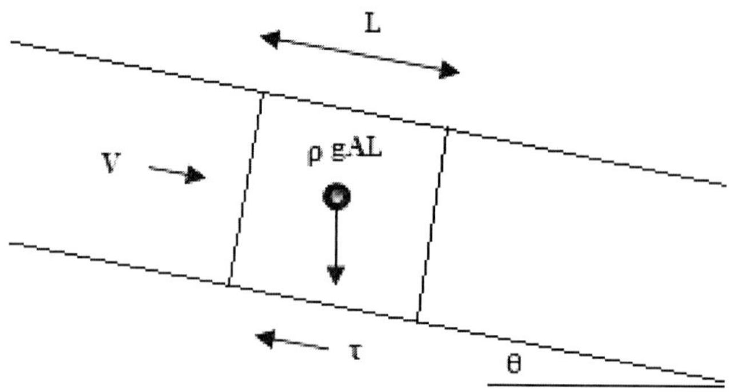

Figure 7.4. Schematic Diagram of an Open Channel Flow Showing Gravity and Frictional Forces.

Eq. (7.31) can also be written as:

$$V = C\sqrt{RS_o} \tag{7.32}$$

Here C is the Chézy coefficient that combines all the constants and parameters whose values are known. Eq. (7.32) is known as the Chézy Equation (as reported in Chézy, 1897). Note that C is not a constant in the strict sense of the word as it depends on the Reynolds Number and the boundary roughness. Its units are $\sqrt{m}/s$ and the values vary from 60 for rough and 120 for the smooth channels.

Assessment of Chézy Coefficient

Although Chézy's equation is quite simple, yet the selection of correct value of C is often a problem. A few of the methods used to estimate the values of C are as follows.

Bazin proposed the following equation to calculate the value of C.

$$C = 157.6/(181+K_1\sqrt{R}) \tag{7.33}$$

K_1 is the Bazin constant. Table (7.3) lists the values of K_1 for various channel surfaces.

Table 7.3. Values of Bazin's Constant for Various Types of Channel Surfaces

Channel Surface Type	*Bazin's Constant, K_1*
Smoothly shaved wooden board, smooth cement coated	0.11
Concrete, brick or non-planed wood	0.21
Smooth rubble masonry or poor brick work	0.83
Earth channels in very good condition	1.54
Earth channels in rough condition	3.17
Dredged earth channel, average conditions	2.36

The Ganguillet-Kutter equation to assess the values of C is:

$$C = (23 + 1/m + 0.00155/S/[1 + \{23 + (0.00155/S)\}(m/\sqrt{R})] \tag{7.34}$$

Here, S is the channel slope. Values of m for various surface conditions are taken from Table 7.4.

Table 7.4. Values of m in the Ganguillet-Kutter Equation for Various Surface Conditions

Surface Condition	*m*
Smoothly shaved wooden board, smooth cement coated	0.010-0.013
Rough wooden board, relatively smooth concrete	0.013-0.017
Brick, coated with mortar	0.01-0.020
Non-finished concrete	0.017-0.030
Concrete with exposed gravel	0.025-0.033
Rough masonry	0.012-0.018
Both sides stone-paved but bottom face irregular earth	0.015-0.018
Deep, sand bed river whose cross-sections are uniform	0.028-0.035
Gravel bed river whose cross-sections are uniform and whose banks are covered with wild grass	0.030-0.040
Bending river with large stones and wild grass	0.035-0.050
Unlined canal, earth straight	0,018-0.025
Lined canal, concrete	0.012-0.017

The Powell formula to calculate values of C seems to be more realistic since it contains the hydraulic radius R, the absolute roughness e, and the Reynolds number Re. However, this formula is implicit and needs trial and error to solve for C.

$$C = -42 \log ((C/4Re) + e/R) \quad (7.35)$$

According to this relationship, C depends on the Reynolds number Re and therefore, on the kinematic viscosity, ν. There is no term that takes care of the effect of the slope S on the coefficient C. The absolute roughness, e appearing in the Powell formula is a measurable parameter. The equation is applicable for the entire domain of turbulent flow.

Relation of C with Darcy-Weisbach friction coefficient

Let us consider Darcy-Weisbach (1840) equation for head loss in pipe flow.

$$H_f = fLV^2/(2Dg)$$

Such that H_f is head loss, f is the Darcy-Weisbach friction coefficient, L is the length and D is the pipe diameter. For a channel, D is replaced with hydraulic radius, R (D/4 = R so D = 4R) and H_f with LS_f *i.e.* drop in head over a length L, we have

$$LS_f = fLV^2/(8Rg)$$

$$R\,S_f = fV^2/(8g)$$

$$V = [(8g/f)\,(RS_f)]^{1/2} \quad (7.36)$$

For uniform flows, S_f is equal to bottom slope, S_o. Comparing eq. (7.32) with eq. (7.36), we get

$$C = \sqrt{(8g/f)} \tag{7.37}$$

Eq. (7.37) relates the Chezy's coefficient with that of Darcy-Weisbach friction coefficient.

Example 7.3

Use Chezy's equation to calculate the head loss due to friction in a pipe of diameter 180 mm and length 50 m. Velocity of water may be taken as 2 m/s and C as 50.

Using eq. (7.32),

$$V = C\sqrt{RS} \text{ such that } S = H_f/L$$

$$2 = 50\,[(R) \times (H_f/50)]^{1/2}$$

For a circular pipe, $R = d/4$

$$(2/50)^2 = (0.18/4) \times (H_f/50)$$

$$H_f = 0.0016 \times 50/0.045 = 1.77 \text{ m}$$

MANNING'S EQUATION

Using data from several experiments, Manning (1891) showed that for a uniform flow C in the Chezy's equation is equal to $R^{1/6}/n$, such that Chézy equation now known as Manning's equation becomes

$$V = (1/n)\,(R^{2/3})\,(S^{1/2}) \tag{7.38}$$

In which n is a constant, known as Manning's coefficient. S is the channel bed slope, m/m. This equation is commonly used in irrigation and drainage works to design open channels. The design comprises finding out the dimensions of the channel cross-section that will suffice to carry the design discharge at a given flow velocity, slope, and shape. It also includes determining the discharge capacity, if cross-section and other dimensional components are given. Following steps describe the procedure to design open channels.

Hydraulic radius, R is calculated for the desired shape of the channel as per procedure outlined in a previous section of this chapter. For this purpose, known or assumed dimensions of the channel along with the guidelines of best hydraulic section are used. The side slopes depend primarily on the engineering properties of the material through which the channel is excavated (Chow, 1959). The commonly used side slopes for different kinds of soil vary from nearly vertical to 3:1 (Table 7.5).

The average slope of the ground shall be determined from the longitudinal section of the ground along the proposed alignment. The maximum average slope of the channel will be limited by this value.

Eq. (7.38) is most sensitive to Manning's coefficient, n. The Manning's n depends upon a number of factors, such as

- ✰ Surface roughness of the channel
- ✰ State of vegetation in the channel

Table 7.5. Recommended Side Slopes for Channels in Various Types of Materials (Chow, 1959)

Material	*Side Slope*
Rock	Nearly vertical
Muck and peat soils	1/4 : 1
Stiff clay or earth with concrete lining	1/2 : 1 to 1 : 1
Earth with stone lining for large channels	1 : 1
Firm clay or earth for small ditches	1 1/2 : 1
Loose, sandy earth	2 : 1
Sandy loam or porous clay	3 : 1

- ☆ Channel alignment and irregularities
- ☆ Silting and scouring
- ☆ Obstructions in the channel, if any
- ☆ Size and shape of the channel
- ☆ Stage and discharge
- ☆ Seasonal change in suspended material and bed load

Thus, careful selection of this parameter is paramount to the use of Manning's equation. Find out the value of n by one of the following ways.

Manning's roughness, n relates to Darcy-Weisbach, f as per the following relation.

$$R^{1/6}/n = (8\,g/f)^{0.5}$$

$$n = R^{1/6}\,(f/8g)^{0.5} \qquad (7.39)$$

The most common approach to select the Manning's roughness coefficient is to use the tabulated values given in various textbooks. Table 7.6 or Table 7.7 may prove handy to select the value for the given condition. In the absence of these tables or if particular conditions are not mentioned in these tables, one can use the following equation to estimate the value of n. It takes care of various factors that affect the values of Manning's n (Cowan, 1956).

$$n = (n_0 + n_1 + n_2 + n_3 + n_4)\,n_5 \qquad (7.40)$$

Here n_0 is the base value of n for a straight, uniform, smooth channel in natural materials. Other factors such as n_1 takes care of surface irregularities, n_2 of variations in shape and size of the channel cross section, n_3 for obstructions, n_4 for vegetation and flow conditions and n_5 for meandering of the channel (Table 7.8). Use an equivalent Manning's roughness coefficient for composite sections *i.e.* the sections composed of different roughnesses along the perimeter.

Table 7.6. Limiting Velocities for Clear and Turbid Water in Straight Channels after Aging and Selected Values of n

Material	*Maximum Permissible Velocities (m/s)*	*n*
Fine sand	0.5	0.020
Sandy loam	0.58	0.020
Silt loam	0.67	0.020
Firm loam	0.83	0.020
Stiff clay	1.25	0.025
Fine gravel	0.83	0.020
Coarse gravel	1.33	0.025
Gravel	1.2	
Disintegrated rock	1.5	
Hard rock	4.0	
Brick masonry with cement pointing	2.5	
Brick masonry with cement plaster	4.0	
Concrete	6.0	

Source: Fortier and Scobey, 1926, Schwab *et al.*, 1993.

Table 7.7. Maximum Allowable Velocities for Non-vegetative Channels of Varying Channel Materials

Channel Materials	*Manning´s, n*	*Maximum Allowable Velocities (m/s)*		
		Clear Water	*Water with Colloidal Silt*	*Water with Sand Gravel/ Fragment*
Fine sand, colloids	0.020	0.46	0.76	0.46
Sandy loam, non-colloids	0.020	0.53	0.76	0.61
Silt loam, non-colloids	0.020	0.61	0.91	0.61
Alluvial silts, non-colloids	0.020	0.61	1.07	0.61
Ordinary firm loam	0.020	0.76	1.07	0.69
Stiff clay, very colloidal	0.025	1.14	1.52	0.91
Alluvial silt, colloidal	0.025	1.14	1.52	0.91
Fine gravel	0.025	0.75	1.52	1.14
Coarse gravel, non-colloidal	0.025	1.22	1.83	1.98
Cobbles and shingles	0.035	1.52	1.68	1.98
Shales and hard pan	-	1.83	1.83	1.52

The capacity of the channel to carry flow is determined by the relation given as

$$Q = AV \tag{7.41}$$

Here, Q is the capacity of the channel, m^3/s, V is the average flow velocity, m/s, and A is the cross-sectional area of channel perpendicular to flow, m^2. The

Table 7.8. Values of n_0 to n_5 of eq. (7.40) for Various Channel Conditions

Material, n_0		*Degree of Irregularity, n_1*		*Variation of Cross-section, n_2*		*Channel Obstructions, n_3*		*Vegetation, n_4*		*Degree of Meandering, n_5*	
Earth	0.020	Smooth	0.0	Gradual	0.00	Negligible	0.0	Low	0.005-0.010	None	1.00
Rock	0.025	Minor	0.005	Alternating occasional	0.005	Minor	0.01 to 0.015	Medium	0.01-0.025	Minor	1.00
Fine gravel	0.024	Moderate	0.01	Alternating frequently	0.01 to 0.015	Appreciable	0.02 to 0.03	High	0.025-0.050	Appreciable	1.15
Coarse gravel	0.028	Severe	0.02			Severe	0.04 to 0.06	Very high	0.050-0.100	Severe	1.70

velocity of flow for use in eq. (7.41) is assessed with the Manning's equation. The channel design should ensure that the flow velocities remain within a range of acceptable minimum and maximum values to avoid scouring or sedimentation (Kumar and Singh, 2005). The minimum permissible velocity is the lowest velocity that prevents sedimentation and vegetative growth (some estimate give 0.6 – 0.9 m/s for sedimentation and 0.75 m/s for vegetation). The maximum velocities to avoid scouring of the channel bed should be restricted as per guidelines reported in Table 7.6 or Table 7.7. Following example illustrates the use of the Manning´s equation (eq. 7.38) to design open channels.

Example 7.3

Design a trapezoidal open ditch to carry a discharge of 0.26 m^3/s. The soil is silt loam and maximum permissible slope of the channel bed is 0.1 per cent. Assume n as 0.04.

Discharge rate Q= 0.26 m^3/s

Side slopes for silt loam is taken as= 1.5:1 (Table 7.5), so that $\tan\theta = 1/1.5=0.666$ or θ=33.66 or $\theta/2$= 16.83

Assuming y= 0.5 m, b=2 y tan $\theta/2$ =2×0.5×0.3025=.30 m ~ 30 cm

$$\text{Length of side} = \sqrt{(1.5\times0.5)^2+(0.5)^2} = 0.901\text{m}$$

P= 2×0.90+0.30= 2.10 m

$$A = \frac{(3\times0.5+0.30)+0.30}{2}\times0.5 = 0.53\,\text{m}^2 \text{ so that } R = \frac{A}{P} = \frac{0.53}{2.1} = 0.252\,\text{m}$$

$$S = \frac{0.1}{100} = 0.001\frac{\text{m}}{\text{m}}$$

$V = (0.252)^{2/3}\times(0.001)^{1/2}/0.04 = 0.315$ m/s

The velocity is within permissible limit (Table 7.6).

Q= 0.53×0.315 = 0.167 m^3/s

Since the discharge is less than the design discharge, assumed dimensions do not suffice. Therefore, increase the value of y by 0.1 m.

Assuming y= 0.6 m, b=2 y tan $\theta/2$ =2×0.6×0.3025=.36 m ~ 36 cm

$$\text{Length of side} = \sqrt{(1.5\times0.6)^2+(0.6)^2} = 1.12\text{m}$$

P= 2×1.12+0.36= 2.6 m

$$A = \frac{(3\times0.6+0.36)+0.36}{2}\times0.6 = 0.756\,\text{m}^2$$

$$R = \frac{0.756}{2.6} = 0.29\,\text{m}$$

$V = (0.29)^{2/3}\times(0.001)^{1/2}/0.04 = 0.346$ m/s

The velocity is still within permissible limit.

$$Q= 0.756\times0.346 = 0.261\ m^3/s$$

The capacity of the channel is higher than the design capacity and hence the dimensions are OK in this respect.

NON-UNIFORM FLOW

Assumption of uniform flow resolves many problems in hydraulics of open channel flow. On the other hand, it may be necessary in some situations to use non-uniform flow principles. This necessitates the use of the concept of super-critical, critical and sub-critical flow regimes. The following section deals with specific energy as a pre-cursor to the concept of super-critical, critical and sub-critical flow regimes. Discussion on hydraulic jump is included in a later section.

Specific Energy

The total energy head at a point in an open channel is the sum of the potential and kinetic energy and internal energy of the flowing water (Figure 7.5). The elevation of the water surface *i.e.* the sum of depth of flow (y) and the elevation of channel bottom (Z) from an arbitrary datum represents the potential energy. The water surface elevation is a measure of the potential work that the flow can do during its movement from a higher elevation to a lower elevation. The kinetic energy is the energy of motion as measured by the flow velocity (v). As water flows down the channel, the flow loses energy because of friction and turbulence. The total energy head between two points in a channel reach can be set equal to one another by adding the losses, H_f between the sections to the downstream total energy head. This equality represents the energy equation expressed as:

$$Z_1 + y_1 + v_1^2/2g = Z_2 + y_2 + v_2^2/2g + H_f \tag{7.42}$$

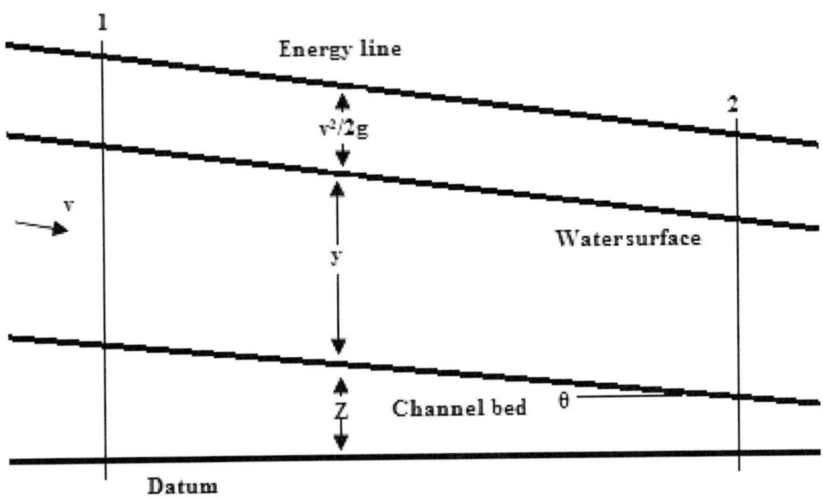

Figure 7.5. Energy Components of an Open Channel Flow.

Here, Z represents the channel elevations above an arbitrary datum, y the depth of flow, and v the average channel velocity. The notations 1 and 2 with each parameter represent these parameters at channel sections 1 and 2, respectively. The H_f is the head or energy loss between channel sections 1 and 2 and g is the acceleration due to gravity. It is a form of Bernoulli's equation and it assumes that the flow is incompressible and steady. Since in open channels, the water surface is open to the atmosphere, the pressure term between two points has the same value and is ignored.

The total head relative to the channel bottom (Z = 0) is called the specific energy E, which expresses the energy per unit weight. If the cross-sectional area of the open channel is A and the discharge Q, (such that v= Q/A), then specific energy after neglecting H_f is:

$$E= y+ Q^2/(2g\, A^2) \tag{7.43}$$

Here, cross-sectional area A depends on water depth y and can be defined as A = A(y). Examination of eq. (7.43) reveals that it can be represented as a functional relation between the three variables as:

$$f\,(E, y, Q) = 0$$

The relationship between any two variables can be obtained by keeping one of the variables as constant.

Specific energy diagram, Q constant

Now, consider a rectangular channel with a steady but non-uniform flow. If Q is the discharge through the channel, b the channel width, and q the discharge per unit width such that q = Q/b, then eq. (7.43) is written as:

$$E= y+ q^2/(2g\, y^2) \tag{7.44}$$

Eq. (7.44) gives the variation of specific energy (E) with the depth of flow (y). Hence, for a given discharge q, it is possible to calculate the corresponding values of E by varying the depth of flow, y. For a channel of known geometry, for the same specific energy there will be two values of y. It means there are two positive roots of the equation. It also means that any particular discharge Q can pass through a given channel at two depths and still maintain the same specific energy, E. A graph between specific energy, E (along X-axis) and depth of flow, y (along Y-axis) plots as a cubic parabola (Figure 7.6). Of the two alternate depths, smaller y_1 has a large velocity head while the other y_2 having larger depth has smaller velocity head. For the same Q, difference between the two alternate depths increases with increase in specific energy or decreases with decreasing E. The two possible depths having the same specific energy are known as alternate depths and correspond to the sub-critical and super-critical flow regions, respectively. At a certain value, E = Ec, and the two depths merge with each other (point C in Figure 7.6). Since, there is no value for y for $E < E_c$, it means that the flow under this condition is not possible in the region. Physically, sub-critical flow is deep and the velocities are slow. This means sub-critical flow has high potential energy and low kinetic energy. Super-critical flow on the other hand tends to be shallow with fast velocities. It has low potential but high kinetic energy. The flow is critical, if the flow conditions for a given discharge, Q, results in minimum specific energy, E_{min}. The depth of flow at this stage is critical depth, y_c and velocity the critical velocity, v_c given by $\sqrt{(g\,y_c)}$.

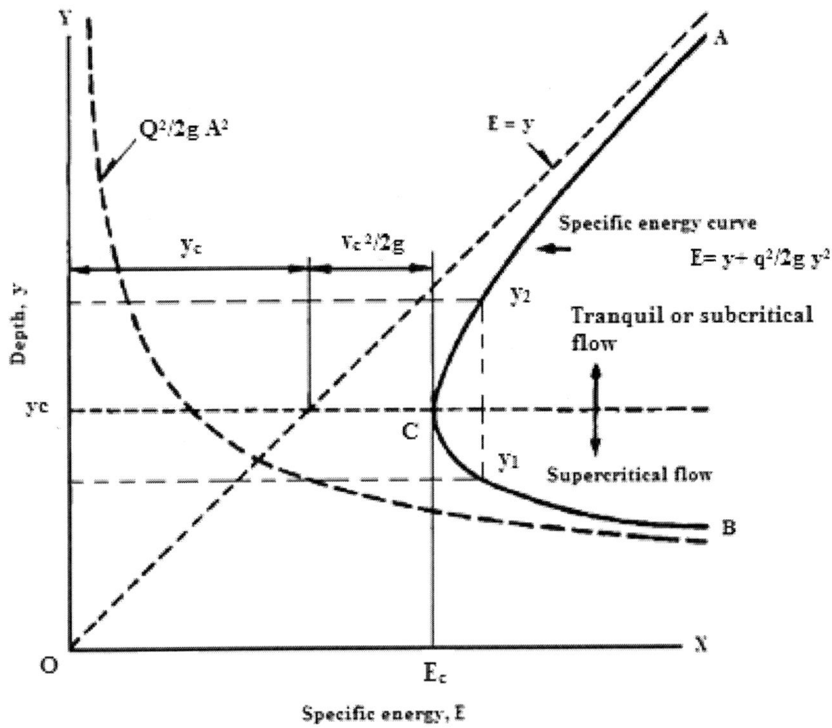

Figure 7.6. Specific Energy Curve.

Now if we draw a family of E-y curves with constant Q, we end up in various curves as shown in Figure 7.7. A line OS drawn from the origin will be an asymptot to the specific energy curve. Similarly, if a line OQ is drawn through the point of minimum E_c, it will have a slope of 2/3 such that 2E/3 = y. Besides, the continuity equation applies to steady state conditions such that:

$$q = y_1\,v_1 = y_2\,v_2$$

In a rectangular channel with specific discharge, q, if one knows one depth (y_1), it is possible to calculate the other depth with the help of following equation:

$$y_2 = 2y_1/[-1 + 8/(Fr_1)^2] \quad (7.45)$$

Here Fr_1 is the Froude number corresponding to the depth of flow, y_1. Section on hydraulic jump includes some more discussion on this issue.

Critical Energy and Critical Depth

For constant discharge Q, Figure 7.6 depicts the relation between the specific energy and the water depth. The critical point of minimum energy occurs where dE/dy = 0. Rewriting eq. (7.44)

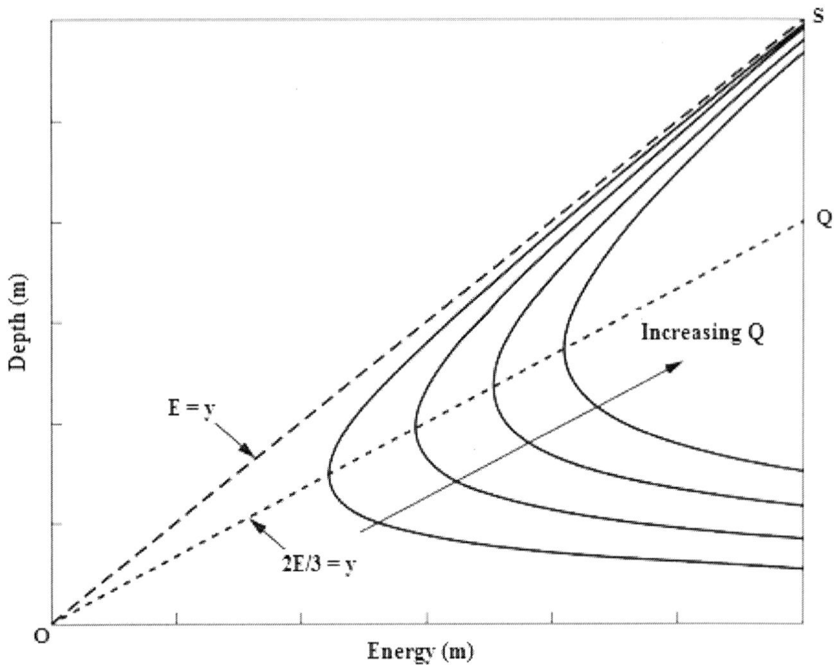

Figure 7.7. A Family of E-y Curves. Note the Dashed Line for 2E/3= *y* Crosses each Curve at its Minimum Energy Value.

$$E = y + q^2/2g\,y^2 \tag{7.46}$$

Such that

$$dE/dy = 1 - 2\,(q^2/2g\,y^3) = 0 \tag{7.47}$$

$$q^2/(g\,y^3) = 1 \tag{7.48}$$

Since y at this stage is y_c, the critical depth at which specific energy is minimum is given as

$$y_c = (q^2/g)^{1/3} \tag{7.49}$$

From eq. (7.44), the energy at the critical water depth is

$$E_c = q^2/2gy_c^{\,2} + y_c \tag{7.50}$$

Since $q^2 = g\,y_c^{\,3}$ (eq. 7.49)

$$E_c = y_c/2 + y_c = 1.5\,y_c \tag{7.51}$$

The specific energy, E_c for the critical flow situation, is 1.5 times the critical water depth, y_c. The statement made by eq. (7.51) is also being depicted in Figure (7.7). It shows that a line passing through the critical values on each successive discharge curve corresponds to y = (2/3) E. The corresponding critical velocity v_c is:

$$v_c = q/y_c = \sqrt{(g\,y_c^{\,3})}/y_c = \sqrt{(g\,y_c)} \tag{7.52}$$

According to the definition, Froude number, Fr is equal to $v_a/\sqrt{gy}$ such that v_a is the average velocity. Thus, minimum energy, E_c corresponds to Fr = 1 (from eq. 7.51). At the critical condition, the flow velocity coincides with the travelling velocity of a wave in a water channel of small depth, the so-called long wave. For depths deeper than y_c, the velocity is smaller than the travelling velocity of the long wave and the flow is tranquil (or sub-critical) flow. For shallower depths than y_c, the velocity is larger than the travelling velocity of the long wave and the flow is rapid (or super-critical) flow.

Discharge-Depth Curve

In the earlier derivation, Q was constant. Now let us keep E as constant so that

$$E = \text{Constant} = E_1 = f(y,Q)$$

For a given specific energy $E_1 = y + Q^2/2gA^2$

$$Q = A\sqrt{2g(E_1 - y)} \text{ or } q = y\sqrt{2g(E_1 - y)} \text{ such that } q = Q/b \quad (7.53)$$

A plot of the variation in discharge with water depth is parabolic in nature, and is known as Koch parabola (Figure 7.8, Moglen, 2015). For a rectangular channel of width b, eq. (7.53) is written as:

$$Q = b\, y\, \sqrt{2g}\, \sqrt{(E_1 - y)} \quad (7.54)$$

The maximum discharge can be obtained by differentiating eq. (7.54) with respect to y and equating it to zero. Also taking q = Q/b

$$dq/dy = \sqrt{2g}\, (\sqrt{(E_1 - y - y/2)}/\sqrt{(E_1 - y)} = 0$$

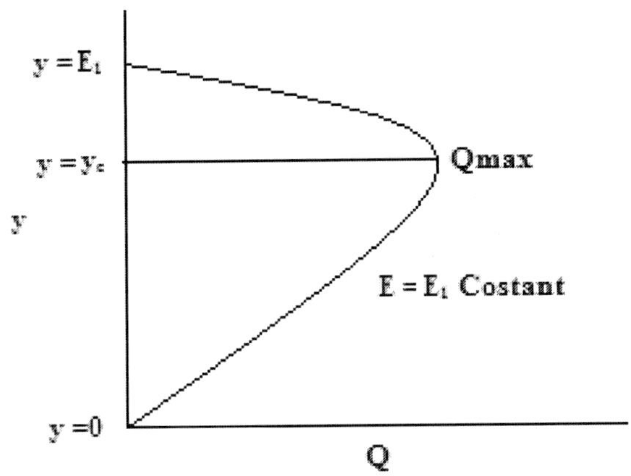

Figure 7.8. A Plot of y-Q at Constant E.

$$\sqrt{2g}\, (E_1 - y - y/2)/\sqrt{(E_1 - y)} = 0 \quad (7.55)$$

Simplifying and replacing y with y_c, we have

$$E_1 = 3/2\, y_c = 1.5\, y_c \quad (7.56)$$

Incorporating the value of E_1 in eq. (7.53), replacing y with y_c and simplifying

$$q_{max} = \sqrt{2g}\, y_c \sqrt{(3/2\, y_c - y_c)} = \sqrt{2g}\, y_c \sqrt{y_c/2}$$

$$q_{max} = y_c^{1.5}\sqrt{g} \qquad (7.57)$$

Example 7.4

A rectangular channel 5 m wide carries a discharge of 15 m^3/s. Find the minimum specific energy of the flowing water.

Minimum specific energy occurs at the critical depth of flow.

$q = 15/5 = 3\ m^3/s$ per m width of the channel

$y_c = (q^2/g)^{1/3} = (9/9.81)^{1/3} = 0.9717$ m

The value of minimum specific energy is given as:

$E_{min} = 1.5\, y_c = 1.5\times0.9717 = 1.46$ m of water

Example 7.5

For a rectangular channel, find the critical water depth if specific energy head is 1.8 m.

$E_c = 1.5\, y_c = 1.8$

$y_c = 1.2$ m

Example 7.6

A rectangular channel 2.50 m wide has a specific energy of 1.50 m when carrying a discharge of 6.48 m^3/s. Calculate the alternate depths and corresponding Froude numbers.

$E = y + v^2/2g$

Since $v = Q/A$, we have

$1.5 = y + [6.48/(2.5\times y)]^2/(2\times9.81)$

$1.5 = y + 0.34243/y^2$

Solving this equation by trial and error, the alternate depths y_1 and y_2 are obtained as, $y_1 = 1.30$ m and $y_2 = 0.63$ m

Froude number $= v/\sqrt{(g\,y)} = Q/[by\times\sqrt{(gy)}]$

$Fr = 6.48/(2.5\, y\times \sqrt{9.81}\, y) = 0.828/y^{1.5}$

For $y_1 = 1.3$, we have $Fr = 0.56$ sub-critical flow

For $y_2 = 0.63$, we have $Fr = 1.67$ super-critical flow

Example 7.7

Calculate the bottom width of a rectangular channel required to carry a discharge of 16.0 m^3/s at critical flow at a depth of 1.25 m.

$q_{max} = y_c^{1.5}\sqrt{g} = (g\, y_c^{3})^{1/2} = (9.81\times1.25^3)^{1/2} = 4.376\ m^3/s$ per m width

Bottom width = 16/4.376 = 3.656 m

MOMENTUM EQUATION

Conservation of energy enunciated by Bernoulli's equation has several useful applications. On the contrary, if energy leaks from the system via frictional head losses then this equation overstates the energy available to the flow. Consequently, the related predictions of velocity and depth are proportionately in error. In such cases it may not be possible to neglect head losses, H_f indicated earlier in eq. (7.42). An alternative to the energy conservation approach is the momentum principle (also known as the momentum-impulse principle). The linear momentum or just the momentum is the mass times the velocity of a body. Conservation of momentum or the momentum principle states that the net force acting on a fluid mass is equal to the change in momentum of flow per unit time in that direction. Conservation of linear momentum, in addition to continuity of mass and conservation of energy, is the third principle to solve open channel flow problems.

The Newton's second law of motion describes the force acting on a fluid mass 'm' as:

$$F = m \times a \qquad (7.58)$$

Here 'a' is the acceleration acting in the same direction as force F. Since a is also equal to dv/dt, eq. (7.58) can also be written as:

$$F = m\, dv/dt \qquad (7.59)$$

Since m is a constant, it can be taken inside the differential such that

$$Fdt = d(mv) \qquad (7.60)$$

Eq. (7.60) is the impulse-momentum equation. It states that the impulse of a force 'F' acting on a fluid mass 'm' in a short interval of time 'dt' is equal to the change of momentum d(mv) in the direction of force. For a controlled volume, integrating both sides of equation (7.60) with respect to time, we have:

$$F = \rho V_0(v_2 - v_1)/t \qquad (7.61)$$

Here V_0 is the volume of water in the control section, v_2 and v_1 are velocities at the two ends of the control section and ρ is the density of water. Since V_0/t is the discharge, Q and γ, the unit weight of water is ρg, we have

$$F = \gamma Q\,(v_2 - v_1)/g \qquad (7.62)$$

Momentum Equation in Open Channel Flow

According to Newton's second law of motion, the change of momentum per unit time is equal to the resultant of all external forces applied to the moving body. Besides the elevation difference, there are three kinds of forces acting on the fluid body (Figure 7.9).

- ☆ The force resulting from the pressure difference at sections 1 and 2: $P_1A_1 - P_2A_2$
- ☆ The force exerted by the downstream component of the weight of the fluid body: $\gamma V_0 \sin\theta$

☆ The force of friction and resistance acting over the boundary area, $F_f = \tau_o A$, such that τ_o is the shear stress offered by the fluid flow.

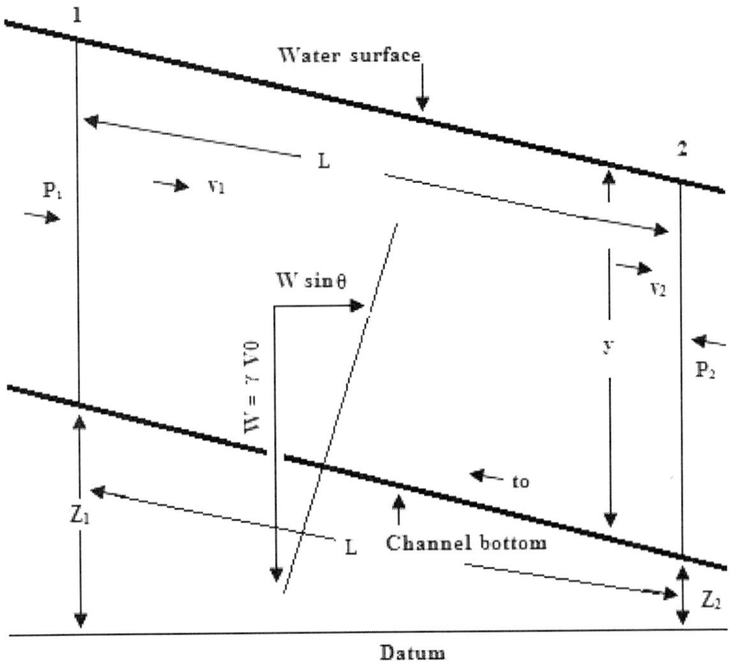

Figure 7.9. Depiction of Various Forces for Application of Momentum Principle in Open Channel Flow.

The momentum of the flow passing the channel section per unit time is expressed as before (eq. 7.62) but to take care of non-uniform velocities at most sections it is multiplied by β, a momentum-flux coefficient. For uniform flows, β is equal to 1.0. Writing β_1 and β_2 for the two sections

$$\text{Momentum} = \gamma Q(\beta_1 v_2 - \beta_2 v_1)/g \qquad (7.63)$$

Since the resultant of three forces acting on the fluid body must be equal to the change in momentum, we have

$$\frac{\gamma Q}{g}(\beta_2 v_2 - \beta_1 v_1) = P_1 - P_2 + W \operatorname{Sin}\theta - F_f \qquad (7.64)$$

Here, P_1 and P_2 are the resultants of pressure acting on two sections; W is the weight of water enclosed between section 1 and 2; F_f is the total external force of friction and resistant action along the surface of contact between the water and the channel. If the slope of the channel is relatively small, then in the short reach of a rectangular channel of width, b

$P_1 = (½)\ \gamma b y_1^2$

$P_2 = (½)\ \gamma\, b y_2^2$

$Q = ½ (v_1 + v_2)\, b\, y_a$

$y_a = ½ (y_1 + y_2)$

$\sin \theta = (Z_1 - Z_2)/L$

$W = \gamma\, b\, y_a\, L$

$F_f = \gamma\, H_f\, b\, y_a$

Substituting all the above expression for the corresponding items in eq. (7.64), dividing throughout by the channel width, b and simplifying, the two dimensional form of the momentum equation is

$$Z_1 + y_1 + \left(\frac{\beta_1 v_1^2}{2g}\right) = z_2 + y_2\left(\frac{\beta_1 v_2^2}{2g}\right) + H_f \tag{7.65}$$

Where, H_f is the loss due to external forces exerted on water by the walls of the channel. Further assuming that θ is small and H_f is negligible and $\beta_1 = \beta_2 = 1$, then

$$y_1 + \left(\frac{v_1^2}{2g}\right) = y_2 + \left(\frac{v_2^2}{2g}\right) \tag{7.66}$$

$$y_1 - y_2 = \left(\frac{v_2^2}{2g}\right) - \left(\frac{v_1^2}{2g}\right) \tag{7.67}$$

Specific Force

Beginning with the momentum equation (eq. 7.64) and assuming that the channel is short, horizontal and prismatic, we can neglect F_f and W sin θ. If $\beta_1 = \beta_2 = 1.0$, then eq. (7.64) is simplified to:

$$\frac{\gamma Q}{g}(v_2 - v_1) = P_1 - P_2 \tag{7.68}$$

The hydrostatic pressure forces P_1 and P_2 are respectively,

$P_1 = \gamma A_1\, y_{1c}$

$P_2 = \gamma A_2\, y_{2c}$

Such that y_{1c} and y_{2c} are the distances to the centroid below the water surface at section 1 and 2.

Also $v_1 = Q_1/A_1$ and $v_2 = Q_2/A_2$

$Q = ½ (Q_1 + Q_2)$

After incorporating these values and taking $Q = Q_1 = Q_2$, the momentum equation reduces to

$$\frac{Q^2}{gA_1} + A_1 y_{1c} = \frac{Q^2}{gA_2} + A_2\, y_{2c} \tag{7.69}$$

It means that the specific forces at sections 1 and 2 are equal provided it is possible to ignore the external forces and the weight effect of water in the reach between the two sections. Moreover, two sides of the above equation are analogous. Therefore, general form of the equation for a short channel section is written as:

$$F = \frac{Q^2}{gA} + A\,y_c \tag{7.70}$$

The first term of the above equation represents rate of change of momentum of the flow passing through the channel section per unit weight of water and the second term the force per unit weight of water. Since both terms are essentially force per unit weight of water, their sum is known as the specific force and equation (7.70) is known as specific force equation

For a given channel section and discharge, a plot of depth of flow against the specific force, is known as specific - force diagram. This curve has two limbs as in the case of energy diagram (Figure 7.6). One limb approaches the horizontal axis asymptotically toward the right. The other limb rises upward and extends indefinitely to the right. For a given value of the specific force, the curve has two possible depths y_1 and y_2. These two depths constitute the initial and sequent depths of a hydraulic jump. The minimum specific force is at the critical depth at point C (Figure 7.6). To obtain minimum value of the specific force, differentiate specific force eq. (7.70) against y and equate it to zero. Thus

$$\frac{dF}{dY} = \frac{Q^2}{gA^2}\frac{dA}{dY} + \frac{d(AY_c)}{dY} \tag{7.71}$$

Since $d(Ay_c) \sim A\,dy$, $dA/dy = b$, Q/A = average velocity, V and $A/b = y$, we get

$$0 = -\frac{bQ^2}{gA^2} + A$$

Here b and y are width and depth of the rectangular channel.

$$V^2/g = y \tag{7.72}$$

It is the same as the criterion for critical depth of flow. Therefore, the depth at the minimum value of specific force is the critical depth.

HYDRAULIC JUMP

A hydraulic jump occurs, if a super-critical flow with higher velocity, v_1 upstream meets a sub-critical flow of sufficient depth with decreased (lower) velocity, v_2 downstream. This local phenomenon known as hydraulic jump, is defined as the rise of water level due to the transformation of unstable shooting flow (super-critical) to the stable streaming flow (sub-critical flow). The flow becomes turbulent until the flow attains a new normal depth downstream. The hydraulic jump does not occur, if Froude number is equal to or less than 1.0. Jump also does not occur if flow changes from sub-critical to super-critical flow. A hydraulic jump frequently occurs under the following situations.

- In canals below a regulating sluice
- At toe of a spillway
- On the downstream of narrow channel
- At places, where a steep channel slope suddenly turns flat

Hydraulic jump has many practical applications. It is used to accomplish one or more than one of the following.

- To dissipate the high kinetic energy of water near the toe of the spillway and to protect the bed and banks of a river near a hydraulic structure
- To raise water level for irrigation and other distribution works
- To increase the discharge of a sluice by holding back tail water
- To mix chemicals in water purification plant
- To remove air pockets from water supply lines and prevents air locks
- To aerate the stream polluted by bio-degradable materials. It is helpful in removing the wastes in the streams causing pollution
- To ensure effectiveness of hydraulic structures like weirs, notches and flumes *etc.* It prevents the scouring action on the downstream side of the structures.

Basic Characteristics of Hydraulic Jump

Bradley and Peterka (1957), after extensive experimental investigations, classified the hydraulic jump into five categories depending upon the initial Froude number Fr_1 (Table 7.9, US Bureau of Reclamation, 1955 as reported in Chin, 2006). According to Chaudhry (1993), the best jumps occur when Fr_1 is more than 4.5 but less than 9.

Because of substantial dissipation of energy owing to the turbulence associated with the jump, the nature of the hydraulic jump cannot be explained by the use of the energy equation. Instead, we need to use the conservation of momentum equation. The total momentum, P, of a flow is simply the sum of the static, F_s and dynamic, F_d forces as quantified in Figure 7.10.

$$F_s + F_d = \gamma b y^2/2 + \rho Q v$$

For simplicity, divide the above equation by b, width of the channel as well as by the specific weight of water, γ. Note that Q/b is the unit discharge, q and that $\gamma = \rho g$, then

$$(F_s + F_d)/b\gamma = y^2/2 + qv/g \tag{7.73}$$

Finally, define M as the open channel flow momentum function, which is equal to the left-hand side of eq. (7.73). Let us also replace v with q/y. The result is the momentum function (Belanger, 1838):

$$M = (F_s + F_d)/b\gamma = y^2/2 + q^2/gy \tag{7.74}$$

Table 7.9. Classification of Hydraulic Jumps

Type of Jump	*Froude Number*	*Remarks*
Critical flow	$Fr_1 = 1$	Wavy surface, celerity c = √gy, no hydraulic jump is formed
Undular jump	$1 < Fr_1 < 1.7$	Slightly ruffled water surface with small undulation, two conjugate depths are close, transition is not abrupt, jump is irregular and is not properly formed, fraction of energy dissipation is <5 per cent
Weak jump	$1.7 < Fr_1 < 2.5$	Eddies and rollers are formed on the surface, energy loss is small, and the ratio of final depth to initial depth ~ between 2.0 and 3.1, fraction of energy dissipation varies from 5 to 15 per cent.
Oscillating jump	$2.5 < Fr_1 < 4.5$	Oscillating jump forms when an oscillating jet enters into super-critical state and particles starts oscillating in either clockwise or anticlockwise direction, generates surface waves that persist beyond the end of the jump, ratio of final depth to initial depth ~ between 3.1 to 5.0, no periodicity, difficult to handle, baffle blocks or appurtenances are of little value, very commonly observed in canals, can travel for kilometres doing unlimited damage to earthen banks and ripraps, fraction of energy dissipation is about 15-45 per cent, this type of jump should be avoided to prevent its destructive effects
Steady jump	$4\ 5 < Fr_1 < 9.0$	Position of jump fixed regardless of downstream conditions, good dissipation of energy, ratio of final to initial depth ~ between 5.9 and 12.0, the bed surface is quite rough so the particles start to tend in one direction with high velocity and turbulence, frictional losses are more, position is sensitive to variation of tail water, fraction of energy dissipation is 45 to 70 per cent
Strong jump	$Fr_1 > 9.0$	Strong jump is a perfect jump formed when frictional losses are more, air pressure division is equal and velocity is very high to result in significant losses, the water changes its state from super-critical to sub-critical in a very short length compared to all other types of jumps, jump becomes increasingly rough, Froude number before the jump, Fr_1 should not be allowed to exceed 12 as otherwise required stilling basins will be large and massive, ability of jump to dissipate energy is massive, ratio of final to initial depth ~ over 12 and may exceed 20, efficiency is about 70-85 per cent

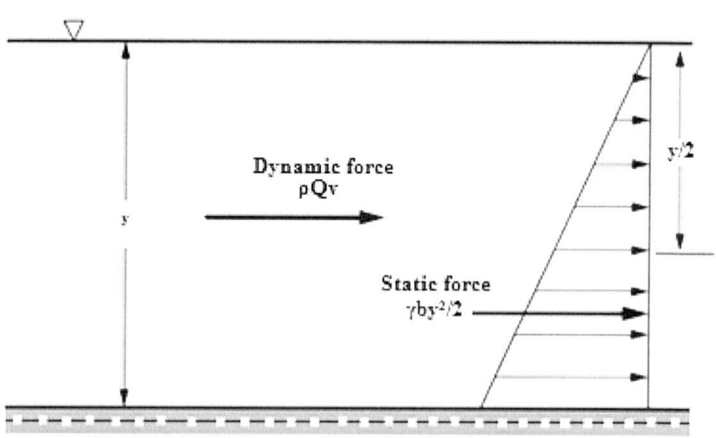

Figure 7.10. Static and Dynamic Forces at a Channel Section.

The units of M in eq. (7.74) are L^2. The units of the momentum function in the context of open channel flow is L^3, reducing to L^2 in the context of a unit width analysis in a rectangular channel.

A cross-section view of the flow from upstream of the hydraulic jump to downstream is shown in Figure 7.11. Look at a block of the flow bounded by imaginary vertical planes at cross sections 1 and 2. Since the momentum function at the two sections should be same, we have $M_1 = M_2$, or

$$y_1^2/2 + q^2/gy_1 - y_2^2/2 - q^2/gy_2 = 0 \tag{7.75}$$

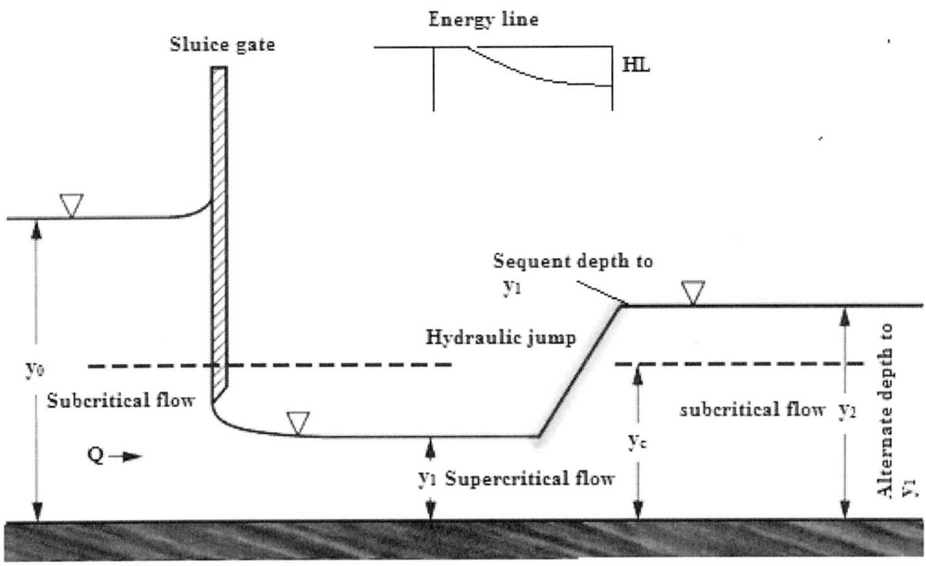

Figure 7.11. Formation of Hydraulic Jump Below a Sluice Gate.

Our goal is to develop an equation that relates the depth on one side of the hydraulic jump to the depth and flow conditions on the other side of the jump. Following assumption are made to derive the expression for the depth of hydraulic jump.

- ☆ Flow is uniform and pressure distribution is due to hydrostatic forces before and after the jump
- ☆ Losses due to friction on the surface of the bed of the channel are negligible
- ☆ The slope of the bed of the channel is small, so that the component of weight of the fluid in the direction of flow is negligibly small.

Eq. (7.75) is written as:

$$0.5\,(y_1^2 - y_2^2) - q^2\,(1/y_2 - 1/y_1)/g$$

Factoring the static force term and grouping a common denominator in the dynamic force term, we get:

$$0.5\,(y_1 - y_2)\,(y_1 + y_2) - q^2\,((y_1 - y_2)/y_1\,y_2)/g$$

Dividing throughout by $(y_1 - y_2)$

$0.5\ (y_1 + y_2) - q^2\ (1/y_1\ y_2)/g$

Multiplying by $2y_1$,

$(y_1^2 + y_2\ y_1) - 2q^2/y_2\ g$

This is a quadratic equation for y_1, with a = 1, b = y_2, c = $-2q^2/gy_2$. Hence,

$$y_1 = \frac{\left[-y_2 \pm \left\{y_2^2 - 4.0\times\left(\frac{-2q^2}{gy_2}\right)\right\}^{0.5}\right]}{2}$$

Multiply the second term under the radical by y_2^2/y_2^2

$$y_1 = \frac{\left[-y_2 \pm y_2\left\{1 + 8.0\times\left(\frac{-q^2}{gy_2^3}\right)\right\}^{0.5}\right]}{2}$$

Taking only the positive term and incorporating $Fr_2^2 = q^2/gy_2^3$

$$y_1 = 0.5\ y_2\ [(1 + 8\ Fr_2^2)^{0.5} - 1] \tag{7.76}$$

Notice that the right-hand side of the equation depends only on the conditions at location 2. Further, since the derivation began from a simple statement of the momentum function value on both the upstream and downstream sides of the jump, it is possible to reverse the subscripts (1 and 2). In other words, if we know the conditions on one side of the jump, we can use eq. (7.76) to solve for the depth on the other side of the jump.

$$y_2 = 0.5\ y_1\ [(1 + 8\ Fr_1^2)^{0.5} - 1] \tag{7.77}$$

In this derivation, the depth y_1 and y_2 is conjugate depth pair.

Example 7.8

Consider a system with a specific discharge of 3 m^3/s. If the upstream depth, y_1, is 0.54 m, determine the upstream Froude number, Fr_1, and specific energy, E_1. If a hydraulic jump spontaneously occurs from this depth, determine the downstream conjugate depth, y_2, and downstream specific energy, E_2. Show that the momentum function at the two points is same.

$Fr_1^2 = q^2/gy_1^3 = 3^2/(9.81\times0.54^3) = 2.41$

$Fr_1 = 1.55$

$E_1 = q^2/2gy_1^2 + y_1 = 9/(2\times9.81\times0.54^2) + 0.54 = 2.11$ m

Using eq. (7.77)

$y_2 = 0.5\times0.54\ [(1 + 8\times2.41^2)^{0.5} - 1] = 0.27[\sqrt{(1+46.46)} - 1] = 0.27\times5.89 = 1.59$ m

$E_2 = q^2/2gy_2^2 + y_2 = 3^2/(2\times9.81\times1.59^2) + 1.59 = 1.77$ m

Momentum function given by eq. (7.74) is

$$M_1 = (0.54)^2/2 + 9/(9.81\times0.54) = 1.84\ m^2$$

$$M_2 = (1.59)^2/2 + 9/(9.81\times1.59) = 1.84\ m^2$$

Energy Dissipation due to Hydraulic Jump

When hydraulic jump take place, energy loss occurs due to formation of eddies and turbulence (Figure 7.11). This loss of energy is equal to the difference in specific energies at sections 1-1 and 2-2. If E_1 and E_2 are the energy at section 1-1 and 2-2 respectively, then loss of energy due to hydraulic jump

$$HL = E_1 - E_2 = (y_1 + v_1^2/2g) - (y_2 + v_2^2/2g) = (v_1^2/2g - v_2^2/2g) - (y_2 - y_1)$$

Since $v_1 = q/y_1$ and $v_2 = q/y_2$, we have

$$(q^2/2gy_1^2 - q^2/2gy_2^2) - (y_2 - y_1) = q^2(y_2^2 - y_1^2)/(2g\,y_1^2y_2^2) - (y_2 - y_1) \qquad (7.78)$$

But as per earlier derivation

$$q^2 = gy_1y_2(y_1+y_2)/2$$

Substituting in eq. (7.78), we have

$$(y_1+y_2)(y_2^2 - y_1^2)/(4y_1y_2) - (y_2 - y_1) = (y_2 - y_1)[(y_2-y_1)^2]/(4y_1y_2)$$

$$HL = (y_2 - y_1)^3/4y_1y_2 \qquad (7.79)$$

Jump Efficiency

The ratio of specific energy after the jump (E_2) to the energy before the jump (E_1) is defined as the efficiency of the jump. It is written as:

$$E_2/E_1 = (y_2 + v_2^2/2g)/(y_1 + v_1^2/2g)$$

$$E_2/E_1 = (y_2/y_1 + v_2^2/2gy_1)/(1 + v_1^2/2gy_1) \qquad (7.80)$$

Incorporating the value of y_2/y_1 from eq. (7.71) and replacing v_2 by $v_1(y_1/y_2)$ and simplifying, we get

$$E_2/E_1 = \{[(1 + 8\,Fr_1^2)^{0.5}]^3 - 4\,Fr_1^2 + 1\}/[8\,Fr_1^2\,(2 + Fr_1^2)] \qquad (7.81)$$

Eq. (7.81) indicates that the jump efficiency is a dimensionless function of Froude number Fr_1 before the jump.

Relative Loss of Energy

Relative loss of energy is the ratio of loss of energy due to hydraulic jump to the specific energy prior to hydraulic jump. One can write

$$\Delta E/E_1 = (E_1 - E_2)/E_1 = [(y_2 - y_1)^3/4y_1y_2]/(y_1 + v_1^2/2g) \qquad (7.82)$$

$$\Delta E/E_1 = (y_2/y_1 - 1)^3\,(y_1^3/4y_1y_2)/[0.5\,y_1\,(2 + v_1^2/2gy_1)]$$

$$\Delta E/E_1 = (y_2/y_1 - 1)^3\,(y_1^2/4y_2)/[0.5\,y_1\,(2 + v_1^2/2gy_1)]$$

$$\Delta E/E_1 = ½\,[-1 + 1 + 8\,Fr_1^2 - 1]^3/(2y_2/y_1)\,(2 + Fr_1^2)]$$

Since y_2/y_1 is known from hydraulic jump and $Fr_1 = v_1/\sqrt{gd_1}$

$$\frac{\Delta E}{E_1}=\frac{1}{8}\frac{(-3+\sqrt{1+8Fr_1^2})^3}{(-1+\sqrt{1+8Fr_1^2})(2+Fr_1^2)} \tag{7.83}$$

Equation (7.83) gives the relative energy loss equation following a hydraulic jump.

Since it is possible to find E_1 and E_2 using the formula

$$E = q^2/(2gy^2) + y \tag{7.84}$$

Energy dissipation, jump efficiency and relative loss of energy, all can be calculated using E_1 and E_2 without the use of the formulae derived in the following sections.

Examples 7.9

Find the depth of flow of water after the hydraulic jump in a rectangular channel of 5 m width having a discharge of 20 m^3/s. The depth of water in the channel before hydraulic jump is 0.5 m.

Width of channel, b = 5 m; Depth of flow before jump, y_1 = 0.5 m; Discharge, Q = 20 m^3/s.

Let the depth of flow after hydraulic jump is y_2, hence from eq. (7.77)

$Fr_1 = v/\sqrt{gy_1} = Q/(by_1\sqrt{gy_1}) = 20/5(0.5\times\sqrt{9.81\times0.5}) = 8/\sqrt{(9.81\times0.5)} = 3.61$

$y_2 = 0.5y_1\,(-1 + \sqrt{(1 + 8\,Fr_1^2)}) = 0.5\times0.5\,(-1 + \sqrt{(1 + 8\,(3.61)^2)}) = 2.32$ m

Example 7.10

For the example of 7.8, determine the energy loss after the hydraulic jump, jump efficiency and the relative loss of energy.

Fr_1^2= 2.41, E_1 = 2.11 m, y_2= 1.59 m, E_2 = 1.77 m

Since, values of E_1 and E_2 are determined in example 7.8, we have HL

HL = 2.11 – 1.77 = 0.34 m

Or we can also use eq. (7.79) to calculate the HL as

$HL = (1.59\text{-}0.54)^3/(4\times1.59\times0.54) = 0.34$

Jump efficiency = E_2/E_1 = 1.77/2.11 = 0.84 or 84 per cent

Since, Fr_1 =2.41, we can also use eq. (7.81) to find out jump efficiency as

$\{[(1 + 46.5)^{0.5}]^3 - 23.23 + 1\}/[46.4\,(2+ 5.81)] = \{[6.89]^3 - 22.23\}/[362.38] = 0.84$

Relative loss of energy then is:

$HL/E_1 = 0.34/2.11 = 0.16$

We can also use eq. (7.83) as follows

$$\frac{\Delta E}{E_1}=\frac{1}{8}\frac{(-3+\sqrt{1+8\times5.8})^3}{(-1+\sqrt{1+8\times5.8})(2+5.8)}=\frac{1}{8}\frac{(-3+\sqrt{1+46.4})^3}{(-1+\sqrt{47.4})(2+5.8)}=\frac{1}{8}\frac{(3.88)^3}{(5.88)(7.8)}=0.16$$

QUESTIONS (THEORY)

1. Show that best hydraulic section for a trapezoidal channel is when the sides makes an external angle of 60° with the horizontal.
2. Show that the triangular channel section is the most economical when each of its sloping side makes an angle of 45° with vertical.
3. Define Froude number and comment upon the state of flow for various values of this number.
4. Derive the Chezy's equation to determine flow velocity in open channels. Derive the relation of Chezy's constant with that of Darcy-Weisbach friction factor.
5. Explain the application of Manning's equation in calculating the channel dimension to carry the given discharge.
6. What do you understand by specific energy? Derive the equation of specific energy for an open channel.
7. Draw a neat diagram showing the relationship of specific energy with depth of flow. Explain the salient features of this diagram.
8. Show that critical specific energy for a rectangular channel is 1.5 times the critical flow depth.
9. Derive the specific force equation for an open channel flow.
10. Discuss the benefits or applications of hydraulic jumps.
11. Discuss the classification of hydraulic jumps as affected by Froude number.
12. Derive the relation between depths of flow before and after the hydraulic jump.
13. Derive the equation for energy dissipation in a hydraulic jump.
14. Differentiate between the following: open channel and pipe flow, rigid and mobile boundary channels, prismatic and non-prismatic channels, specific energy and specific force, and jump efficiency and relative loss of energy in a hydraulic jump

Chapter 8

Pipe Flow

Pipes, natural (as veins and arteries) or artificial (as water pipes supplying water in the communities), are closed conduits through which fluids flow. Artificial pipe systems consist of inlets, the pipe, bends, valves, and outlets. Most common shapes of pipes are circular or square although these can be of any other shape as well. A square cross section is called a duct *e.g.* heating and air-conditioning ducts. Circular cross section is able to withstand higher-pressure differentials without distortion compared to the ducts, which are capable to withstand low-pressure differential across walls.

Pressure pipe flow refers to full water flow in a closed conduit of circular cross sections under a certain pressure gradient. For a given discharge (Q), pipe flow at any location is described by the pipe's cross section, the elevation, and the pressure and flow velocity in the pipe. Whereas elevation of a pipe section is measured in relation to a horizontal reference datum, the mean velocity of flow (V) is given by the discharge (Q) divided by the cross-sectional area (A). The total energy at any point in a pipe system is given by

Total Energy = Kinetic Energy + Potential Energy + Pressure Energy

In terms of head, the above statement takes the following form

$$H = P/\rho g + Z + V^2/2g \tag{8.1}$$

Here H is the total energy head, V is the flow velocity, P is the pressure, Z is the elevation above the datum, g is the acceleration due to gravity and ρ is the density of water. Based on the above description, the concept of energy applied to pipe flow has the following three energy components:

- ☆ Energy due to motion (Kinetic Energy)
- ☆ Energy due to elevation (Gravitational Potential Energy)
- ☆ Energy due to pressure

Note that the energy can be transformed from one form to the other during fluid flow, or it is used to perform work.

LAMINAR AND TURBULENT FLOW

Inject a dye in the middle of the stream of free flowing water in a pipe. Depending upon the type of flow, different scenes will emerge (Figure 8.1). For small enough flow velocities, the dye streak (a streak line) remains a well-defined line as it flows along. Slight blurring, if any may be due to molecular diffusion of the dye into the surrounding water. The motion of the fluid particles is very orderly with all particles moving in straight lines parallel to the pipe walls. For an intermediate flow rate, the dye streak fluctuates in time and space, and intermittent bursts of irregular behaviour appear along the streak. On the other hand, when flow rates are large, the dye streak is blurred immediately. It spreads across the entire pipe in a random fashion. These three characteristics flows are termed as laminar, transitional, and turbulent flow, respectively (Figure 8.1).

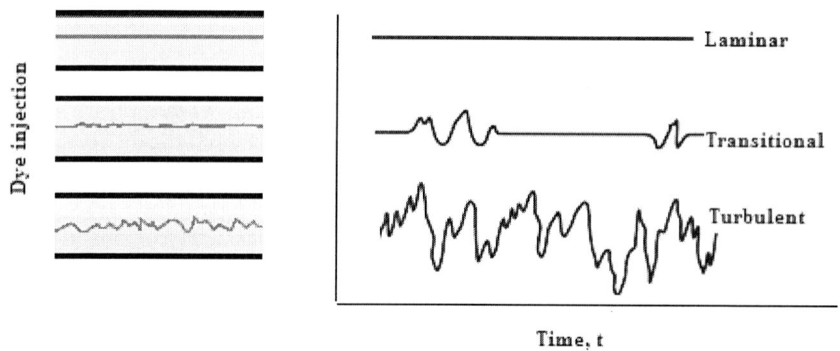

Figure 8.1. A View of Streak Line in Laminar, Transitional and Turbulent Flows.

Osbourne Reynolds first investigated the phenomenon of laminar and turbulent flows in the 1880s in an experiment, which has become a classical experiment in fluid mechanics. The Reynolds number, Re primarily governs whether a flow will be laminar or turbulent (See Chapter 4). In piped water systems, laminar flow is rare. On the other hand, laminar flow is easy to analyze using simple mathematical procedures. Therefore, researchers make appropriate assumptions so that flow can be treated as laminar. For example, for a well-defined streak line, it is assumed that velocity component in the direction of flow is enough to describe the problem.

$$V = u\,i \tag{8.2}$$

Here V is the velocity vector function, and u is the scalar function of (x,y,z,t) and i the unit vectors in the x-direction. On the other hand, most common kind of flow in most field situations is turbulent. Velocity along the pipe is unsteady and accompanied by random components normal to pipe.

$$V = ui + vj + wk$$

Here again u, v, and w are the scalar functions of (x,y,z,t) and i, j and k are the unit vectors in the x, y and z directions. Note that

$$\nabla = \frac{\partial}{\partial x} i + \frac{\partial}{\partial y} j + \frac{\partial}{\partial z} k \tag{8.3}$$

Apparently, mathematical analysis of turbulent flow may pose many insurmountable difficulties.

ENTRANCE REGION FLOW

Any flowing fluid enters the pipe at some point. The region of flow near the entry point is termed as the entrance (entry) region or developing flow region. The fluid enters the pipe with a nearly uniform velocity profile (see section 1, Figure 8.2). As the fluid moves further, viscous effects cause it to stick to the pipe wall, described as no-slip condition. A boundary layer in which viscous effects are important is produced along the pipe wall such that the initial velocity profile changes with distance along the pipe length, x, until the fluid reaches the end of the entrance length (section 2, Figure 8.2). Beyond this point velocity profile does not change with x. Viscous effects are of considerable significance within the boundary layer. Outside the boundary layer, the viscous effects are negligible. The shape of the velocity profile in the pipe depends on whether the flow is laminar or turbulent, as does the length of the entrance region, L_1.

For laminar flow $L_1/D = 0.06\ Re$ (8.4)

For turbulent flow $L_1/D = 4.4\ Re^{1/6}$ (8.5)

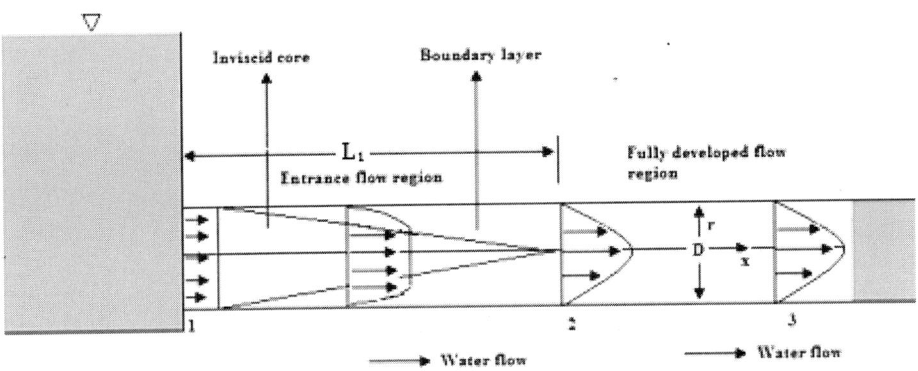

Figure 8.2. Fluid Flow in a Pipe in the Entrance Region and Beyond.

L_1/D is known as dimensionless entrance length such that D is the diameter of the pipe. For practical engineering problems where Re is in the range of 10000 to 100000, L_1 is in the range of 20D to 30D (eq. 8.5). As the fluid reaches the end of the entrance region, the flow is simpler to describe because the velocity is only a function of the distance from the pipe center line, r, and independent of x. The flow between sections (2) and (3) is termed fully developed. The pressure drops linearly with length only in the fully developed flow region (Figure 8.3). Three main physical attributes of fully developed flow are:

- ✰ Boundary layers from opposite sides of the pipe no longer grow as they get merged
- ✰ The stream wise velocity component satisfies $u_x = 0$
- ✰ The radial (or in the case of, *e.g.*, square ducts, the wall-normal) component of velocity is zero, *i.e.*, v = 0.

These flow properties, especially the latter two, are quite important and will be used in the later sections of this chapter.

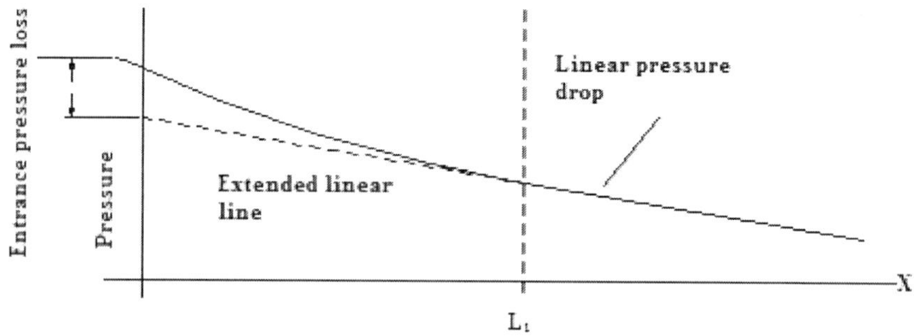

Figure 8.3. Pressure Distribution along the Horizontal Pipe.

POISEUILLE AND COUETTE FLOW

The laminar viscous flow is classified in two broad classes namely the Poiseuille and the Couette flow.

Poiseuille Flow: In this type of flow, applied pressure difference between the stationary surfaces causes the fluid motion.

Couette Flow: In this kind of flow, a moving surface drags adjacent fluid along with it and thereby imparts a motion to the rest of the fluid.

Occasionally, both types of motion may occur simultaneously, as in the screw extruder.

FLOW THROUGH PARALLEL PLATES

Consider two stationary plates kept at a distance, d apart. A viscous fluid is flowing between two plates from left to right. Following assumptions are made to simplify the problem.

- ✰ We consider the flow of real fluids, the fluids possessing viscosity. This implies shear stresses or the frictional forces develop between the fluid layers as well as between the fluid and solid surfaces. The mechanical energy at the upstream section has to be more than the downstream section because the fluid flows by virtue of the energy gradient.
- ✰ Liquid is incompressible.

- ☆ The flow is assumed to be bi-dimensional ($\partial/\partial z = 0$), and viscous Newtonian flow with negligible gravitational effects. It implies flow velocity parallel to the plates u is not 0 but v, w are 0, if the infinite plates assumption is made.
- ☆ The flow is fully developed and stationary ($\partial/\partial t = 0$).

Following two approaches are adopted to solve problems of this nature.

- ☆ Direct derivation with the help of Navier-Stokes equations
- ☆ Shell Balance approach (Conservation of momentum)

Former approach is more elegant. More often, the latter is used because it is simple to apply. The following stepwise procedure describes the application of the Navier-Stokes equations.

Application of Navier-Stokes Equations

- ☆ Make reasonable simplifying assumptions. Almost all of the cases treated involve steady incompressible flow of a Newtonian fluid in a single coordinate direction. Further, gravity may or may not be important, and a certain amount of symmetry may be apparent
- ☆ Write down the equations of motion (both mass continuity and momentum balances) and simplify them according to the assumptions made in the previous step. Only a few terms will remain in the final equation
- ☆ The simplified continuity equation yields information that can be used to simplify the momentum equations
- ☆ Integrate the simplified equations in order to obtain expressions for the dependent variables such as velocities and pressure
- ☆ Invoke the boundary conditions to evaluate the constants appearing in the previous step
- ☆ At this stage, the problem is essentially solved for the pressure and velocities. If desired, shear-stress distributions can be derived by differentiating the velocities yielding velocity gradients.

Simple Poiseuille Flow: Flow between Two Parallel Stationary Plates

Let us consider an element located between two stationary plates having dimensions of $\Delta x \times \Delta y \times \Delta z$ in x, y and z Cartesian coordinate system (Figure 8.4). The pressure varies only in the x-direction, because the flow is steady and incompressible and fluid flows only in the x-direction. It is a two-dimensional problem with the flow occurring only in the x-direction. Using the basic equations of continuity and Navier-Stokes equations, the components of velocity u, v, w in x, y and z coordinates are

$$u = u, v = 0, \text{ and } w = 0$$

Without writing the full equations, we write only the simplified equation considering the flow conditions described in the previous paragraph. From continuity equation, we have

$\partial u/\partial x + \partial v/\partial y = 0$

Since v=0

$\partial u/\partial x = 0$ (8.6)

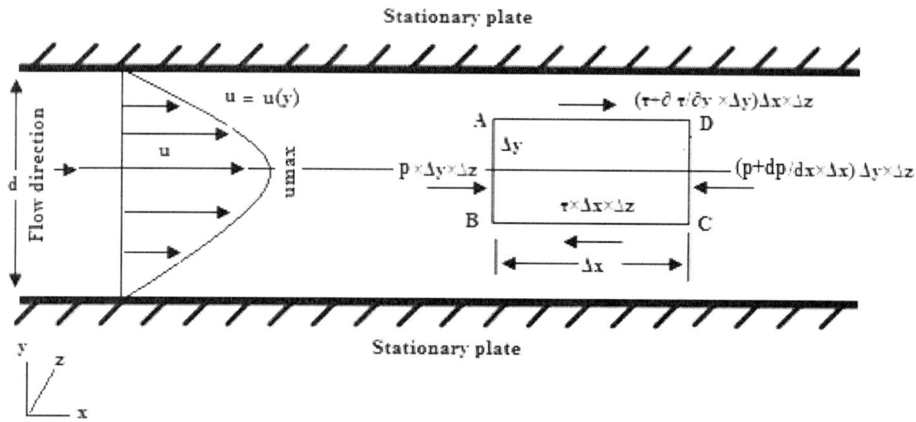

Figure 8.4. Fluid Flow between Two Stationary Plates and Shell Balance Diagram.

As such, u is independent of the distance from the inlet such that same velocity profile will appear for all values of x. Since $\partial u/\partial z = 0$ (In comparison with their spacing, d, the plates extend for a very long distance in the z direction, all locations in this direction appear essentially identical to one another. It follows that $u = u(y)$ is a function of y only. Thus, partial derivative change to ordinary differential equation

$du/dx = 0$ or $u = u(y)$ (8.7)

The x-momentum equation is

$$\rho\left(u\frac{\partial u}{\partial t}+v\frac{\partial u}{\partial x}\right)=-\frac{\partial p}{\partial x}+\mu\left(\frac{\partial^2 u}{\partial x^2}+\frac{\partial^2 u}{\partial y^2}\right)+\rho gx \quad (8.8)$$

Since ρ=0 and if gravity acts vertically downwards; $gy = -g$ and $gx = gz = 0$, then.

$$\frac{\partial p}{\partial x}=\mu\frac{\partial^2 u}{\partial y^2} \quad (8.9)$$

The y-momentum equation is

$$\rho\left(u\frac{\partial v}{\partial t}+v\frac{\partial v}{\partial x}\right)=-\frac{\partial p}{\partial y}+\mu\left(\frac{\partial^2 v}{\partial x^2}+\frac{\partial^2 v}{\partial y^2}\right)+\rho gy \quad (8.10)$$

Since ρ=0 and v =0

$$\frac{\partial p}{\partial x}=-\rho g$$

If we neglect the gravity term, then $\frac{\partial p}{\partial y} = 0$ (8.11)

The z-momentum equation is

$$\rho\left(u\frac{\partial w}{\partial t}+v\frac{\partial w}{\partial x}+w\frac{\partial w}{\partial z}\right)=-\frac{\partial p}{\partial x}+\mu\left(\frac{\partial^2 w}{\partial x^2}+\frac{\partial^2 w}{\partial yv}\right)+\rho gz \quad (8.12)$$

$$\frac{\partial p}{\partial z} = 0 \quad \text{(Since } \int \text{and w are zero)} \quad (8.13)$$

From eqs. (8.9), (8.11) and (8.13), it is clear that:

$$\frac{\partial p}{\partial x} = \mu\frac{\partial^2 u}{\partial y^2}$$

$$\frac{\partial p}{\partial y} = 0 \text{ and } \frac{\partial p}{\partial z} = 0$$

It means, p is a function of x only and its partial derivative can be taken as ordinary derivative such that

$$\frac{dp}{dx} = \mu\frac{d^2 u}{dy^2} = \text{Constant} \quad (8.14)$$

It flows from the principle of calculus, which states that we can equate the differentials only if the relation is a constant. Further, derivations will follow as in the second approach following eq. (8.17). Note that this constant is negative, because pressure must decrease in the flow direction to overcome the resisting wall shear stress.

Shell Balance Approach (Conservation of Momentum)

In this approach, we assume a small shell and use the following balance equation.

Rate of momentum in - rate of momentum out + sum of all forces = 0

Following basic steps describe the shell balance approach.

1. (Momentum from shear stress into system) - (Momentum from shear stress out of system) is calculated. Momentum from shear stress goes into the shell at y and leaves the system at $y + \Delta y$. Assess the momentum from shear stress, which is given as: Shear stress×Area= $\tau_{yx} \times A$
2. Momentum from flow- Momentum flows into the system at $x = 0$ and out at $x = L$. Since we deal with the steady flow, the momentum flow at $x = 0$ is equal to the momentum flow at $x = L$. Therefore, the two cancel out.
3. Gravity force on the shell (As per assumptions, it is negligible)
4. Assess pressure forces
5. Plug into conservation of momentum equation and solve for τ_{yx}

6. Apply Newton's law of viscosity for a Newtonian fluid such that $\tau_{yx} = -\mu(du/dy)$
7. Integrate to find an equation for the velocity and use boundary conditions to find constants of integration

Now let us derive the basic equation with the free body diagram of a shell (Figure 8.4). The forces acting on the elements are:

- ✰ Pressure force, $p\times\Delta y\times\Delta z$ on face AB
- ✰ Pressure force, $(p+\partial p/\partial x\times\Delta x)\Delta y\times\Delta z$ on face CD
- ✰ Shear force, $\tau\times\Delta x\times\Delta z$ on face BC
- ✰ Shear force, $(\tau+\partial\tau/\partial y\times\Delta y)\Delta x\times\Delta z$ on face AD

For steady and uniform flow, there is no acceleration; therefore, resultant force in the direction of flow is zero.

$$p\times\Delta y\times\Delta z - (p+\partial p/\partial x\times\Delta x)\Delta y\times\Delta z - \tau\times\Delta x\times\Delta z + (\tau+\partial\tau/\partial y\times\Delta y)\Delta x\times\Delta z = 0 \qquad (8.15)$$

Note that $\partial p/\partial x$ is negative as the pressure deceases in the direction of flow. Opening the brackets and cancelling out the terms, eq. (8.15) reduces to

$$(-\partial p/\partial x\times\Delta x)\Delta y\times\Delta z + (\partial\tau/\partial y\times\Delta y)\Delta x\times\Delta z = 0$$

Dividing by $\Delta x\times\Delta y\times\Delta z$ throughout, we have

$$-\partial p/\partial x + \partial\tau/\partial y = 0 \qquad (8.16)$$

For a laminar flow of a Newtonian fluid, the shear stress is simply proportional to the velocity gradient, $\tau = \mu\partial u/\partial y$. So the above equation becomes

$$-\partial p/\partial x + \mu\partial^2 u/\partial y^2 = 0$$

As discussed before, partial derivatives can be taken as ordinary derivatives, such that

$$\mu d^2u/dy^2 = dp/dx \qquad (8.17)$$

It is the same equation as eq. (8.14) derived from the basic principles.

For the case illustrated in Figure (8.4), we integrate eq. (8.17) two times and evaluate constants using the boundary conditions to arrive at the final solution.

$$\frac{du}{dy} = \frac{1}{\mu}\frac{dp}{dx}y + c1 \qquad (8.18)$$

$$u = \frac{1}{\mu}\frac{dp}{dx}\frac{y^2}{2} + c1y + c2 \qquad (8.19)$$

Since plates are stationary, we have $y = 0$, $u = 0$ so $c2 = 0$

$y = d$, $u = 0$

$$0 = \frac{1}{\mu}\frac{dp}{dx}\frac{d^2}{2} + c1d \qquad (8.20)$$

$$c1 = -\frac{1}{\mu}\frac{dp}{dx}\frac{d}{2}$$

$$u = \frac{1}{\mu}\frac{dp}{dx}\frac{y^2}{2} - \frac{1}{\mu}\frac{dp}{dx}\frac{yd}{2} \quad (8.21)$$

$$u = -\frac{1}{2\mu}\frac{dp}{dx}(d-y)y$$

Note that negative sign will be taken care of when we incorporated dp/dx, which is negative.

It is clear that velocity distribution forms a parabola with u_{max} at d/2 as du/dy is zero at y =d/2.

$$\frac{du}{dy} = -\frac{1}{2\mu}\frac{dp}{dx}(d-2y) = 0 \quad (8.22)$$

Incorporating y =d/2 in the velocity equation

$$u_{max} = -\frac{1}{2\mu}\frac{dp}{dx}\left(d - \frac{d}{2}\right)\frac{d}{2} = -\frac{1}{8\mu}\frac{dp}{dx}d^2 \quad (8.23)$$

Integrate the velocity eq. (8.21) to obtain the volumetric flow rate.

$$\int_0^d u\,dy = Q = -\frac{1}{12\mu}\frac{dp}{dx}d^3 \quad (8.24)$$

Since, u_{avg} =Q/d, we have

$$u_{avg} = \frac{1}{1.5}u_{max} \quad (8.25)$$

Shear stress, $\tau = -\mu du/dy = ½ (d - 2y)\, dp/dx$ (8.26)

So far, we noticed negative signs for the equations of discharge and velocity. It is because dp/dx is negative. Putting L as the length of the plate in the flow direction and Δp as the pressure difference, and integrating in the x-direction, we have

$-dp/dx = \Delta p/L$

Incorporating in the discharge equation, we get

$Q = \Delta p d^3/(12\mu L)$ (8.27)

This expression states that the flow rate is

- Directly proportional to pressure gradient
- Directly proportional to the cube of gap width (the distance between the two plates)
- Inversely proportional to the viscosity

It may be mentioned that in Figure 8.4, we can also take the origin at the center of the two plates with a distance between the plate as 2h (top plate at +h and the bottom plate at –h such that h=d/2). In that event, the velocity distribution u is:

$$u = -\frac{1}{2\mu}\frac{dp}{dx}(h^2 - y^2)$$

Integration $\int_{-d}^{d} u dy$ will yield the discharge equation as before (eq. 8.24)

$$Q = -\frac{2}{3\mu}\frac{dp}{dx}h^3$$

Note that in this case h is ½ of the d in eq. (8.24). Students may derive the equation from the basics beginning with eq. (8.15).

Couette Flow

Let us first consider the Couette- Poiseuille flow, where flow is due to both pressure gradient and the moving plate. Beginning with eq. (8.19), we rewrite the equation as follows.

$$u = \frac{1}{\mu}\frac{dp}{dx}\frac{y^2}{2} + c1y + c2 \tag{8.28}$$

Since in this case, upper plate is moving with a velocity, U and the lower plate is stationary (Figure 8.5), we have y = 0, u =0 so c 2 =0

y = d, u =U

$$U = \frac{1}{\mu}\frac{dp}{dx}\frac{d^2}{2} + c1d \tag{8.29}$$

$$c1 = \frac{U}{d} - \frac{1}{\mu}\frac{dp}{dx}\frac{d}{2} \tag{8.30}$$

$$u = \frac{1}{\mu}\frac{dp}{dx}\frac{y^2}{2} + y\frac{U}{d} - \frac{yd}{2\mu}\frac{dp}{dx} \tag{8.31}$$

$$u = \frac{U}{d}y - \frac{yd}{2\mu}\frac{dp}{dx}\left(1 - \frac{y}{d}\right) \tag{8.32}$$

We can express the above equation as

$$\frac{u}{U} = \frac{t}{d} - \frac{py}{d}\left(1 - \frac{y}{d}\right) \tag{8.33}$$

$$\text{such that } p = \frac{d^2}{2\mu U}\frac{dp}{dx} \tag{8.34}$$

Eq. (8.33) describes the velocity distribution in non-dimensional form across the channel with P as a parameter known as the non-dimensional pressure gradient. It is the case of Couette- Poiseuille flow between the two parallel plates. When p=0, then

$$\frac{u}{U} = \frac{y}{d} \tag{8.35}$$

Eq. (8.35) is applicable to a planer or simple Couette flow. As already stated, it is a shear-driven flow. The shearing force is produced only by motion of the upper plate travelling in the x direction at a constant speed U, provided we ignore all other body forces (McDonough, 2009). The velocity profile for such a case shown in Figure (8.5) reveal that it linearly increases from 0 at the stationary plate to U at the moving plate. Velocity profile for the Couette- Poiseuille flow varies with pressure gradient.

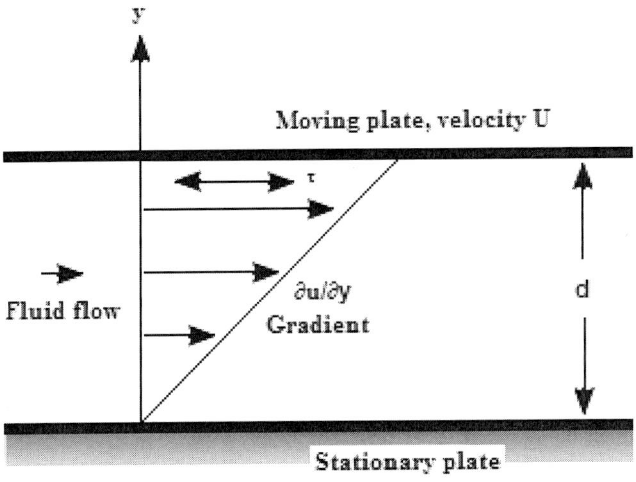

Figure 8.5. Velocity Profile between a Stationary and a Moving Plate.

LAMINAR FLOW IN PIPES

One of the most important practical problems in fluid engineering is the analysis of pipe flow. Let us consider the flow in a long, straight, constant diameter section of a pipe as a fully developed laminar flow. The gravitational effects (mass forces) are neglected. The velocity profile is the same at any cross section of the pipe, being outside the entrance region. We propose to work out the velocity profile u(r) in a pipe of radius R with a pressure gradient $\partial p/\partial x$. Consider a fluid element of length Δx and radius r centred on the axis of a horizontal pipe of diameter D in a fully developed laminar flow (Figure 8.6).

From the free body diagram, the force balance is written as (Figure 8.6)

$$p_1 \pi r^2 - (p_1 + (\partial p/\partial x)\,\Delta x)\,\pi r^2 - 2\pi r \Delta x\,\tau = 0 \tag{8.36}$$

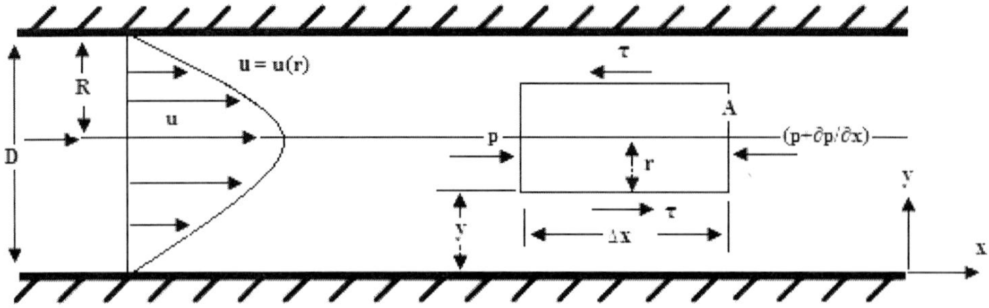

Figure 8.6. Schematic Illustration of Shell Balance Approach for Laminar Flow in a Pipe Section.

Simplifying the equation, we have

$$(\partial p/\partial x)\ r - 2\tau = 0$$

$$2\tau = r\ \partial p/\partial x \tag{8.37}$$

We assume that u and τ vary only in the y-direction and that p varies only in the x-direction. It also means that our partial derivatives collapse to ordinary derivatives (Matthew, 2015). Moreover, eq. (8.37) can be integrated between x_1 and x_2 where pressures in the pipes are p_1 and p_2 such that $\Delta p = p_2 - p_1 < 0$, and $x_1 - x_2 =$ L, the pipe length. By doing so, we have

$$-\Delta p/L = 2\tau/r \tag{8.38}$$

For a fully developed flow, Δp/L is constant and hence, the right hand side is a constant. It happens if

$$\tau = C\ r$$

Here C is a constant. At r = 0, τ = 0. Thus, there is no shear stress at the center line. At r = D/2, τ is maximum and is known as wall shear stress τ_w. Hence C = 2 τ_w/D. Therefore

$$\tau = 2\tau_w r/D = \tau_w r/R \tag{8.39}$$

In Poiseuille flow, the shear stress varies linearly from center line to the wall (Figure 8.7). It holds true whether the flow is laminar or turbulent.

By definition, the wall shear stress for a laminar flow of a Newtonian fluid in a pipe is given by

$$\tau = \mu\ \partial u/\partial y \tag{8.40}$$

If y is the distance of the fluid element from the pipe wall, then y =R- r or dy = -dr. The negative sign is due to the fact that the velocity decreases from the pipe center line to the pipe wall. Thus eq. (8.40) becomes

$$\tau = -\ \mu\ \partial u/\partial r$$

Incorporating in eq. (8.38), it becomes

$$(\Delta p/2L\mu)\ r = \partial u/\partial r \tag{8.41}$$

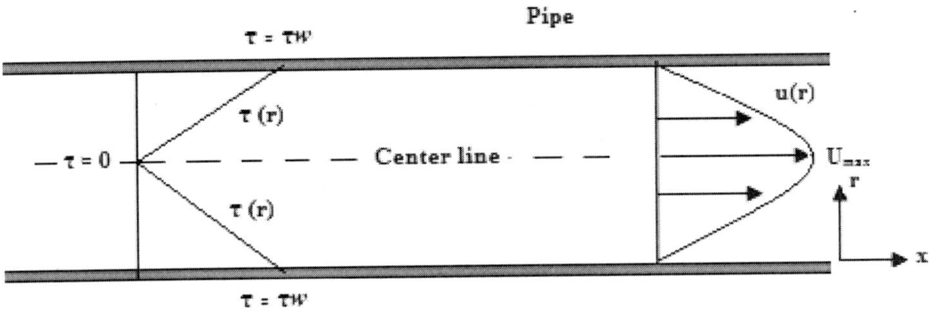

Figure 8.7. Distribution of Shear Stress and Laminar Flow Velocity Profile in a Circular Pipe.

Integrating w.r.t. r, we get

$$u = \frac{1}{4\mu}\frac{\Delta p}{L}r^2 + c_1 \tag{8.42}$$

Here c_1 is a constant. Since u= 0 at r =R, c_1 is obtained as

$$c_1 = -\frac{1}{4\mu}\frac{\Delta p}{L}R^2$$

$$u = \frac{1}{4\mu}\frac{\Delta p}{L}r^2 - \frac{1}{4\mu}\frac{\Delta p}{L}R^2 \tag{8.43}$$

$$u = \frac{\left[\left(-\frac{\Delta p}{L}\right)R^2\left\{1-\left(\frac{r}{R}\right)^2\right\}\right]}{4\mu} = -\frac{\Delta p[R^2 - r^2]}{4L\mu} \tag{8.44}$$

From this equation, it is clear that the velocity distribution is parabolic with a maximum at the center of the pipe, becoming zero at the pipe wall.

$$u_{\max} = -\frac{1}{4\mu}\frac{\Delta p}{L}R^2 \tag{8.45}$$

To obtain the volume flow rate, Q through any cross section of radius R, we integrate the velocity profile across the pipe. Since the flow is axi-symmetric about the center line, the velocity is constant on small area elements consisting of rings of radius r and thickness dr. Thus, the volumetric flow rate through the pipe is given as

$$Q = \int u dA = \int_0^R u(r)2\pi r dr$$

$$Q = \int_0^R 2\pi r u dr$$

$$Q = 2\pi u_{\max}\int_0^R \left[1-\left(\frac{r}{R}\right)^2\right] r dr \tag{8.46}$$

$$Q = 2\pi u_{max}\left(\frac{r^2}{2} - \frac{r^4}{4R^2}\right) \text{ (Take limits from 0 to R)} \tag{8.47}$$

$$Q = \frac{\pi R^2 u_{max}}{2} = \frac{\pi R^4}{8\mu}\left(-\frac{\Delta p}{L}\right) \tag{8.48}$$

By definition, the average velocity is the flow rate divided by the cross-sectional area, $u_{avg} = V = Q/\pi R^2$

$$V = \left(-\frac{\Delta p}{L}\right)\frac{R^2}{8\mu} = \frac{u_{max}}{2} \tag{8.49}$$

To find out the location of the average velocity, we incorporate eq. (8.49) in eq. (8.44)

$$u = 2V\left[1 - \left(\frac{r}{R}\right)^2\right]$$

For u = V

$$r = \frac{R}{\sqrt{2}} \tag{8.50}$$

It means the average velocity in a pipe of radius R is experienced along a ring $R/\sqrt{2}$ distance away from the walls of the pipe.

Loss of Pressure Head (H_f)

For flow of real fluids through a horizontal pipe section, the head loss due to friction is

$$H_f = -\Delta p/\rho g \tag{8.51}$$

From eq. (8.49)

$$\Delta p = -8V\mu L/R^2 \tag{8.52}$$

Here V is the mean velocity of flow. Combining eq. (8.51) and (8.52)

$$H_f = 8V\mu L/R^2\rho g = 32V\mu L/\rho g D^2 = 128Q\mu L/\rho\pi g D^4 \quad (Q = V/A) \tag{8.53}$$

$$Q = \pi D^4 \rho g H_f/128\mu L = \pi D^4(-\Delta p)/128\mu L \tag{8.54}$$

Eq. (8.53) or eq. (8.54) is the Hagen Poiseuille equation. It is why; many times pipe flow is also known as Hagen-Poiseuille's flow. This law for horizontal pipe states that the flow rate is:

- ☆ Directly proportional to the pressure drop
- ☆ Directly proportional to the fourth power of the pipe radius
- ☆ Inversely proportional to the viscosity
- ☆ Inversely proportional to the pipe length

This equation calculates pressure-drop in a horizontal pipe of Length L and inside-diameter D, for a viscous incompressible fluid flowing under steady state fully developed laminar flow condition. One can also calculate the viscosity of the fluid by measuring the pressure drop.

From Darcy-Weisbach equation, we can write

$$H_f = f\,L\,V^2/2g\,D \tag{8.54}$$

Comparing the Darcy-Weisbach equation with eq. (8.53), we have

$$32V\mu L/\rho g D^2 = f\,L\,V^2/2g\,D \tag{8.55}$$

$$f = 64\,\mu/\rho DV = 64/Re \tag{8.56}$$

This relation finds application in calculating the Darcy-Weisbach friction factor for laminar flows in circular pipes.

From the above analysis, it is possible to draw the following important inferences.

- ☆ The nature of velocity profile in a laminar pipe flow is parabolic with zero at the wall and maximum at the center line
- ☆ The shear stress varies linearly from center line to the wall, maximum shear stress being at the wall and zero at the center line. It is true for both laminar as well as turbulent flow
- ☆ The maximum velocity in a laminar pipe flow is twice that of average velocity
- ☆ In a laminar pipe flow, the Darcy-Weisbach friction factor drops with increase in flow Reynolds number

Example 8.1

Oil with a dynamic viscosity of 0.40 Ns/m^2 flows in a 10 m long pipe of 20 mm diameter. Calculate the discharge, if the pressure drop is 20 kPa? Check whether the flow is laminar or not. Take the value of density of the oil as 890 kg/m^3.

Since $Q = \pi R^4 \Delta p/8\mu L$

$$Q = 3.142\times.01^4\times 20000/8\times10\times0.40 = 1.96\times10^{-5}\ m^3/s$$

$Re = \rho VD/\mu$

Since $V = Q/(\pi r^2)$, we have

$$Re = \rho QD/(\mu\,\pi r^2)$$

$$Re = 890\times1.96\times10^{-5}\times0.02/(0.40\times3.142\times0.01^2) = 2.77$$

Flow is laminar as Re is much less than 2000.

Example 8.2

Water with a dynamic viscosity of 0.001 Ns/m^2 flows in a 20 m pipe having the radius of 10 mm. What should be the pressure drop to result in a discharge of 2.0×10^{-5}? Check whether the flow is laminar or not. Take the density of the water as 1000 kg/m^3. What is the Darcy-Weisbach friction factor for this set-up?

Since $\Delta P = 8Q\mu L/(\pi R^4)$

$\Delta p = 8 \times 2.0 \times 10^{-5} \times 0.001 \times 20/(3.142 \times 0.01^4) = 101.85$ Pa

$Re = \rho QD/(\mu \, \pi r^2)$

$Re = 1000 \times 2.0 \times 10^{-5} \times 0.02/(0.001 \times 3.142 \times 0.01^2) = 1273.1$

Since Re is less than 2000, flow is laminar.

Friction factor f = 64/Re = 64/1273.1 = 0.0503

Example 8.3

Oil of density 900 kg/m^{3}flows in a pipe of 100 mm diameter. Its dynamic viscosity is 0.018 Ns/m^2. Determine the pressure drop per meter length. Also determine the velocity of flow at a radius of 35 mm. Take the Reynolds number of the flow as 260.

$Re = \rho V D/\mu$

$V = (260 \times 0.018)/(900 \times 0.1) = 0.052$ m/s

$\Delta p = 32\mu L\, V/D^2 = 32 \times 0.018 \times 1 \times 0.052/0.1^2 = 3.0$ Pa

Using eq. (8.44), velocity at a radius of r, V_r is

$V_r = \{\Delta p/4L\mu\}(R^2 - r^2) = [3.0/(4 \times 1 \times 0.018)]\,(0.05^2 - 0.035^2) = 0.053$ m/s

PIPE CAPACITY: MANNING'S EQUATION

Although typically applied for open-channel flow, Manning's equation is also used to calculate the velocity of flow or the discharge through closed-conduits. Manning's equation for fluid velocity in SI units is reproduced as:

$$V = (1/n)\, R^{2/3}\, S^{1/2} \tag{8.57}$$

Here V is the velocity, m/s, R is the hydraulic radius, m, S is the slope of the energy grade line and n is the Manning's roughness coefficient. Application of this equation for open channel flow has been dealt with in a previous chapter. Since the procedure is essentially similar, readers may refer Chapter 7 of this book for detailed discussions. Examples 8.4 and 8.5 illustrate the application of this equation for pipe flow. The values of n for various kinds of pipes are reported in Table 8.1 (US Govt. Printing Press, 1961 and other sources).

Table 8.1. Manning's n for Various Pipe/Channel Materials

Nature of Surface	*Manning's n (Range)*
Brass	0.009 -0.013
Cast iron	0.011 – 0.015
Clay (drainage pipe)	0.011 – 0.017
Concrete	0.011 – 0.013
Copper	0.009 -0.013
HDPE or PVC	0.009 – 0.011
Vitrified clay	0.012 – 0.014
Steel	0.009 – 0.011
Monolithic concrete	0.012 – 0.017

Nature of Surface	Manning's n (Range)
Cement rubble	0.017 – 0.025
Brick	0.014 – 0.017
Laminated treated wood	0.015 – 0.017
Corrugated metals	0.019 – 0.030
Vitrified sewer pipe	0.010 – 0.017
Open channels	
Lined with concrete	0.013 – 0.022
Earth, clean, after weathering	0.018 – 0.020

Example 8.4

Determine the head loss in a 45 cm concrete pipe 30 m long flowing full with an average velocity of 1.0 m/s.

Manning's n from Table 8.1 = 0.012

For a pipe flowing full:

$$R = d/4 = 0.45/4 = 0.1125 \text{ m}$$

$$V = (1/n)\,(R^{2/3})\,(S^{1/2})$$

$$S = [nV/(R^{2/3})]^2$$

$$S = [0.012\times 1/(0.1125)^{2/3}]^2 = 0.00265$$

Since slope of energy line is head loss/length of pipe, head loss is

Head Loss = Slope× Pipe length

Head loss = 0.00265× 30 = 0.08 m

Example 8.5

Determine the discharge of water through a vitrified clay pipe of 3 cm diameter laid at a slope of 2 per cent.

Manning's n from Table 8.1 = 0.013

For a pipe flowing full: $R = d/4 = 0.03/4 = 0.0075$ m

$$V = (1/n)\,(R^{2/3)}\,(S^{1/2})$$

$$V = (1/0.013)\,(0.0075^{0.667})\,(0.02^{0.5})$$

$$V = 76.92 \times 0.0382 \times 0.1414 = 0.415 \text{ m/s}$$

Discharge = VA = $0.415 \times (\pi \times 0.03^2/4) = 0.415 \times 0.00071 = 0.0003 \text{ m}^3/\text{s}$ or 0.3 L/s

QUESTIONS (THEORY)

1. Briefly discuss the characteristics of flow in the entrance region.
2. Show that the maximum velocity in a simple Poiseuille's flow is 1.5 times the average velocity of flow between two stationary plates. Use the shell balance approach.

3. Use the Shell balance approach to show that maximum velocity is 2.0 times the average velocity for laminar flow in pipes.

4. For a simple Couette flow show that velocity at any distance, y from the stationary plate is given as u=Uy/d such that u is the velocity, U is the velocity of the moving plate and d is the distance between the stationary and moving plates.

5. Using the Hagen-Poiseuille's and Darcy-Weisbach equations show that Darcy-Weisbach friction factor, f for laminar flow is equal to 64/Re. Here Re is the Reynolds number.

6. Differentiate between the Poiseuille and the Couette flows.

Chapter 9

Hydraulic Losses in Pipe Flow

INTRODUCTION

Liquids flow in the pipes either by gravity or because of pressure difference in the direction of flow. In the later case, a piezometer connected to the pipe shows a rise in the level of the liquid in the piezometer. The pressure, P in that case is equal to ρgh such that ρ is the density of the fluid, g is acceleration due to gravity and h the fluid level in the piezometer. For an ideal fluid, the two piezometers at L distance apart will show the same level or pressure. On the other hand, when a viscous fluid flows through the pipe, some of its energy is lost because of the resistance experienced due to:

- Fiction between the fluid and the walls of the pipe
- Friction between adjacent fluid particles as they move relative to one another; and
- Turbulence caused on the way as the flow is redirected or affected in some manner by piping entrances and exits, or in components such as piping valves, flow reducers, and other fittings.

The loss is reflected in the reduced pressure or the total head (comprising of sum of elevation head, velocity head and pressure head) of the fluid as it moves through the pipe system. The head loss is directly proportional to the length of pipe, the square of the fluid velocity, and a term accounting for fluid friction called the friction factor. For similar conditions, the head loss is inversely proportional to the diameter of the pipe. The total head loss comprises major and minor head losses.

$$\text{Total head loss} = \text{Major head loss} + \text{Minor head loss} \qquad (9.1)$$

This chapter describes the theory and practices of assessing these losses, and discusses the implications of head loss in fluid flows.

MAJOR AND MINOR HEAD LOSSES

Major head loss is attributed to friction in the conveying pipelines and characteristics of the flowing fluid. While the major head loss occurs continuously along the conveying pipeline/duct, its extent largely depends upon roughness of the pipeline. Minor head losses on the other hand, result from one or a combination of more than one of the following:

- Head loss at the entrance of pipe
- Head loss at the exit of pipe
- Head loss due to obstructions in pipe
- Head loss due to sudden contraction in pipe
- Head loss due to sudden expansion in pipe
- Head loss due to pipe fittings

Head loss because of minor losses is small compared to the major head losses. It is why, we call these losses as minor losses.

ESTIMATION OF MAJOR HEAD LOSSES

Following methods are used to estimate the major head losses.

Darcy-Weisbach Formula

The equation derived by Henry Darcy (1803 to 1858), Julius Weisbach (1806 to 1871), and others about the middle of the nineteenth century, popularly known as the Darcy-Weisbach equation describes the pipe flow. This equation is valid for fully developed, steady state and incompressible flow. Consider a uniform pipeline through which water flows at a uniform rate. If there is no friction resistance, pressure P_1 and P_2 at any two points along the length should be equal. On the other hand, because of the friction losses following inequality holds true (Figure 9.1).

$$V_1^2/2g + P_1/\rho g + Z_1 > V_2^2/2g + P_2/\rho g + Z_2 \qquad (9.2)$$

Here, P, V and Z are the pressure, velocity and elevation heads of fluid with 1 and 2 indicating the sections 1 and 2 in the flow path. To restore equality, one adds a scalar quantity to the right side of this inequality, such that

$$V_1^2/2g + P_1/\rho g + Z_1 = V_2^2/2g + P_2/\rho g + Z_2 + H_f \qquad (9.3)$$

Here H_f is head loss due to friction (Figure 9.1) and is given as

$$H_f = H_1 - H_2 \qquad (9.4)$$

Note that H_1 is always greater than H_2. In a horizontal pipe when $Z_1 = Z_2$ and diameter of pipe is constant such that $V_1 = V_2$, hydraulic loss is equal to the head of r pressure drop or head loss

$$H_f = (P_1 - P_2)/\rho g \qquad (9.5)$$

To derive the Darcy-Weisbach equation, let us consider once again Figure (9.1). The fluid moves from point 1 to point 2 because of the pressure force given as P_1A and P_2A at points 1 and 2 respectively, such that

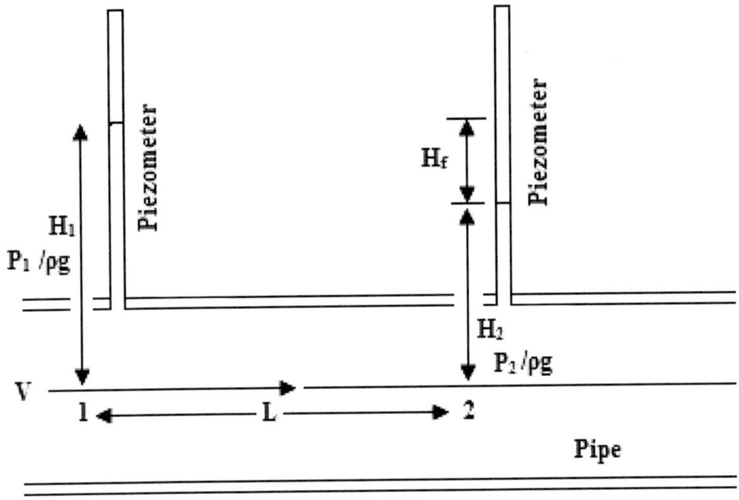

Figure 9.1. Schematic Diagram Showing Friction Loss in Pipe Flows.

$$\text{Force to move the fluid} = P_1 A - P_2 A \tag{9.6}$$

Here A is the area of the pipe sections. The frictional force opposes this movement. It has been established from Froude's experiments that frictional resistance force is proportional to length of the pipe, wetted perimeter, p (for circular pipes, it is π d, such that d is the diameter of the pipe) and V^n (for commercial pipes n =2) such that

$$\text{Frictional resistance force} = f' \times p \times L \times V^2 \tag{9.7}$$

Here L is the length of the pipe section and f' is the friction factor depending upon the type of the pipe material.

At equilibrium,

$$P_1 A - P_2 A = f' \times p \times L \times V^2 \tag{9.8}$$

If we, divide both sides by specific weight $\gamma = \rho g$ (For water γ is 9807 N/m³since ρ is 1000 kg/m³, and g is 9.807 m/s²)

$$(P_1 A - P_2 A)/\gamma = (f' \times \pi d \times L \times V^2)/\gamma \tag{9.9}$$

$$H_f = f' \times \pi d \times L \times V^2/A\gamma \tag{9.10}$$

Dividing and multiplying eq. (9.10) by 2 g

$$H_f = (f'2g/\gamma) \times p \times L \times V^2/(2g A) \tag{9.11}$$

If we incorporate $f_f = f' \, 2g/\gamma$ (A dimensionless number)

$$H_f = f_f \, p \, LV^2/(2g A) \tag{9.12}$$

$$H_f = f_f L \, V^2/(2g \, m) \tag{9.13}$$

Here m is the hydraulic radius (A/p), and f_f is the Fanning's co-efficient of friction. The friction factor or coefficient, f_f depends on the flow type *i.e.* whether laminar, transient or turbulent (the Reynolds Number) and the roughness of the tube or duct. For circular pipes, m = d/4, therefore, for circular pipes eq. (9.13) is written as:

$$H_f = 4\, f_f\, L\, V^2/(2g\, d) \tag{9.14}$$

Here H_f is in m if L is in m,V is in m/s, d the diameter of pipe in m, ρ is in kg/m^3,and g is in m/s^2.

In terms of pressure drop, Δp, eq. (9.14) can be written as (using relation h = P/ρg)

$$\Delta p = 4\, f_f\, L\, \rho V^2/(2d) \tag{9.15}$$

For units used in eq. (9.14) Δp is the pressure loss in Pa (N/m^2). For non-circular pipes, we use hydraulic diameter, d_h in eq. (9.12), eq. (9.12) takes the form

$$H_f = 4\, f_f\, LV^2/(2g\, d_h) \tag{9.16}$$

Here hydraulic diameter is defined as

$$d_h = 4 \times \text{area of flow/perimeter of pipe in contact with fluid} \tag{9.17}$$

Note that the hydraulic diameter is not the same as the equivalent diameter discussed in some later section of this chapter. Whereas the hydraulic diameter is used to calculate pressure loss in ducts or pipes or as the characteristic length to calculate e/d in the Moody's diagram, the equivalent diameter is the diameter of a circular duct or pipe that gives the same pressure loss as a rectangular duct or pipe.

By replacing f_f to Darcy-Weisbach friction factor f_d, eq. (9.14) is commonly written as

$$H_f = f_d\, L\, V^2/(2g\, d) \tag{9.18}$$

Her f_d is the Darcy–Weisbach coefficient, which is 4 times the f_f (Cowan, 1956). Thus, the Darcy-Weisbach friction factor is obtained by multiplying the Fanning friction factor by four or vice-versa. In the following sections, we will use eq. (9.18) and D-W coefficients to calculate head losses or discharge. This equation (eq. 9.18) can also be expressed in terms of discharge through the pipe. The discharge, Q of a full-flowing circular pipe of diameter, d is

$$Q = (\pi\, d^2/4)\, V \tag{9.19}$$

Therefore, the Darcy–Weisbach equation (eq. 9.18) in terms of Q is

$$H_f = 8\, f_d\, L\, (1/(\pi^2 g))\, (Q^2/d^5) \tag{9.20}$$

We use eq. (9.18) or eq. (9.20) to solve the following three types of problems in the design and analysis of piping systems.

- ✰ To determine the pressure drop for a known pipe length and diameter for a specified flow rate (or velocity)
- ✰ To determine the flow rate for a known pipe length and diameter for a specified pressure drop
- ✰ To determine the pipe diameter for a known pipe length and flow rate for a specified pressure drop.

Normally, one uses eq. (9.18) to find out velocity to calculate discharge. In that case, it should be possible to have a check on the flow velocity as per its guidelines in pipe flow. Recommended maximum velocities in low-pressure pipelines are in the range of 1.3 to 1.5 m/s. One may prefer high velocities to minimize the pipe diameter to reduce the cost. At the same time, it may result in higher frictional losses and higher cost of protection against water hammer. Minimum flow velocities to prevent sedimentation of fine sand are around 0.5 m/s.

Example 9.1

Air in a duct of *32 cm* diameter is flowing at a velocity *6 m/s*. The density of air is *1.2 kg/m³*. The friction coefficient f_d is estimated to be *0.02* and the length of the duct is 2 *m*. Calculate the friction loss per unit length of the pipe. Neglect the minor losses.

Knowing that $H_f = \Delta p/\rho g$, we use eq. (9.18) such that

$\Delta p = f_d \, L \, \rho V^2/(2d)$

$\Delta p = 0.02 \times 1 \times 1.2 \times 6^2/(2 \times 0.32) = 1.35 \; Pa$

Assessment of Friction Coefficient

The Darcy-Weisbach, f_d is a complex function of the Reynolds number and relative roughness. There are several options to select an appropriate friction coefficient for a given system. Let us look only at some of these options.

Laminar flow

In the laminar flow regime (Reynolds number, Re < 2000), the friction factor can be obtained by any of the following two methods. The easiest option is to use the following equations.

$f_d = 64/Re$ (Re is in the range of 500 to 2000). (9.21)

To calculate f_f, use the following equation.

$f_f = 16/Re$ (Re is in the range of 500 to 2000). (9.22)

The other option is to use Moody's graphical method illustrated in a following section on turbulent flow.

Transitional flow

Inconsistency of the flow patterns in the transitional flow regime makes the assessment or prediction of friction factor almost impossible. No appropriate relation is available to adequately describe this flow regime. Because of this reason, Darcy-Weisbach equation is applied almost exclusively to calculate head loss in turbulent flow regimes.

Turbulent flow

In the turbulent flow regime, the Colebrook-White equation is widely accepted for assessing the friction factor (Colebrook and White, 1937). The only drawback is that it is implicit. Thus, one is forced to use an iterative procedure to solve for f_d.

$$1/\sqrt{f_d} = -\text{Log}\,[(e/D)/3.7 + 2.51/(Re\sqrt{f_d})] \tag{9.23}$$

Here e is roughness height in the pipe, mm and D the diameter of the pipe, mm. Surface roughness e is due to the projections of the wall, which protrude out of the laminar sub-layer. Thus, e/D represents the relative roughness of pipes. Table 9.1 lists the values of e for various kinds of pipes. The values of f_f for concrete pipes of various diameters and flow velocities are reported in Table 9.2.

Table 9.1. Roughness Height in Various Kinds of Pipes and their State of Use

Material	*State of Use*	*Value or Range (mm)*
Aluminium, drawn/pressed	New	0.0013 - 0.0015
Aluminium, drawn/pressed	Used	0.0015 - 0.03
Asbestos-cement	New, smooth	0.03 - 0.1
Brass, drawn/pressed	New	0.0013 - 0.0014
Brass, drawn/pressed	Used	0.0014 - 0.03
Cast iron	Average city sewerage	1.2
Cast iron	Incrusted	1.2 - 3.0
Cast iron	New, bituminized	0.10 - 0.13
Cast iron	New, with skin	0.2 - 0.6
Cast iron	Operated for several years, cleaned	1.5
Cast iron	Slightly rusty	1.0 - 1.5
Clay	New, clay tile	9.0
Clay, drainage-pipe	New, calcined	0.7
Concrete	New, medium rough	1.0 - 2.0
Concrete	New, rough	2.0 - 3.0
Concrete	New, smooth	0.3 - 0.8
Concrete	Operated for several years	0.2 - 0.3
Concrete, centrifugal-	New, smooth plastered	0.1 - 0.15
Concrete, centrifugal-	New, without plaster	0.2 - 0.8
Concrete, steel-	New, smooth	01. - 0.15
Copper, drawn/pressed	New	0.0013 - 0.0015
Copper, drawn/pressed	Used	0.0015 - 0.03
Glass, drawn/pressed	New	0.0013 - 0.0015
Glass, drawn/pressed	Used	0.0015 - 0.03
Plastic, drawn/pressed	New	0.0013 - 0.0015
Plastic, drawn/pressed	Used	0.0015 - 0.03
Rubber	New, smooth	0.0016
Steel	After long operation, cleaned	0.15 - 0.20
Steel	Homogeneous corrosion pits	0.15
Steel	Intensely incrusted	2.0 - 4.0
Steel	Slightly rusty and incrusted	0.15 - 0.40
Steel, longitudinal welded	New, bituminized	0.01 - 0.05
Steel, longitudinal welded	New, galvanized	0.008

Material	*State of Use*	*Value or Range (mm)*
Steel, longitudinal welded	New, rolling skin	0.04 - 0.1
Steel, weld less	New, galvanized	0.10 - 0.16
Steel, weld less	New, neatly galvanized	0.07 - 0.10
Steel, weld less	New, pickled	0.03 - 0.04
Steel, weld less	New, rolling skin	0.02 - 0.06
Steel, weld less	New, un-pickled	0.03 - 0.06
Stoneware		0.25
Wood	After long operation	0.1
Wood	New	0.2 - 1.0

Table 9.2. Fanning's Coefficients, f_f for Concrete Pipes (Darcy-Weisbach factor = $4f_f$)

Diameter (cm)	*Velocity (cm/s)*				
	30	*60*	*90*	*120*	*150*
15	0.0102	0.0094	0.0090	0.0086	0.0084
22.5	0.0080	0.0074	0.0070	0.0067	0.0064
30	0.0067	0.0062	0.0059	0.0056	0.0054
45	0.0056	0.0051	0.0048	0.0046	0.0044
60	0.0050	0.0046	0.0044	0.0042	0.0040
90	0.0044	0.0040	0.0038	0.0036	0.0034
120	0.0039	0.0037	0.0035	0.0033	0.0030

Serghide's equation, an approximation of the Colebrook equation, calculates the f_d explicitly with a very low error (less than 0.003 per cent). It can be applied across the entire range of the Moody diagram for fluids flowing in a circular pipe.

$$1/\sqrt{f_d} = A - (B-A)^2/(C - 2B + A) \quad (9.24)$$

$$A = -2\,\mathrm{Log}[(e/D)/3.7 + 12/Re] \quad (9.25)$$

$$B = -2\,\mathrm{Log}[(e/D)/3.7 + (2.51 \times A/Re)] \quad (9.26)$$

$$C = -2\,\mathrm{Log}[(e/D)/3.7 + (2.51 \times B/Re)] \quad (9.27)$$

The Swamee–Jain equation applies to full-flowing circular pipes to determine the Darcy–Weisbach friction factor (Swamee and Jain, 1976). It is also an approximation of the implicit Colebrook–White equation for Re in the range of 100 to 10^6.

$$f_d = 0.25/\{\mathrm{Log}\,[(e/D)/3.7 + (5.74/Re\ 0.9)]\}^2 \quad (9.28)$$

Besides, it is possible to approximate the friction factor by few simple equations for some prescribed ranges. For example, for Re greater than 4000 but less than 10^6, f_d is given as:

$$f_d = 0.0791/Re^{1/4} \quad (9.29)$$

Alternatively, the Darcy friction factor is related to Manning's n through the following relationship:

$$n = R^{1/6} (\sqrt{f_d}/8g) \qquad (9.30)$$

Here n is the Manning's friction factor and R the hydraulic radius.

Example 9.2

Water is flowing in a pipe of 50 cm diameter. If the dynamic viscosity of water is 0.55×10^{-3} Ns/m², calculate the maximum velocity for which flow in the pipe remains laminar. Assume appropriate values of any parameter, if required.

Since $Re = \rho V d/\mu$

Let us take density of water, $\rho = 1000$ kg/m³. Also assuming that flow remains laminar for a maximum value of Re as 2000 (literature reports values of Re up to 2300). Then limiting value of V is

$$V = Re\ \mu/\rho\ d = 2000\times 0.55\times10^{-3}/(0.5\times1000) = 0.0022 \text{ m/s}$$

Graphical solution: Moody's diagram

For laminar flows, it is easy to use the Moody diagram, which is a graphical representation of eq. (9.21). No iterations are required for this case. On the other hand, Moody's diagram is commonly employed to obtain f_d for rough pipes with turbulent flows (Figure 9.2, Moody, 1947). The use of this diagram in such cases requires an iterative process wherein flow rate (velocity) is guessed to estimate f_d. With this f_d, a new flow rate is calculated. This new rate is used to further estimate f_d and a new flow rate. This process continues until the calculated flow rate nearly matches the last flow rate. Use the following steps to determine f_d.

- ✰ Calculate the Reynolds number. If the flow velocity or discharge is unknown then assume a reasonable value to calculate the Reynolds number
- ✰ Calculate the relative roughness of pipe of known diameter with the help of Table 9.1. For non-circular ducts, calculate the hydraulic diameter, d_h with the help of eq. (9.17)
- ✰ Begin on y-axis on the right hand side of the diagram with known relative roughness. Move parallel to the nearest curved line of the relative roughness. Proceed until you reach a line vertical to Reynolds number. Mark this point on the chart
- ✰ Move parallel to x-axis and reach y-axis on the left. Read the friction factor. If assumed velocity has been used then calculate the friction losses and velocity of flow using Darcy-Weisbach equation. Calculate the Reynolds number with this velocity and check with the previous Reynolds number. If the new value differs from the assumed value, take this as new value of Reynolds number and proceed with previous steps once again. Repeat until the values match
- ✰ The last calculated value of the f is the actual value of f_d.

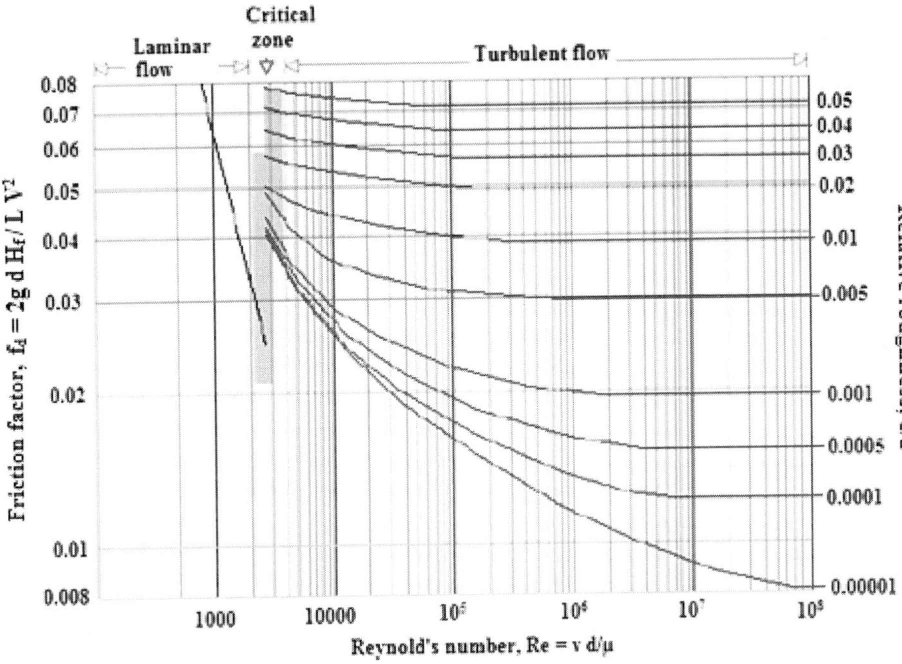

Figure 9.2. Moody's Diagram to Assess Fraction Factor, f_d.

Example 9.3

Find out the velocity of water flow in a concrete pipe of 15.4 cm diameter given that the energy loss is 6 m of head over a length of 300 m. Compute the volume flow rate at this velocity.

The nearest value of f_f is selected from Table 9.2 as 0.0084 for a velocity of about 150 cm/s. Thus, $\mathbf{f_d} = 4\times0.0084 = 0.0336 \sim 0.34$.

Using eq. (9.18)

$$V = (2Hf\ d\ g/f_d\ L)^{1/2} = [2\times 6\times0.154\times9.81/(0.034\times300)]^{1/2} = 1.33\ m/s$$

The area of the pipe is $(\pi/4)\ (15.4/100)^2 = 0.0186\ m^2 \sim 0.019\ m^2$

Since Q = VA, we have

$$Q = 1.33 \times 0.019 = 0.025\ m^3/s$$

Hazen-Williams Equation

Hazen-Williams equation is an empirical equation developed for water flows in larger pipes (diameter ≥5 cm) within a moderate range of velocity ($V \leq 3$ m/s). It is simple and easy to use. Unlike Darcy-Weisbach friction factor, Hazen-Williams coefficient does not depend upon the Reynolds number. The equation in its most widely used form is:

$$V = 0.85\ C\ R^{0.63}\ S^{0.54} \tag{9.33}$$

The discharge for a circular pipe can be written as:

$$Q = 0.278\ C\ d^{2.63}\ S^{0.54} \tag{9.34}$$

Here C is the Hazen William's coefficient, dimensionless, depends upon the pipe material and its age - the higher the factor, the smoother the pipe or tube. Also R is the hydraulic radius, m, d is inside diameter, m, and S is the slope of energy grade line *i.e.* head loss/pipe length = h/L, dimensionless. Although equation is valid for water, it is suggested that it can be used for some other fluids as well provided right hand side of eq. (9.33) or (9.34) is multiplied by SG, the specific gravity of fluid (for water it is 1.0). Some common values of C are given in Table 9.3. For other materials, the tabulated values are available in the literature. To calculate the pressure drop, use any of the above equations to assess head loss, h in m/m. Multiply the head loss by the specific weight of the fluid to give the pressure drop.

$$p = h\,\gamma \tag{9.35}$$

Such that p is the pressure loss (N/m^2, Pa) and γ is the specific weight (N/m^3).

Table 9.3. Values of the Hazen-Williams Coefficient C for Various Pipe Materials

Pipe Material	*C*
Brass	130-140
Brick	100
Cast iron (New, old)	130, 100
Concrete or concrete lined (smooth, rough)	140,100
Copper	130-140
Ductile iron (Cement mortar lined)	140
Glass	140
Iron (worn/pitted)	60 – 80
Polyvinyl chloride (PVC) or high density polyvinyl chloride (HDPE)	150
Smooth wood	120
Smooth masonry	120
Smooth pipes (all metals)	130 – 140
Steel, commercial	140-150
Corrugated steel	60
Vitrified clay	110

Example 9.4

What will be the velocity of flow of water in Example 9.3 if the pipe is made of plastic? Compute the volume of flow rate at this velocity. Use the Hazen-Williams equation for solving the problem.

$$S = h/L = 6/300 = 0.02$$

$$R = d/4 = (0.154)/4 = 0.0385\ m$$

C= 150 (Table 9.3)

Then, using eq. (9.33)

$$V = 0.85\ C\ R^{0.63}\ s^{0.54} = 0.85(150)(0.0385)^{0.63}(0.02)^{0.54} = 1.98\ m/s$$

$$Q = A\ V = (0.019 \times 1.98) = 0.038\ m^3/s$$

Example 9.5

Determine the head loss in a 45 cm concrete pipe flowing full with an average velocity of 1.0 m/s having the length of 30 m. Make use of Manning's equation to solve the problem.

Manning's n from Table 8.1 = 0.012

For a pipe flowing full:

$$R = d/4 = 0.45/4 = 0.1125\ m$$

Slope of energy line = Head loss/Length of pipe

$$V = (1/n)\ (R^{2/3})\ (S^{1/2})$$

$$S = [nV/(R^{2/3})]^2$$

$$S = [0.012 \times 1/(0.1125)^{2/3}]^2 = 0.00265$$

Head Loss = h× L = Slope× Pipe length

Head loss = 0.00265× 30 = 0.08 m

MINOR HEAD LOSSES

The fluid flow in a typical piping system encounters inlets, exits, enlargements, contractions, various fittings, bends and elbows, *etc.* These components interrupt the smooth flow of fluid causing head losses because of mixing and flow separation (Chanson, 2004 and May, 1999). In typical systems with long pipes, the total losses involve major and minor head losses. Note that the minor loss for a pipe system having short length but many bends and valves can be even higher than major losses. Besides, a partially closed valve can cause a greater pressure drop than a long pipe.

A commonly used method to express minor losses is to introduce an equivalent length (L_e) of a straight pipe to represents the head loss for any of the system component discussed in the beginning of this chapter. It stems from the fact that the minor losses are also proportional to the velocity head ($V^2/2g$) as follows.

$$H_f = f_d\ L_e\ V^2/(2g\ d)$$

Here L_e is the equivalent length for a given component of the pipe system. In practice, equivalent length method simply increases the multiplying factor in the Darcy-Weisbach equation (*i.e.* f L/d) by a length of straight pipe (L_e), which would give rise to a pressure loss equivalent to the loss in the pipe fitting. Adding the major and minor head losses, we have

$$H_f = f_d\ (L+L_e)\ V^2/(2g\ d) \qquad (9.36)$$

Thus, the factor f_dL/d is changed to $f_d(L+L_e)/d$ for a given fitting. For more than one fitting, we sum up the L_e over all the components/fittings such that equation takes the following form.

$$H_f = f_d (L+\Sigma L_e) V^2/(2g\ d) \tag{9.37}$$

Table 9.4 presents ratios of equivalent length (L_e) to fitting diameter. Readers may also refer https://neutrium.net/fluid_flow/pressure-loss-from-fittings-equivalent-length-method/ for values of additional pipe fittings. Equivalent length can be calculated for the known diameter of the fitting. Note that L_e/d represents K_L/f_d. Table 9.5 gives similar values for some fittings in ft of straight pipe as a function of fitting size.

Table 9.4. Values of L_e/d for Various Fittings (eq. 9.37)

Fitting Type	*L_e/d*
Gate valve, fully open	8
Check valve, swing type	110
Ball valve, fully open	3
Ball valve, reduced bore	25
Globe valve, fully open	320
90° screwed elbow	30
45° screwed elbow	16
Long radius bend	13
Standard Tee- flow-through run	20
Standard Tee- flow-through branch	60
Welded Tee, through run	10
Welded Tee, through branch	60

Table 9.5. Friction Losses as Equivalent Lengths in Feet of Straight Pipe (PVC) for Various Fittings

Fitting	*Nominal Pipe Size (inches)*												
	1/2	*3/4*	*1*	*1.25*	*1.5*	*2*	*2.5*	*3*	*4*	*6*	*8*	*10*	*12*
90° Elbow	1.5	2.0	2.5	3.8	4.0	5.7	6.9	7.9	12.0	18.0	22.0	26	32
45° Elbow	0.8	1.1	1.4	1.8	2.1	2.6	3.1	4.0	5.1	8.0	10.6	13.5	15.5
Gate valve	0.3	0.4	0.6	0.8	1.0	1.5	2.0	3.0					
Tee - flow through - run	1.0	1.4	1.7	2.3	2.7	4.3	5.1	6.2	8.3	12.5	16.5	17.5	20.0
Tee flow through- branch	4.0	5.0	6.0	7.0	8.0	12.0	15.0	16.0	22.0	32.7	49.0	57.0	67.0
Male/Female adapter	1.0	1.5	2.0	2.8	3.5	4.5	5.5	6.5	9.0	14			

The other option to express minor losses in fittings and other appurtenances, H_{fmi} is as follows.

$$H_{f\,mi} = f_d\, L\, V^2/(2g\ d) = f_d\, (L_e/d)\, (V^2/2g) = K_m\, (V^2/2g) \tag{9.38}$$

Such that $K_m = f_d (L_e/d)$ or $L_e = K_m d/f_d$ (9.39)

The total head losses, H_t for all the fittings and appurtenances including major losses is written as:

$$H_t = H_{ma} + \Sigma H_{mi} = \{V^2/2g\} [fL/d + \Sigma K_m] \quad (9.40)$$

Here H_{ma} is major head losses and ΣK_m is the sum of the resistance coefficients for various fixtures used in the pipeline. Note that the losses need to be added separately if the pipe sizes and the average velocities in various components differ.

Following sections deal with this issue in a detailed manner.

Head Loss at the Entrance of Pipe (h_{en})

Let us assume that the fluid enters the piping system from the right hand side as shown by the arrow mark (Figure 9.3). The entrance head loss is:

$$h_{en} = K_{en} V^2/2g \quad (9.41)$$

Here h_{en} is head loss at the entrance and K_{en} is the entrance head loss coefficient. K_{en} values vary with entrance geometries. If the area at the entrance tends to infinity, it is a situation that corresponds to the flow from a large reservoir into a sharp edged pipe (r/D =0). Here r is the radius of rounded pipe inlet. The value of K_{en} in this case tends to 0.5 for a flush/square-edged entrance (Figure 9.3 left). The end of the pipe should not protrude into the reservoir. For rounded entrance (Figure 9.3 middle), values of K_{en} vary from 0.28 to 0.04 as r/D varies from 0.02 to 0.15+. A protruding pipe (Figure 9.3 right) results in higher head loss such that K_{en} values are around or equal to 1.0.

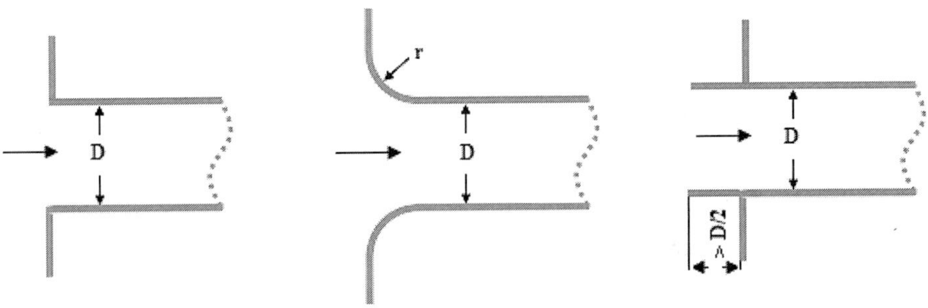

Figure 9.3. Entrance Geometry Governing the Value of K_{en}.

Head Loss at Exit of Pipe (h_{ex})

When a fluid exits a pipe into a much large body of the same fluid, all of the kinetic energy is dissipated because the velocity reduces to zero. Thus, the losses in the system are one velocity head regardless of the exit geometry (Figure 9.4).

$$h_{ex} = K_{ex} V_1^2/2g \quad (9.42)$$

Here V_1 is the mean velocity of flow in the pipe just before the exit. K_{ex} is equal to 1.0.

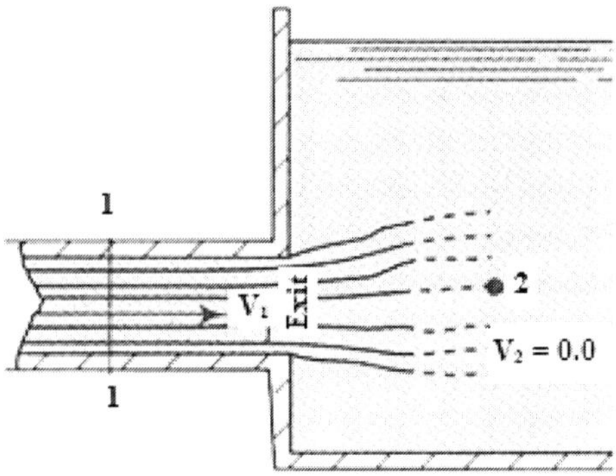

Figure 9.4. Head Loss at the Pipe Eexit Terminating in a Tank.

Similarly, for a liquid freely discharging into the atmosphere, the head loss at the pipe exit is given by eq. (9.42). The K_{ex} in this case irrespective of the exit geometry is equal to one.

Head Loss due to Gradual Enlargement or Contraction

The loss of energy in a gradual enlargement is much less than sudden enlargement because the gradual decrease in the velocity eliminates eddies. The head loss mainly depends on the ratio of the pipe diameters and the angle of enlargement. Minor losses increase with increasing angle (in pipe enlargement). Although, minor losses decrease with decreasing angle, but one will always need a longer pipe to make the transition – that means higher major losses. Therefore, there is a trade-off between the angle and the head loss. Minimum loss including minor and friction losses occur for an angle of 7°. Head loss due to gradual enlargement, h_{gl} is calculated as per the following equation.

$$h_{gl} = kV_1^2/2g \tag{9.43}$$

Here V_1 is the velocity of flow in the smaller pipe. The values of k depend on the diameters of two pipes and angle of divergence. Similarly, when a gradual contraction is provided in the pipe, calculate the head loss, h_{gc} as per the following relation.

$$h_{gc} = kV_2^2/2g \tag{9.44}$$

Here V_2 is the velocity of flow in the smaller pipe. The values of k depend on the diameters of two pipes and angle of convergence also known as angle of cone.

Head Loss Due to Sudden Contraction in Pipe

For an abrupt contraction (Figure 9.5), streamlines are unable to follow the abrupt change in geometry. As such, stremlines gradually converge from an upstream section of the larger tube. At immediate downstream of the junction, the cross-sectional area of the stream tube becomes minimum, being less than that of the smaller pipe. This section of the stream tube is known as vena contracta. Thereafter, stream widens again to fill the pipe. The velocity of flow in the converging part of the stream tube from section 1-1 to section 2-2 (vena contracta) increases due to continuity while the pressure decreases in the direction of flow in compliance with the Bernoulli's theorem. In an accelerating flow under a favourable pressure gradient, losses due to separation will not occur. But in the decelerating part of the flow from section 2-2 to section 3-3, where the stream tube expands to fill the pipe, losses take place in a similar fashion as in case of a sudden geometrical enlargement. Thus, the loss due to contraction is not really for the contraction, but due to the expansion following contraction. If the diameters, areas and flow velocities at the 3-sections are given as follows (Figure 9.5), then the head loss is given as:

	At Section 1-1	*At Section 2-2 Vena Contracta*	*At Section 3-3*
Diameter of pipe	d_1	d_2	d_2
Area of pipe	A_1	A_c	A_2
Flow velocities	V_1	V_c	V_2

Head loss due to sudden contraction = head loss up to vena contracta + head loss due to sudden enlargement

$$H_{sc} = 0 + (V_c - V_2)^2/2g = (V_c - V_2)^2/2g \quad (9.45)$$

Since $V_c A_c = V_2 A_2$ (9.46)

$$V_c = V_2/(A_c/A_2) = V_2/C_c \quad (9.47)$$

$$H_{sc} = (V_2/C_c - V_2)^2/2g \quad (9.48)$$

$$H_{sc} = (V_2^2/2g)\,(1/C_c -1)^2 = K_{sc}\,(V_2^2/2g) \quad (9.59)$$

So that K_{sc}, the k-value for sudden contraction is equal to

$$K_{sc} = (1/C_c -1)^2 \quad (9.50)$$

In general, value of K_{sc} is 0.5.

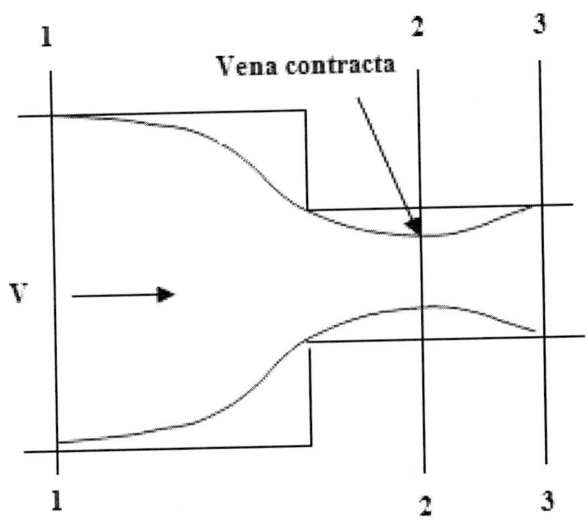

Figure 9.5. Formation of Vena Contracta in a Sudden Contraction in the Flow.

For coaxial circular pipes with flows at high Reynolds numbers, Table 9.6 gives representative values of K_{sc}. As A_1 tends to infinity (A_1/A_2 tends to 0), the value of K_{sc} in eq. (9.50) tends to 0.5. This limiting situation corresponds to the flow from a large reservoir into a sharp edged pipe (entry loss), provided the end of the pipe does not protrude into the reservoir.

Table 9.6. Values of K_{sc} for Various Ratios of A_2/A_1

A_2/A_1	0	0.04	0.16	0.36	0.64	1.0
K_{sc}	0.5	0.45	0.38	0.28	0.14	0

Example 9.7

A 20 cm diameter pipe reduces suddenly to 10 cm. The pipe is carrying water at the rate of 50 L/s. Calculate a) Head loss due to contraction b) Pressure difference between two sections. Take C_c=0.6

d_1=0.2 m, A_1 =0.0314, V_1 = 1.592

d_2= 0.1 m, A_2 = 0.00785, V_2 = 6.369

Q=50 L/s= 50×10^{-3} m^3/s

$$H_{sc} = (6.369)^2/(2 \times 9.81)[1/0.6 - 1]^2 = 2.0675 \times 0.4444 = 0.919 \text{ m}$$

Applying Bernoulli equation for $Z_1 = Z_2$

It will appear as

$$P_1 - P_2 = (V_2^2/2 - V_1^2/2 + g\,H_{sc}) \tag{9.51}$$

$$\Delta\rho = (20.28 - 1.267 + 9.81 \times 0.919)1000 = 28028 \text{ N/m}^2 \text{ or } 28.03 \text{ KN/m}^2$$

Losses Due to Sudden Enlargement

If the cross-section of a pipe with fluid flowing through it is abruptly enlarged (Figure 9.6) at a certain place, fluid emerging from the smaller pipe is unable to follow the abrupt deviation of the boundary. The streamlines take a typical diverging pattern (Figure 9.6). This creates pockets of turbulent eddies in the corners where mechanical energy dissipates into intermolecular energy. It happens because the fluid flows against an adverse pressure gradient, the upstream pressure p_1 at section a-b being lower than the downstream pressure p_2 at section e-f. Note that upstream velocity V_1 is higher than the downstream velocity V_2 because of continuity. The fluid particles near the wall due to their low kinetic energy cannot overcome the adverse pressure hill in the direction of flow and hence follow up the reverse path under the favourable pressure gradient (from p_2 to p_1). For high values of Reynolds number, one can assume the velocity in the smaller pipe to be sensibly uniform over the cross section. Due to the vigorous mixing caused by the turbulence, the velocity is again uniform at a downstream section e-f far away from the enlargement (approximately 8 times the larger diameter).

Applying Bernoulli's theorem across section a-b and e-f, we have

$$H_{sc} = (p_1 - p_2)/\rho g + (V_1^2 - V_2^2)/2g \tag{9.52}$$

In the control volume abcdefgh (Figure 9.6), the momentum theorem can be written as

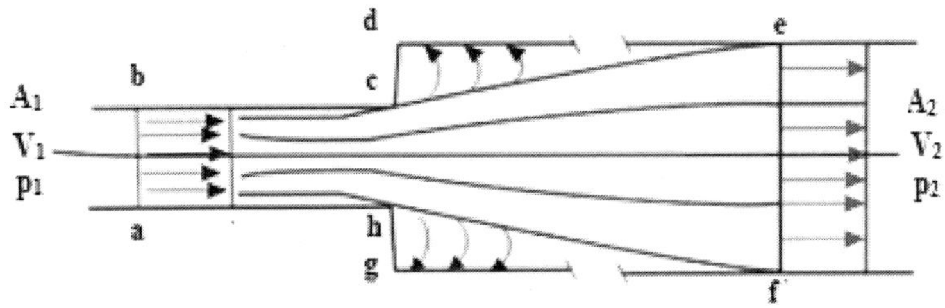

Figure 9.6. Flow Behaviour through an Abrupt Finite Enlargement.

$$p_1 A_1 + p' (A_2 - A_1) - p_2 A_2 = \rho Q (V_2 - V_1) \quad (9.53)$$

Here A_1, A_2 are the cross-sectional areas of the smaller and larger parts of the pipe respectively, and p' is the mean pressure of the eddying fluid over the annular face, gd. From experimental evidences, it is known that $p' \sim p_1$.

Hence, eq. (9.53) becomes

$$A_2 (p_2 - p_1) = \rho Q (V_1 - V_2) \quad (9.54)$$

$$p_2 - p_1 = \rho V_2 (V_1 - V_2) \quad \text{(After incorporating } Q = V_2 A_2) \quad (9.55)$$

Substituting $(p_2 - p_1)$ in the head loss eq. (9.52), we have

$$H_{sc} = -V_2 (V_1 - V_2)/g + (V_1^2 - V_2^2)/2g \quad (9.56)$$

$$H_{sc} = [-2V_2 V_1 + 2V_2^2 + V_1^2 - V_2^2]/2g \quad (9.57)$$

$$H_{sc} = [(V_1 - V_2)^2]/2g = [V_1^2/2g] [1 - A_1/A_2]^2 \quad (9.58)$$

$$H_{sc} = K_{sc} [V_1^2/2g] \quad (9.59)$$

Such that K_{sc} is equal to $[1 - A_1/A_2]^2$

In view of the assumptions made, eq. (9.58) is subject to some inaccuracies, but experiments show that for coaxial pipes differences are within few per cent of the actual values. If, in eq. (9.58), A_2 tends to infinity, then the head loss at an abrupt enlargement tends to $V_1^2/2g$ (Exit as shown in Figure 9.4). It is because the fluid velocities are arrested in the large reservoir, the entire kinetic energy at the outlet of the pipe is dissipated into intermolecular energy of the reservoir through the creation of turbulent eddies. In such circumstances, the loss is usually termed as the exit loss for the pipe. It equals the velocity head at the discharge end of the pipe.

Example 9.8

Calculate the head loss, if a pipe of diameter 20 cm suddenly enlarges to a diameter of 40 cm. The rate of flow of water through the pipe is 260 L/s.

$d_1 = 0.2\text{m}, d_2 = 0.4 \text{ m}, A_1 = 0.0314 \text{ m}^2, A_2 = 0.1256 \text{ m}^2$

$Q = 260 \text{ L/s} = 0.26 \text{ m}^3/\text{s}$

$V_1 = 0.26/0.0314 = 8.28 \text{ m/s}, V_2 = 0.26/0.1256 = 2.07 \text{ m/s}$

Applying eq. (9.58)

$$H_{sc} = (V_1 - V_2)^2/2g = 1.97 \text{ m}$$

Example 9.9

Determine the head loss when water flow of 100 L/min moves from 25.3 mm tube to 73.8 mm tube. Take $K_{sc} = 0.72$.

$$d_1 = 25.3 \text{ mm}, \quad A_1 = 0.0005017 \text{ m}^2$$

$$d_2 = 73.8 \text{ mm}, \quad A_2 = 0.004282 \text{ m}^2$$

$$V_1 = Q_1/A_1 = (100/60{,}000)/0.0005017 = 3.32 \text{ m/s}$$

Applying eq. (9.59)

Head loss = $0.72\,(3.32)^2/(2 \times 9.81) = 0.40$ m

Head Loss Due to Obstruction in Pipe (H_{obs})

The loss of energy due to an obstruction in the pipe occurs because the cross sectional area of the pipe is reduced due to the obstruction. An abrupt enlargement of the stream beyond the obstruction follows. Let us consider an irregular object placed in the path of flow (Figure 9.7). If A is the area of the pipe and maximum area of obstruction is a, then the area of flow at section 1-1 will be (A-a). There is a formation of vena contracta as the fluid has just passed through the obstruction. It again enlarges and fills the pipe having the same velocity as before the obstruction. If V is the velocity of liquid in pipe, V_c is the velocity of liquid at vena contracta and A_c is the area of cross-sectional area at vena contracta, then

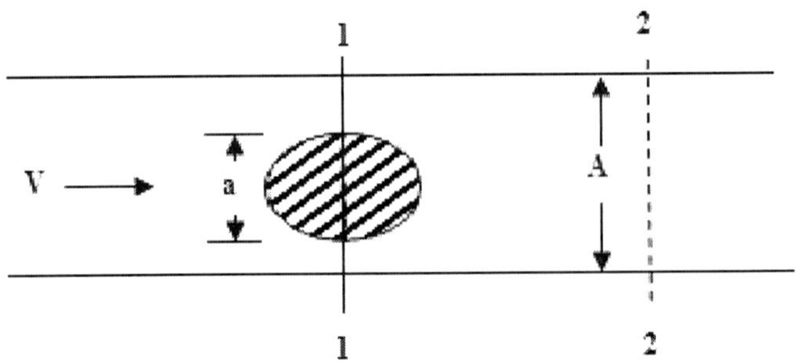

Figure 9.7. View of an Obstruction in the Flow Path in a Pipe.

Loss of head due to obstruction = Loss of head due to enlargement from vena contracta

$$H_{obs} = [(V_c - V)^2/2g] \tag{9.60}$$

As per continuity equation

$$A_c V_c = A\,V \tag{9.61}$$

If we define C_c as coefficient of contraction = Area of Vena contraction/Area of obstruction

$$C_c = A_c/(A - a) \quad (9.62)$$

Substituting A_c in the continuity equation (9.61), we have

$$C_c\,(A - a)\,V_c = A\,V \quad (9.63)$$

$$V_c = AV/C_c(A - a) \quad (9.64)$$

Substituting V_c in the head loss equation (eq. 9.60)

$$H_{obs} = [(V_c - V)^2/2g] = [(AV/C_c\,(A - a) - V)^2/2g] \quad (9.65)$$

$$H_{obs} = [V^2/2g]\,[(A/C_c\,(A - a) - 1)^2] \quad (9.66)$$

$$H_{obs} = K_{obs}\,(V^2/2g) \quad (9.67)$$

Such that $K_{obs} = [(A/C_c\,(A - a) - 1)^2]$

Example 9.10

The velocity of water flowing through a 12 cm diameter pipe is 3.0 m/s. An iron plate of 6 cm diameter obstructed the flow path. Calculate, the loss of head due to obstruction, if co-efficient of contraction $C_c = 0.70$.

Calculate areas from the given diameters of pipe and obstruction

$$A = \pi\,(0.12)^2/4 = 0.0113\ m^2$$

$$a = \pi\,(0.06)^2/4 = 0.0028\ m^2$$

Using eq. (9.66)

$$h_{obs} = [3^2/(2\times9.81)]\,[0.0113/0.7\,(0.0113\text{-}0.0028)\ \text{-}1]^2 = 0.459 \times 0.81 = 0.372\ m$$

Head Loss Due to Pipe Fittings (H_{fi})

Components in a piping system normally comprise various types of fitting such as valves, elbows, tees, bends, joints *etc.* The loss coefficients in these cases strongly depend on the shape of these components, besides the size and type of the components. The head loss due to pipe fittings is given by the following equation:

$$H_{fi} = K_L\,V^2/2g \quad (9.68)$$

Here K_L is the friction loss coefficient for a given fitting. Mostly, manufacturers supply the values of K_L for the pipe fittings. Many reference books also report the values of K_L. As an illustration, some of these values are given in Table 9.7.

Head Loss in the System

Head loss through the pipe system is determined by summing-up the head loss from individual component along the flow path. For example, the head loss for the system shown in Figure 9.8 can be written (moving from 1 to 2) as follows.

Head loss = Entrance head loss ($0.5\,V^2/2g$) +Head loss in pipe section L_1 ($f_d\,L_1\,V_1^2/(2g\,d)$) + Head loss in Elbow ($0.9\,V^2/2g$) + Head loss in pipe section L_2 ($f_d\,L_2\,V^2/(2g\,d)$) + Head loss in Elbow ($0.9\,V^2/2g$) + Head loss in pipe section L_3 ($f_d\,L_3\,V^2/(2g\,d)$) + Head loss in globe valve ($10.0\,V^2/2g$) + head loss at exit ($1.0\,V^2/2g$)

Table 9.7. The Values of K_L for Various Pipe Fittings

Fitting	*Type*	K_L
Gate valve	Fully open	0.17
	3/4 open	0.9
	1/2 open	4.5
	1/4 open	24
Plug disk	Fully open	9
	3/4 open	13
	1/2 open	36
	1/4 open	112
Butterfly valve	$\Theta = 5^0$	0.24
	$\Theta = 10^0$	0.52
	$\Theta = 20^0$	1.54
	$\Theta = 40^0$	10.8
	$\Theta = 60^0$	118
45° elbow	Standard (R/D = 1)	0.35
	Long radius (R/D =1.5)	0.2
90° elbow	Standard (R/D = 1)	0.75
	Long radius (R/D =1.5)	0.45
90° elbow square		1.3
180° bend	Close return	1.5
Tee, run through	Branch blanked	0.4
Tee, as elbow	Entering in run	1
Tee, as elbow	Entering in branch	1
Tee, branching flow		1

R: Center to end distance; D: Outer diameter.

The above additions assume that the pipe material and diameter remains same throughout so that flow velocity and friction factor in pipe remains the same.

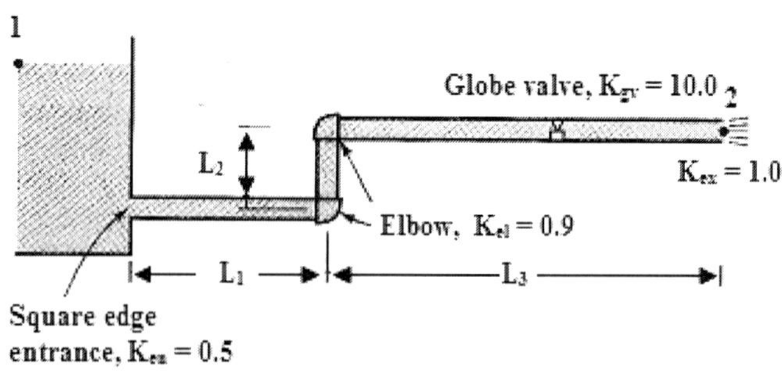

Figure 9.8. Schematic Diagram Showing Various Components of Head Loss.

FLOW THROUGH PIPING SYSTEMS

Pipes in Series and Parallel

When pipes are connected end to end (in continuation with each other) so that the fluid flows in a continuous line without any branching are said to be in series (Figure 9.9). The volume rate of flow through the pipes in series is the same throughout.

$$Q = Q_1 = Q_2 = Q_3 = \ldots \tag{9.69}$$

Here Q is the discharge from the system and Q_1, Q_2 and Q_3 the discharge at various sections in the pipeline. The head loss then is:

$$\text{Head loss, } h_{tot} = h_1 + h_2 + h_3 \ldots \tag{9.70}$$

Here h_{tot} is the total head loss in the system and h_1, h_2 and h_3 are the head loss in various sections of the pipeline.

Pipes connected in a manner that the flow from a pipe branches or divides into two or more separate pipes and then reunite into a single pipe is said to be in parallel (Figure 9.9). The discharge relation for pipes connected in parallel is:

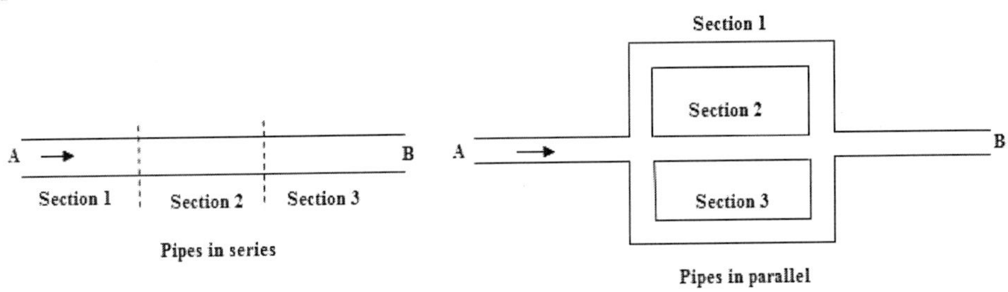

Figure 9.9. Schematics of Pipes Connected in Series and Parallel.

$$Q_a = Q_b = Q_1 + Q_2 + Q_3 \ldots \tag{9.71}$$

Here Q_a and Q_b are the discharge in the system at A and B and Q_1, Q_2 and Q_3 the discharge at various sections in the pipeline. Since the total head at junctions must have single value, therefore the head loss in each branch must be the same irrespective of the path followed by the flow. Head loss equation is written as:

$$H_a - H_b = h_1 = h_2 = h_3 \tag{9.72}$$

Here H_a and H_b are heads at A and B respectively and h_1, h_2 and h_3 are the head loss in various sections.

Dupuit's equation for equivalent pipe: pipe in series

Suppose a pipeline consists of a number of pipes of different sizes and lengths (Figure 9.10). Let d_1, d_2, d_3 be the diameters of lengths L_1, L_2, L_3 of these component pipes. Let V_1, V_2, V_3 be the velocities in these sections.

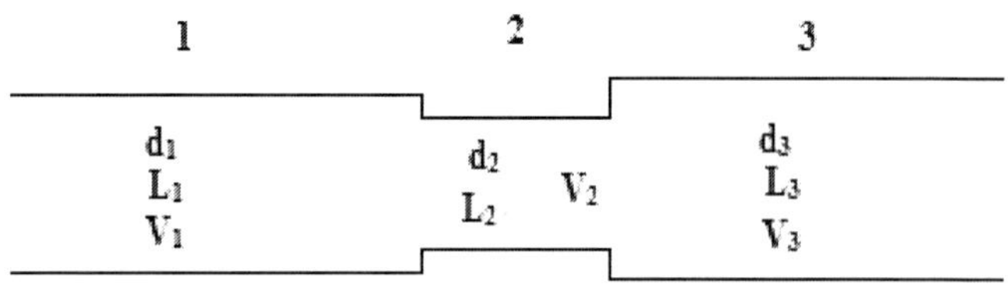

Figure 9.10. Schematic Diagram for Derivation of an Equation of Equivalent Length of Pipes Joined in Series.

For pipes connected in series, rate of discharge Q is the same in all the pipes. Ignoring minor losses, the total loss of head is equal to the sum of the friction losses in the individual pipes. If H_{f1}, H_{f2} and H_{f3} are the head losses in the individual pipe sections, then

$$H_{tot} = H_{f1} + H_{f2} + H_{f3} = [f_{d1} L_1 V_1^2/(2g d_1)] + [f_{d2} L_2 V_2^2/(2g d_2)] + [f_{d3} L_3 V_3^2/(2g d_3)] \quad (9.73)$$

If we incorporate V =Q/A and A =π d²/4, we have

$$H_{tot} = f_{d1}L_1Q^2/(C d_1^5) + f_{d2}L_2Q^2/(C d_2^5) + f_{d3}L_3Q^2/(C d_3^5) \quad (9.74)$$

$$C = 4^2/(2\pi^2 g) = 0.0826$$

If $f_d = f_{d1} = f_{d2} = f_{d3}$

$$f_d L_{eq} Q^2/(C d_{eq}^5) = (f_d Q^2/C) L_1/d_1^5 + (f_d Q^2/C) L_2/d_2^5 + (f_d Q^2/C) L_3/d_3^5$$

Here L_{eq} is the length of equivalent pipe, which has the same head loss and discharge that of series or compound pipe.

Thus, the Dupuit's equation for equivalent length of the pipe can be written as:.

$$L_{eq}/d_{eq}^5 = L_1/d_1^5 + L_2/d_2^5 + L_3/d_3^5 \quad (9.75)$$

Normally, equivalent length is calculated for equivalent diameter, which is taken as the average diameter of the various pipes used. Alternatively, equivalent diameter is calculated for equivalent length, which is taken as the sum of the lengths of various sections.

If there are intermediate fittings, calculate additional equivalent lengths for each fitting with the help of Table 9.4. This length is added to the actual length of the pipe to determine the overall head loss.

Equation for equivalent pipe: pipe in parallel

For Figure 9.11, the head loss from point A to B is

$$H_a - H_b = f_{d1}L1V1^2/(2g\, d1) = f_{d2}L_2V_2^2/(2\, g\, d_2) \quad (9.76)$$

In terms of Q, it is written as:

$$H_a - H_b = 8\, f_{d1}L_1Q_1^2/(\pi^2\, g\, d_1^5) = f_{d2}L_2Q_2^2/(\pi^2\, g\, d_2^5) \quad (9.77)$$

$$Q_1^2 = Q_2^2 (R_b/R_a)$$

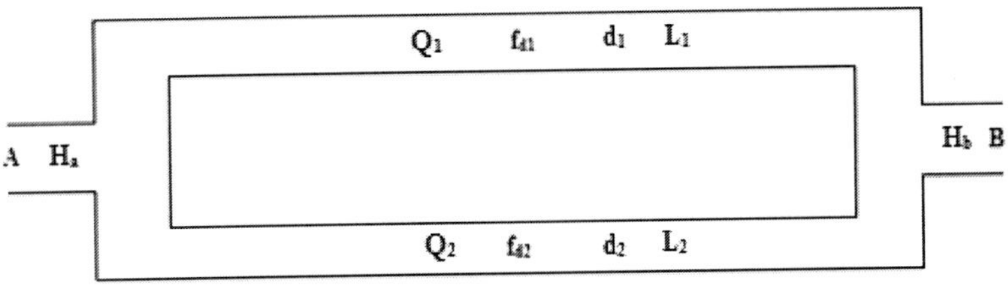

Figure 9.11. Schematic Diagram for Derivation of an Equation of Equivalent Length of Pipes Joined in Parallel.

Such that $R_a = 8\ f_{d1} L_1/(\pi^2\ g\ d_1^5)$ and $R_b = 8\ f_{d2} L_2/(\pi^2\ g\ d_2^5)$

$$Q_1 = K\ Q_2$$

Here $K = \sqrt{(R_b/R_a)}$

Since $Q = Q_1 + Q_2$

$$Q = KQ_2 + Q_2$$

$$Q_2 = Q/(1+K)$$

Similarly $Q_1 = Q\ K/(1+K)$

We can also write the equation in terms of Q as follows:

$$Q = [H_f\ g\ \pi^2\ d^5/(8\ f_d\ L)]^{1/2} \text{ such that} \tag{9.78}$$

$$Q_a = [H_f\ g\ \pi^2\ d_1^5/(8\ f_{d1}\ L_1)]^{1/2}$$

$$Q_b = [H_f\ g\ \pi^2\ d_2^5/(8\ f_{d2}\ L_2)]^{1/2}$$

Since $Q = Q_a + Q_b$, we have

$$Q = [H_f\ g\ \pi^2\ d/(8\ f_d\ L)]^{1/2} = [H_f\ g\ \pi^2\ d_1^5/(8\ f_{d1}\ L_1)]^{1/2} + [H_f\ g\ \pi^2\ d_2/(8\ f_{d2}\ L_2)]^{1/2}$$

Since H_f is same and writing d and L as d_{eq} and L_{eq}, we get

$$(d_{eq}^5/(f_d\ L_{eq}))^{1/2} = (d_1^5/(f_{d1}L_1))^{/2} + (d_2^5/(f_{d2}L_2))^{1/2} + \ldots. \tag{9.79}$$

If friction factors are also same, then

$$(d_{eq}^5/L_{eq})^{1/2} = (d_1^5/L_1)^{1/2} + (d_2^5/L_2)^{1/2} + \ldots. \tag{9.80}$$

Example 9.11

Three pipes having internal diameters of 50, 40 and 30 cm are joined in series. The length of these pipes respectively is 900, 800 and 700 m. Find out the equivalent length of the pipe, if the equivalent diameter is taken as the average diameter of 3 pipes. Also, calculate the equivalent diameter for an equivalent length of pipe.

Equivalent diameter = (50 +40 + 30)/3 = 120/3 = 40 cm

Then using eq. (9.75)

$$L_{eq} = D_{eq}^5\ (L_1/d_1^5 + L_2/d_2^5 + L_3/d_3^5)$$

$$L_{eq} = 0.4^5\ (900/0.5^5 + 800/0.4^5 + 700/0.3^5)$$

$$L_{eq} = 0.01024\ (28800 + 78125 + 288065) = 0.01024 \times 394990 = 4044 \text{ m}$$

For equivalent diameter, equivalent length = 900 + 800 + 700 = 2400 m

From previous calculations, we can write

$2400/d_{eq}^{5} = 394990$

$d_{eq}^{5} = 2400/394990 = 0.006076$

$d_{eq} = 0.36$ m

Example 9.12

Two pipes are joined in parallel. The length and diameter of the first parallel pipe are 2000 m and 1.5 m respectively, while these for the second pipe are 2000 m and 1.25 m. Find rate of flow in each parallel pipe if total discharge in the main pipe is 4m^3/s.Coefficient of Fanning's friction for each parallel pipe is 0.0083.

L_1= 2000 m, d_1= 1.5 m, Q= 4 m^3/s

L_2= 2000 m, d_2= 1.25 m, f_1= f_2= f_3=0.0083

$Q= Q_1 + Q_2 = 4$ m^3/S

$4\ f_1 L_1 V_1^2/(2g\ d_1) = 4\ f_2 L_2 V_2^2/(2\ g\ d_2)$

Therefore,

$V_1^2/1.5 = V_2^2/1.25$

$V_1 = 1.09\ V_2$

Also

$Q_1 = A_1\ V_1 = 1.77\ V_1$

$Q_2 = A_2\ V_2 = 1.23\ V_2$

Since $Q = Q_1 + Q_2$

We have

$1.77\ V_1 + 1.23\ V_2 = 4$

$1.77 \times 1.09\ V_2 + 1.23\ V_2 = 4$

$V_2 = 4/(1.93 + 1.23) = 1.27$ m/s

$Q_2 = 1.27 \times 1.23 = 1.56$ m^3/s

$V_1 = 1.27 \times 1.09 = 1.38$ m/s

$Q_1 = 1.38 \times 1.77 = 2.44$ m^3/s

Example 9.13

Three pipes with the following details are joined in parallel between two points. Calculate the total head loss between the junctions, if flow through the pipes is 0.3 m^3/s.

Pipe	*Length (m)*	*Diameter (cm)*	f_d
1	1000	20	0.02
2	1200	30	0.015
3	800	15	0.02

Let us first find out the equivalent length of pipe assuming d_{eq} as 30 cm and f_{de} as 0.02. Use eq. (9.79)

$$(d_{eq}^5/f_{de}L_{eq})^{1/2} = (d_1^5/f_{d1}L_1)^{1/2} + (d_2^5/f_{d2}L_2)^{1/2} + (d_3^5/f_{d3}L_3)^{1/2}$$

$$(0.3^5/(0.02 \times L_{eq})^{1/2} = (0.2^5/0.02\times 1000)^{1/2} + (0.30^5/0.015 \times 1200)^{1/2} + (0.15^5/0.02 \times 800)^{1/2}$$

$$(0.3^5/(0.02 \times L_{eq})^{1/2} = 0.004 + 0.0116 + 0.00218 = 0.0178$$

$$0.3^5 = 0.000317\times 0.02 \times L_{eq}$$

$$L_{eq} = 383.3 \text{ m}$$

For a discharge of 0.3 m^3/s, the velocity in the pipe of equivalent diameter is 4.24 m/s and therefore, head loss can be given as:

$$H_f = f_d L_1 V_1^2/(2g\, d_1) = 0.02\times 383.3\times 4.24^2/(2 \times 9.81\times 0.3) = 23.43 \text{ m}$$

Example 9.14

Two pipes are connected in parallel between two reservoirs having a difference in level of 2.5 m. The length, diameter, and D-W friction factor for one pipe are 2000 m, 1.2 m, and 0.025. The respective values for other pipe are 2200 m, 1 m, and 0.020 for the second pipe. Find out the flow rate between the two reservoirs.

$$H_f = 8\, f_1\, L_1\, Q_1^2/\pi^2\, g\, d_1^5$$

$$2.5 = 8 \times 0.025\times 2400\times Q_1^2/(\pi^2\times 9.81\times 1.2^5)$$

$$Q_1^2 = 0.0070 \times (\pi^2\times 9.81\times 1.2^5)$$

$$Q_1 = 1.50 \text{ m}^3/\text{s}$$

Similarly, for the other pipe, discharge is 0.69 m^3/s

Therefore, Q = 1.50 + 0.69 = 2.19 m^3/s

QUESTIONS (THEORY)

1. What do you understand by major and minor head losses in pipe flow?
2. Derive the Darcy-Weisbach formula to calculate major friction head losses in pipe flow.
3. Derive the equation for minor head loss due to sudden contraction.
4. Derive the equation for minor head loss due to sudden enlargement.
5. Derive the equation for minor head loss due to an obstruction in the pipe.
6. What do you understand by pipes in series and pipes in parallel?
7. Derive the Dupuit's equation for equivalent pipe when pipes are joined in series.
8. Derive the equation for equivalent pipe when pipes are joined in parallel.

Chapter 10

Dimensional Analysis and Similitude

Dimensional analysis is an important scientific tool especially in fluid mechanics. It is applicable both in mathematical formulations and in experimental quantitative studies for the design of small-scale models (similitude). Dimensional analysis eventually leads to design and simplification of models, reduction in number of parameters, and an extension of the domain of application. Basic introduction to dimensions and units has already been made in Chapter 1. This chapter deals with dimensional analysis, explains its working, and makes some remarks on its utility.

DIMENSIONAL ANALYSIS

A derived quantity is dimensionless if its numerical value remains invariant when the base units are changed. An example is Vt/L, where V is velocity, dx/dt, t is time and L is the length. The dimension of a dimensionless quantity is unity (See eq. 10.1), the factor by which its numerical value changes when base unit's sizes are changed. In other words, quantities whose numerical values are identical in all systems of units within a class are dimensionless. All other quantities are dimensional. Dimensional analysis is a systematic procedure of identifying the variables in a physical phenomena and correlating them to form a set of dimensionless group (Pritchard and Leylegian, 2011). In other words, dimensional analysis is a process of arranging various factors in a manner that they form dimensionless number. Dimensional analysis is commonly used for the following prominent reasons.

- ☆ To test the consistency of a dimensional equation
- ☆ To determine the dimensions and thereby the units of any quantity
- ☆ Derive relation between physical quantities involved in physical phenomena
- ☆ To reduce variables/equations by arranging variables in dimensionless forms

- ☆ To change units from one system to another
- ☆ Analyzing complex flow phenomena by use of scaled models

Dimensional Homogeneity

An equation is dimensionally homogeneous if the dimensions of various terms on both sides of the equation are identical. A dimensionally homogeneous equation is independent of the fundamental units of measurement and is applicable to all systems of units. If an equation truly expresses a proper relationship among variables in a physical process, then it will be dimensionally homogeneous. Let us consider the following equation that states that the period of a pendulum is proportional to the square root of the length of the string divided by g. Using dimensional analysis, we show that it is dimensionally homogeneous.

$$t = 2\pi \sqrt{(l/g)} \tag{10.1}$$

Here t is the period of a pendulum, l is the length of the string and g is acceleration due to gravity. Using the MLT system for eq. (10.1), we have

$$[T] = \sqrt{L/LT^{-2}} = [T] \tag{10.2}$$

Dimensions of LHS and RHS are equal so the equation is dimensionally homogeneous. Other examples of dimensional homogeneity are as follows.

Discharge, Q= Velocity,V × Area, A

$$L^3T^{-1} = LT^{-1}\,L^2 = L^3T^{-1} \tag{10.3}$$

Velocity, V = √2 (acceleration due to gravity x head)

$$V = \sqrt{(2gh)} \tag{10.4}$$

$$LT^{-1} = \sqrt{2\,LT^{-2}.L} = LT^{-1} \tag{10.5}$$

For a known homogeneous equation, it should be possible to determine some unknown value in the equation. For example, Einstein's equation, which states that energy, E equals mass, m times the speed of light, c squared. If we cannot recall that it is c^2, then the equation can be written as:

$$E = mc^N \tag{10.6}$$

Now we can use the dimensions of each term to work out the value of N.

$$ML^2T^{-2} = M\,(LT^{-1})^N \tag{10.7}$$

Since the equation is homogeneous, clearly the value of N should be 2.0.

To ensure that an equation is dimensionally homogeneous is the first step to dimensional analysis. Dimensional homogeneity imposes the following constraints on any mathematical representation or a relationship.

- ☆ Both sides of the equation must have the same dimensions
- ☆ Wherever a sum of quantities appears, all the terms in the sum must have the same dimension (consistency)
- ☆ All arguments of any exponential, logarithmic, trigonometric or other special functions that appear must be dimensionless.

Consistency of the Equation

We can add or subtract physical quantities only if they have the same dimensions. The similarity of dimensions of physical quantities is called consistency of a dimensional equation. For example, we cannot add or subtract mass and velocity as they are of different dimensions. Let us check the consistency of the equation for the displacement of a free falling body given as:

$$x = x_o + v_o t + (1/2)\, gt^2 \tag{10.8}$$

Here, v_o is the initial velocity, g is the acceleration due to gravity, t is the time, x_o is the initial distance, and x the distance travelled at any time t respectively. To check if the above equation is dimensionally correct, we have to prove that dimensions of physical quantities on both the sides are same.

x = distance = [L]

x_o = distance = [L]

$v_o t$ = Velocity × time = $[LT^{-1}] \times [T] = [L]$

at^2 = Acceleration × time^2 = $[LT^{-2}] \times [T^2] = [L]$

$$[L] = [L] + [LT^{-1}] \times [T] + [LT^{-2}] \times [T^2] \tag{10.9}$$

Since dimensions on left hand side equals dimension on right hand side, equation is consistent and dimensionally homogeneous.

Dimensional variables

Dimensional variables are the quantities which have dimensions and whose numerical value may change. We can plot them against each other. In eq. (10.8) x and t are the dimensional variables.

Dimensional constants

Dimensional constants are quantities having dimensions but having a constant value, *e.g.*, gravitation constant (g), Planck's constant (H), Stefan's constant, Boltzmann's constant, and many others. These are known as universal physical constants. Their magnitudes are the same but values vary with the chosen system of units. Some specific constants may also appear in an equation but may vary from case to case. In eq. (10.8), $x_{0,}$ v_0 besides g are the dimensional constants. Normally, these arise in an equation because of initial or boundary conditions.

Pure constants

A constant is a number on its own, or sometimes a letter such as a, b or c to stand for a fixed number. They have no dimensions, but arise while performing mathematical manipulations or during experimental verifications. In eq. (10.8), ½ arises due to mathematical manipulation and is a pure constant.

Non-dimensionlization of Equations

Non-dimensionalization is the process of removal of units from physical quantities by a suitable substitution of variables. Eq. (10.8) can be non-dimensionalized by the following substitutions.

$$x^* = x/x_o$$

$$t^* = v_o t/x_o$$

Such that

$$x'' = 1 + t^{*\prime\prime} + (1/\alpha)\, t^{*2\prime\prime 2}$$

Here $\alpha = v_o^2/gx_o$

Example 10.1

Check whether the equation $W = 0.5\, mV^2 - mgh$ is dimensionally correct. Here W is work done, m is the mass, g is acceleration due to gravity, V the velocity and h the height.

To check whether the equation is dimensionally correct, we first write dimensions of all the physical quantities mentioned in the equation.

W = Work done = Force × Displacement = $[MLT^{-2}] \times [L] = [ML^2T^{-2}]$

$0.5\, mV^2$ = Kinetic energy = $[M] \times [L^2T^{-2}] = [ML^2T^{-2}]$

mgh = Potential energy = $[M] \times [LT^{-2}] \times [L] = [ML^2T^{-2}]$

Since all the dimensions on left and right sides are equal, it is a dimensionally correct equation.

STEPS FOR DIMENSIONAL ANALYSIS

The dimensional analysis for an engineering problem that can be mathematically described consists of following steps.

- Formulate differential equations and/or algebraic expressions, which together with the required boundary conditions adequately describe the problem. It means identify a complete set of independent quantities that determine the dependent variable
- List the dimensions of dependent and independent variables
- Substitute the new dimensionless variables into each differential equation and boundary condition to normalize them
- Group all dimensionless variables and parameters generated into functional form
- Reduce the functionality to the minimum possible number of independent groups; and
- Drop any groups in which arbitrary terms cannot be divided out using other groups of the functionality.

Thus, with the above steps, we finally arrive at an equation, which relates all the physical factors of a problem to each other. Commonly used methods of dimensional analysis are:

- Rayleigh's method
- Buckingham Pi theorem

RAYLEIGH'S METHOD

Rayleigh's method, named after Lord Rayleigh, is a conceptual tool used for dimensional analysis in physics, chemistry, and engineering (Agrawal, 2006). It expresses a functional relationship of variables in the form of an exponential equation, and consists of the following steps:

1. Gather all the independent variables that are likely to influence the dependent variable
2. If R is a variable that depends upon numbers of independent variables $R_1, R_2, R_3, .., R_n$, then the functional equation can be written as $R = F(R_1, R_2, R_3, .., R_n)$
3. Write the above equation in the form $R = C\, R_1^{\,a}\, R_2^{\,b}\, R_3^{\,c}.\, R_n^{\,m}$, where C is a dimensionless constant and a, b, c,., m are arbitrary exponents
4. Express each of the quantities in the equation in some base units in which the solution is required
5. By using dimensional homogeneity, obtain a set of simultaneous equations involving the exponents a, b, c,., m
6. Solve these equations to obtain the value of exponents a, b, c,., m
7. Substitute the values of exponents in the main equation, and form the non-dimensional parameters by grouping the variables with like exponents

Drawback – Two main limitations of the Rayleigh method are:

- ☆ It assumes that an exponential relationship exists between the variables
- ☆ It does not provide any information regarding number of dimensionless groups emerging at the end of dimensional analysis.

Following two examples illustrate the application of Rayleigh's method.

Example 10.1

Show that the period of a pendulum is proportional to the square root of the length of the string divided by acceleration due to gravity, g.

Period of a pendulum is a function of length, l and acceleration due to gravity, g, such that we can write

$$t = f\,(l, g) \tag{10.10}$$

Here f is some function. Using Rayleigh assumption of an exponential function, we have

$$t = C\, l^a\, g^b \tag{10.11}$$

$$T = C \times L^a \times (L\, T^{-2})^b \tag{10.12}$$

Comparing the exponents in eq. (10.12), we get

$$-2b = 1$$

$$a + b = 0$$

Solving the two equations, we get

$b = -1/2$ and $a = 1/2$

Thus, eq. (10.11) transforms to

$$t = C\, l^{1/2}\, g^{-1/2} \qquad (10.13)$$

(Students may use the Buckingham theorem also to prove this)

The value of C is determined from experiments or mathematical manipulations. We know that $C = 2\,\pi$. Thus

$$t = 2\pi\,\sqrt{(l/g)} \qquad (10.14)$$

Example 10.2

The drag on a body in a fluid flow is a function of the body size (expressed via its characteristic length, L) fluid velocity V, density ρ, viscosity μ and acceleration due to gravity, g. Perform dimensional analysis for this problem to find out drag, D, acting on the ship.

$$D = f\,(L,\rho,\mu,V,g)$$

With the Rayleigh Method, we express it as

$$D = C\,L^a\,\rho^b\,\mu^c\,V^d\,g^e \qquad (10.15)$$

Here C is a dimensionless constant, and a, b, c, d, and e are exponents, whose values are not yet known.

Note that the dimensions on the left side must equal those on the right side.

Step 1: Setting up the equation

Write the equation in terms of dimensions only, For the tabulated dimensions equation becomes

Quantity	*Symbol*	*Dimension*
Size	L	L
Viscosity	μ	M/LT
Density	ρ	M/L^3
Velocity	V	L/T
Gravity	g	L/T^2

$$ML/T^2 = L^a\,(M/L^3)^b\,(M/LT)^c\,(L/T)^d\,(L/T^2)^e \qquad (10.16)$$

On the left side, we have $M^1L^1T^{-2}$, which should be equal to the dimensions on the right side. Therefore, the exponents of the right side must be such that the units are $M^1L^1T^{-2}$

Step 2: Solving for the exponents

Equate the exponents of each term of the respective fundamental units in eq. (10.16):

M: $1 = b + c$ (since $M^1 = M^bM^c$)

Similarly

$L: 1 = a - 3b - c + d + e$

$T: -2 = -c - d - 2e$

Herein we have only three equations for 5 unknown variables. Therefore, it is not possible to obtain a complete solution. Thus, we solve a, b, and d in terms of c and e. This choice is mostly based on experience. Therefore,

From M:	$b = 1 - c$	(i)
From T:	$d = 2 - c - 2e$	(ii)
From L:	$a = 1 + 3b + c - d - e$	(iii)

Solving (i), (ii), and (iii) simultaneously, we obtain

$a = 2 + e - c$

Substituting the exponents back into the original eq. (10.15), we obtain

$D = C\, L^{2+e-c} \rho^{1-c} \mu^{c} V^{2-c-2e} g^{e}$

Collecting like exponents together

$$D = C\,(V^2/Lg)^{-e}\,(VL\rho/\mu)^{-c}\,\rho\, L^2\, V^2 \qquad (10.17)$$

Comparing the terms for different exponents

Terms with exponent of 1: $C\,\rho$

Terms with exponent 2: L^2V^2

$$\text{Terms with exponent e: } L^{e}V^{-2e}g^{e} = (V^2/Lg)^{-e} \qquad (10.19)$$

$$\text{Terms with exponent c: } L^{-c}\rho^{-c}\mu^{c}V^{-c} = (VL\rho/\mu)^{-c} \qquad (10.20)$$

Note that the right sides of eqs. (10.19) and (10.20) are the dimensionless groups.

Step 3: Determining the exponents of dimensionless groups

To determine unknown e and c, let us consider the following cases:

If $e = 1$ then eq. (10.19) becomes (Lg/V^2)

If $e = -1$ then eq. (10.19) becomes (V^2/Lg)

If $c = 1$ then eq. (10.20) becomes $(\mu/VL\rho)$

If $c = -1$ then eq. (10.20) becomes $(VL\rho/\mu)$

Similarly, we can put other values as well. For the present problem, for c = -1, we get Reynolds number on right hand side. Similarly, for e = - 1, we get square of the Froude number on the right hand side. Therefore, choose exponents c as -1 and e as -½, since these values gives Reynolds and Froude numbers respectively. Incorporating in eq. (10.17), we obtain

$$D = g(Fr, Re)\,\rho\, L^2 V^2 \qquad (10.21)$$

Here g (Fr, Re) is a dimensionless function. Eq. (10.21) can also be written as:

$$D/\rho L^2 V^2 = g\,(Fr, Re) \tag{10.22}$$

Now the right hand side is also a dimensionless quantity, the drag coefficient, C_D. Such that

$$C_D = D/\rho\, l^2 V^2$$

Thus, drag coefficient is now a function of only 2 dimensionless variables instead of 5.

BUCKINGHAM Π THEOREM

The dimensional analysis of the experimental data of unknown flow problems leads to some non-dimensional parameters. These dimensionless products are often referred as pi (Π) terms. Based on the concept of dimensional homogeneity, these dimensionless parameters can be grouped and expressed in functional forms (White, 2011). This idea explored by the famous scientist Edgar Buckingham (1867-1940) led to the development of Buckingham Π theorem. It states that, if there is physically meaningful equation involving a certain number of variables say n, and these variables contain m primary dimensions (for example M, L, T), the equation relating all the variables will have (n-m) dimensionless groups. Normally it is n-3 groups but for some systems, even an angle is a primary group. In that case, one can have (n-4) independent dimensionless groups. Buckingham referred these groups as Π groups. The final Buckingham equation is of the form:

$$\Pi_1 = f(\Pi_2, \Pi_3, \ldots \Pi_{n-m}).$$

Determination of Π Terms

Several methods are used to form dimensionless products or pi terms in dimensional analysis. A systematic procedure known as method of repeating variables is used to decide and form dimensionless and independent pi terms. Following steps explain the application of this procedure (Buckingham, 1914).

Step I: List all the variables including the dependent variable involved in the problem. All the variables must be independent in nature to minimize the number of variables required to describe the complete system

Step II: Express each variable in terms of basic dimensions

Step III: Decide the required number of pi terms. It can be determined by using Buckingham pi theorem, which states that the number of pi terms is equal to n-m, where n is the number of variables including the dependent variable in the problem (determined from Step I) and m the number of reference dimensions required to describe these variables (determined from Step II)

Step IV: Amongst the original list of variables, select variables that can be combined to form pi terms. These are repeating variables; the required number is equal to the number of reference dimensions. When selecting repeat variables, take following points into account.

- ☆ The dependent variable should not be selected as repeating variable
- ☆ Select the repeating variables in such a way that one variable contains geometric property, other variable contains flow property and third variable contains fluid property. For example
  - ❒ Geometric properties: length, diameter, height
  - ❒ Flow properties: velocity, acceleration
  - ❒ Fluid properties: viscosity, density
- ☆ The repeating variable should not form a dimensionless group
- ☆ No two repeating variables should have the same fundamental dimensions.

Step V: To form pi terms, multiply one of the non-repeating variables by the product of the repeating variables each raised to an exponent that will make the combination dimensionless

Step VI: Repeat 'Step V' for each of the remaining non-repeating variables. The resulting set of pi terms will correspond to the required number obtained as indicated in Step III

Step VII: After obtaining the required number of pi terms, make sure that all the pi terms are dimensionless. To check, simply substitute the basic dimension of the variables into the pi terms

Step VIII: The final relationship is of the following form

$$\Pi_1 = \Phi[\Pi_2, \Pi_3]$$

Here, Π_1 contain the dependent variable in the numerator. If the difference in the number of variables for a given problem and number of reference dimensions is 1, then only one pi term describes the phenomena. The functional relationship for one pi term is a constant quantity.

$$\Pi_1 = \text{Constant} \tag{10.23}$$

The actual functional relationship among pi terms or the constant of eq. (10.23) is determined from experiments.

Following examples illustrate the application of Buckingham Π theorem.

Example 10.3

Obtain a dimensionless expression for R, if the resistance, R experienced by a partially submerged body is a function of flow velocity V, length of the body l, viscosity of the fluid μ, density of the fluid ρ and gravitational acceleration g. (Note that this example is the same as 10.2)

Mathematically, the functional relationship between the parameters is of the following form:

$$R = f(V, L, \mu, \rho, g) \tag{10.24}$$

$$0 = f(R, V, L, \mu, \rho, g)$$

Number of variables n = 6

$R = MLT^{-2}$, $V = LT^{-1}$, $L = L$, $\mu = M L^{-1} T^{-1}$, $\rho = M L^{-3}$ and $g = LT^{-2}$

Number of fundamental dimensions, m = 3, [M], [L], [T]

Therefore, number of Π terms = n-m = 6-3 = 3

Choosing length L, velocity V, and density ρ as the 3 repeating variables, we form Π terms as follows.

a.

$$\Pi_1 = L^a V^b \rho^c R \tag{10.25}$$

$$M^o L^o T^o = L^a [LT^{-1}]^b [ML^{-3}]^c MLT^{-2} \tag{10.26}$$

Equating powers of M, L and T on both sides of eq. (10.26)

$0 = c + 1$, $0 = a + b - 3c + 1$, $0 = -b - 2$

Such that c =-1, b=-2 and a =-2

Since $\Pi_1 = l^a V^b \rho^c R$, we have

$$\Pi_1 = L^{-2} V^{-2} \rho^{-1} R = R/L^2 V^2 \rho$$

b.

$$\Pi_2 = L^x V^y \rho^z \mu$$

$$M^o L^o T^o = L^x [LT^{-1}]^y [ML^{-3}]^z ML^{-1}T^{-1} \tag{10.27}$$

Following the earlier procedure, it can be shown that

z = -1, y =-1 and x = -1

$$\Pi_2 = L^{-1} V^{-1} \rho^{-1} \mu = \mu/L V \rho$$

c.

$$\Pi_3 = L^l V^m \rho^n g$$

$$M^o L^o T^o = L^l [LT^{-1}]^m [ML^{-3}]^n LT^{-2}$$

Equating powers of M, L and T we have the values of l, m and n as

l =1, m =-2 and n = 0

$$\Pi_3 = L^l V^{-2} \rho^o g = Lg/V^2$$

The functional relationship can be given as:

$$\Phi[\Pi_1, \Pi_2, \Pi_3] = 0$$

$$\Phi[R/L^2 V^2 \rho, \mu/L V \rho, Lg/V^2] = 0 \tag{10.28}$$

Or

$$R/L^2 V^2 \rho = \Phi[\mu/L V \rho, Lg/V^2]$$

Since reciprocal of Π term and its square root is non-dimensional.

$$R/L^2 V^2 \rho = \Phi[L V \rho/\mu, V/\sqrt{Lg}] \tag{10.29}$$

Note that term $LV\rho/\mu$ is Reynolds number and $V/\sqrt{Lg}$ is Froude's number.

Therefore, it can be stated that non-dimensional resistance term, $R/L^2 V^2 \rho$ is a function of the two non-dimensional numbers. It is comparable to example 10.2.

Example 10.4

For a steady flow of an incompressible Newtonian fluid through a long, smooth walled, horizontal circular pipe, it is required to measure the pressure drop per unit length of the pipe. Find the number of non-dimensional parameters involved in the problem under consideration and establish the functional relation among these dimensionless parameters.

Let us express all the pertinent variables. The pressure drop per unit length of the pipe, Δp depends upon diameter of the pipe D, pressure p, viscosity μ and flow velocity v such that

$$\Delta P = f(D, \rho, \mu, v) \tag{10.30}$$

Or

$$g(\Delta P, D, \rho, \mu, v) = 0$$

No. of variables, n = 5

No. of fundamental dimensions, m = 3

By Buckingham's theorem, number of dimensionless groups = n - m = 5-3 = 2

As per guidelines, we cannot have ΔP, ρ, and v as repeating variables because $\Delta P/pv^2$ is dimensionless. Moreover, ΔP is also a dependent variable. Therefore, variables D, v and ρ are chosen as the repeating variables. Writing the dimensions of all the variables:

$$\Delta P = [ML^{-2}T^{-2}],\ D = [L],\ v = [LT^{-1}],\ \rho = [ML^{-3}] \text{ and } \mu = M L^{-1} T^{-1}$$

$$\Pi_1 = D^a v^b \rho^c \Delta P$$

$$M^o L^o T^o = L^a [LT^{-1}]^b [ML^{-3}]^c ML^{-2}T^{-2}$$

Equating powers of M, L and T on both sides

For M: 1+ c = 0

For L: -2+ a +b -3c =0

For T :-2 – b =0

Such that c = -1, b= -2 and a =1

Thus $\Pi_1 = = D^1 v^{-2} \rho^{-1} \Delta P = \Delta P\, D/v^2 \rho$

Repeating the process for the remaining non-repeating variables, we have

$$\Pi_2 = D^m v^n \rho^p \mu$$

$$M^o L^o T^o = L^m [LT^{-1}]^n [ML^{-3}]^p M L^{-1} T^{-1} \tag{10.31}$$

$$p + 1 = 0$$

$$m + n - 3p - 1 = 0$$

$$-n - 1 = 0$$

The solution of these algebraic equation gives m =-1, n=-1 and p =-1. Therefore,

$\Pi_2 = \mu / D\ v\ \rho$

Since $\Pi_1 = f(\Pi_2)$, we have

$\Delta P\ D/v^2\ \rho = f(\mu/D\ v\ \rho)$

Because $\mu/D\ v\ \rho$ is the inverse of Reynolds number,

$$\Delta P\ D/v^2\ \rho = f(1/Re) \qquad (10.32)$$

Since the equation is dimensionally homogeneous, Buckingham's theorem asserts that, f is not actually a function of five separate variables, but rather a function of a complex set of dimensionless products of the variables comprising of $P = \Delta PD/\mathbf{v}^2\rho$, and the inverse of Reynolds number, $1/R = \mu/VD\rho$. This relationship may be indicated in the explicit form, $P = f(1/R)$. Practically speaking, to understand and establish the relationships between ΔP and other four independent variables, one needs to conduct a number of experiments by varying one variable in a particular range keeping the other 3 variables as constant. Repeat this step for all other variables within the range in which each of these variables are likely to vary. But for the dimensionless form of the equation (eq. 10.32), only one experiment will suffice to establish the relationship between the two dimensionless variables one on the right and other on the left hand side of the equation.

DIMENSIONLESS NUMBERS IN FLUID DYNAMICS

In dimensional analysis, a dimensionless quantity is a quantity without an associated physical dimension. Dimensionless quantities are often defined as products or ratios of quantities that are not dimensionless, but whose dimensions cancel out when their powers are multiplied. For instance, the engineering strain is a dimensionless quantity. Being a measure of deformation, it is defined as change in length over initial length. Length dimension in the numerator and denominator cancel out.

Importance

1. Dimensionless numbers in engineering and physics helps to reduce variables in the functional relation. They reduce the amount of experimental data required to make correlations of physical phenomena to scalable systems.It means lesser experimental runs.
2. Dimensionless forms allow presenting the solution in a compact way.
3. Dimensionless numbers often correlate with some performance parameter and greatly aid engineering analysis and design.
4. The values of the dimensionless numbers often reflect certain properties. For example, a flow problem with a low Reynolds number will be laminar, a larger value will imply turbulent behaviour.
5. The numbers of dimensionless parameters determine the dimensionality of the space of solutions. For example, if a problem has two dimensionless numbers, the problem behaviour can be inferred by varying these numbers alone.

6. A dimensionless number can be used in the analysis of prototype models, to predict the behaviour of a similar full-scale system. Dimensionless numbers help to compare two systems that are vastly different by combining the parameters of interest. For example, the Reynolds number, Re = velocity × length/kinematic viscosity. If an airfoil has to be tested using a scaled-down model with a particular Re, one could increase fluid velocity or lower kinematic viscosity (change fluids) or vary both for the same Re. It ensures that the system is being tested under comparable circumstances.

Dimensionless Numbers

Dimensionless numbers in fluid mechanics are a set of dimensionless quantities, which describe as ratios of the relative magnitude of fluid and physical system characteristics. Although a large number of dimensionless numbers are listed in various texts (Salih, 2017), but none of these may be the complete list of such numbers. On the other hand, in fluid mechanics inertia force always exists if there is any mass in motion. Dividing this inertia force with other forces like viscous force, gravity force, surface tension, elastic force, pressure force, or vice-versa in some cases gives us some dimensionless numbers. Five such numbers that have inertia force either in the numerator or in the denominator are:

- Reynolds number: Inertia force/viscous force
- Froude number: Inertia force/gravitation force
- Weber number: Inertia force/surface tension force
- Mach number: Inertia force/compressible force
- Euler number: Pressure force/inertia force
- Strouhal number: Inertia forces (local acceleration) to inertia forces (convective acceleration)

This section discusses these numbers to illustrate the importance of dimensionless numbers.

Reynolds number (R_e)

Reynolds number, Re is a measure of the ratio of inertial forces to viscous forces. It quantifies the relative importance of these two forces for a given flow conditions.

Inertia force, F_i = Mass × acceleration = density × volume × velocity/time

Inertia force F_i = density × (Area × Length) × velocity/time

Since length/time = velocity

Inertia force $F_i = \rho A V^2$

Viscous force (F_v) = Shear stress × area = μ × (velocity/length) × Area

$$F_v = \mu V A/L$$

Thus

$$Re = \text{Inertia force/Viscous force} = \rho\ A\ V^2/(\mu\ V\ A/L)$$

$$Re = \rho\ V\ L/\mu = V\ L/\nu \tag{10.33}$$

Here, V is the flow velocity, L is a characteristic linear dimension, (travelled length of the fluid or hydraulic diameter *etc.*), ρ fluid density (kg/m^3), μ dynamic viscosity (Pa-s), ν kinematic viscosity (m^2/s); $\nu = \mu/\rho$.

For pipe flow, characteristic linear dimension, L is the diameter of the pipe, D. For fluids flowing through non-circular closed conduits and open channels, L is replaced with D_h, the hydraulic diameter. Small values of Re indicates that the viscous forces are dominant compared to inertia forces. Such flows are creeping/ viscous flows. Conversely, for large values of Re, inertia forces are dominant the flows are known as inviscid flows. Readers may refer Chapter 4 and Chapter 8 for additional information on this parameter including guidelines to characterize the flow as laminar, transitional and turbulent for pipe and open channel flows.

Froude's number (Fr)

Froude's number (Fr) is the square root of the ratio of inertia force to the force due to gravity. Inertia force derived earlier after incorporating $A = L^2$ is used here.

Inertia force = = $\rho\ L^2\ V^2$

Gravity force = mass × gravitational acceleration = $m \times g = \rho \times \text{Volume} \times g = \rho \times L^3 \times g$

$$\text{Froude number, } Fr = [\rho\ L^2\ V^2/(\rho \times L^3 \times g)]^{1/2} = V/\sqrt{Lg} \tag{10.34}$$

This number is significant for flows with free surface effects such as open-channel flows. In such flows, the characteristics length is the depth of water. Froude number governs the dynamic similarity under flow situations where gravitational force dominates and other forces are relatively small and negligible. Examples are:

- Flow through open channels
- Flow of liquid jets from orifices
- Flow over notches and weirs
- Flow over the spillway of a dam.

Readers may refer Chapter 8 on guidelines to characterize flow on the basis of Froude number and detailed discussion on hydraulic jump.

Weber number (W)

Weber number (W) is square root ratio of the inertia force to the force of surface tension.

Inertia force = $\rho\ L^2\ V^2$

Surface tension is the attractive force in liquids that pulls surface molecules into the rest of the liquid, minimizing the surface area (see Chapter 1). The general formula for measuring surface tension is: Surface tension = (surface force)/(length on which force acts). It is expressed as

$$\gamma = F/L$$

Such that γ is the surface tension per unit length, F is the force applied on the liquid and L is the length where the force acts. Accordingly, surface tension force is given as γ L. Weber number, W is

$$W = [\rho L^2 V^2/(\gamma L)]^{1/2} = V/\sqrt{(\gamma/\rho L)} \quad (10.35)$$

Weber number finds applications in the following types of flows.

- ☆ Flow in capillary tubes
- ☆ Thin sheet flow
- ☆ Liquid atomization

This number is an index of droplet formation and flow of thin film liquids with an interface between two fluids. Flow of water in a river is a case where the inertia force is dominant compared to surface tension force.

Mach number (M_a)

The Mach number is the ratio of flow velocity after a certain limit of the sound speed. In simple words, it is the ratio of the speed of a body to the speed of sound in the surrounding medium. It is the key parameter characterizing the compressibility effects in a fluid flow. Mathematically

$$M_a = V/c = V/(dp/d\rho)^{1/2} = V/(E_v/\rho)^{1/2} \quad (10.37)$$

Here V is the velocity of the flow, c is the local sonic speed or speed of sound in that medium, ρ is the density of the fluid and E_v is the bulk modulus. To explain it further, the speed of sound is equated to Mach 1 speed, which is the sonic speed in that medium. Thus, Mach 0.75 will be 75 per cent of the speed of sound, called as subsonic. Mach 1.65 is 65 per cent faster than the speed of light, and called supersonic. The other commonly used terms are as follows.

Regime	*Subsonic*	*Transonic*	*Sonic*	*Supersonic*	*Hypersonic*	*Hyper velocity*
Mach	< 0.8	0.8–1.3	1.0	1.3–5.0	5.0–10.0	>10.0

Sometimes, square of the Mach number is called Cauchy number, C_a. It is derived as follows.

$$C_a = \text{Inertia force/Compressibility force} = m\,a/\rho c^2 L^2 = \rho L^2 V^2/\rho c^2 L^2 = V^2/c^2.$$

$$C_a = M_a^2 = \rho V^2/E_v \quad (10.38)$$

Both the numbers are predominantly used in problems where fluid compressibility is important. Flow is incompressible as long as the Mach number is less than 0.3 regardless of the fluid at hand. Compressibility of the fluid in that case can be neglected. Most common use of this number is in compressible fluid flow problems, particularly in the field of gas dynamics and aerodynamics.

Euler number (E_u)

Euler's number, E_u is the ratio of pressure force to inertia force. It is used to analyze flow dynamics problems where pressure difference between two points is important. In most aerodynamic model testing, the pressure data are mathematically expressed as:

$$E_u = \Delta p/(\rho V^2/2) \tag{10.36}$$

Here E_u is the dimensionless Euler number, Δp is the difference in local pressure, say drop caused by an obstruction in the flow and free stream pressure. Other parameters are same as earlier described. A perfect frictionless flow corresponds to an Euler number 1.0. The denominator in Eq. (10.36) is the dynamic pressure. Therefore, E_u is also termed as the pressure coefficient, C_p. In cavitation studies, Δp is the difference in liquid stream pressure and liquid-vapour pressure. The dimensional parameter in that case is the known as the cavitation number.

Strouhal number (S_t)

It is a dimensionless parameter having applications in unsteady, oscillating flow problems. It is defined as

$$S_t = \omega L/V \tag{10.39}$$

Here L is the characteristics length descriptive of the flow field and ω is the frequency of oscillation. This number is the measure of the ratio of the inertial forces due to unsteadiness of the flow (local acceleration) to inertia forces resulting from changes in velocity from one point to another point in the flow field (convective acceleration). This kind of unsteady flow develops for fluids flowing past a solid body placed in the moving stream.

HYDRAULIC SIMILITUDE

Prototype and Model

Prototype is a physical structure for which engineering design is required. Model on the other hand, is the scaled down representation of the physical structure. Models are commonly used to predict as to how the prototype is likely to perform under actual field conditions. A model, usually smaller than the prototype, less expensive and easy to construct and operate, is used to conduct laboratory studies. In some cases, when amplified and more focused studies are undertaken, the size of model can even be larger than the prototype.

Kinds of Models

Commonly used models are of two types.

Undistorted models

These models are geometrically similar to their prototype. It means the scale ratio for the linear dimensions of the model and its prototype are the same.

Distorted models

These models are not geometrically similar to their prototypes. In other words, the scale ratio for the linear dimensions of the model and its prototype are not the same. These models are used where horizontal and vertical dimensions vary widely *i.e.* length is much larger than width and depth of flow, say a river. The depth of the water in the undistorted model in such cases becomes too small to accurately measure its variations. In such case, distorted models prove quite useful.

Importance of Model Studies

- ☆ Most of the equipment and hydraulic structures like dams, canals *etc.* require lot of investments and time to construct. Scale-down model tests are performed to understand their performance. It saves significant time and resources. If the model for some or the other reason fails to perform as per anticipation, it will not have financial implications for investors. The model can be easily re-designed for testing, if required
- ☆ In many fields, new designs have high uncertainties because of some unexpected problems. Model studies as a part of the product design allow engineers and designers to explore several design alternatives, test new theories and confirm the performance before going into production of the new product
- ☆ Sometimes mathematical relationships and equations are not enough to design an engineering product. A model provides valuable data on its geometrical appearance, force and pressure distributions, capacity and performance, *etc.* Any problem encountered during tests can be rectified before actual design and commissioning of the prototype.
- ☆ It is possible to test alternate plans and modifications within a relatively short time simulating all expected and unexpected situations. The design engineers can observe conditions resulting from a particular arrangement and satisfy themselves as to the adequacy of the plan.
- ☆ The cost of model studies varies with area of study, characteristics of the streams, nature of the problem, and number of plans and alternate plans to be tested before an acceptable solution is developed. The cost of model studies may usually be less than 0.10 per cent of the cost of the project, a small price to pay for the assurance that the most practical and economical design has been developed.

Examples of model studies

Hydraulic structures like dams, reservoirs, spillways *etc.* require model testing before finalizing their designs. Similarly, automobiles, planes, and rockets require model testing in wind tunnels.

SIMILITUDE

Similitude in a general sense is the indication of a known relationship between a model and prototype. The model tests must yield data that can be scaled-up to obtain design parameters for the prototype (Buckingham, 1915). The results obtained from model experiments can be scaled to the prototype only if a complete similarity exists between the model and the prototype. For that to happen, the model and the prototype must be similar in the following senses.

- Geometrically similar
- Kinematically similar
- Dynamically similar

Geometric Similarity

A model and a prototype is geometric similar if and only if all body dimensions in all the three coordinates have the same linear-scale ratio. All linear dimensions of the model should be related to corresponding dimensions of the prototype by a constant scale factor. It requires that the model and the prototype be of the same shape. Usually, one or more of these terms will involve ratios of important lengths, which are purely geometrical in nature. Thus, for geometric similarity

$$l_m/l_p = b_m/b_p = h_m/h_p = d_m/d_p = L_r \text{ (scale ratio)}$$

Here, l_m, b_m, h_m, and d_m are the respective dimensions of the model length, breadth, height and diameter respectively and l_p, b_p, h_p and d_p are the corresponding linear dimensions of the prototype. L_r is a constant known as scale ratio or the scale factor.

It emerges that

$$A_r = A_m/A_p = (l_m \times b_m)/(l_p \times b_p) = L_r^2$$

Such that A_r is the area scale ratio and A_m and A_p are the respective areas of the model and the prototype.

Similarly, the volume scale ratio can be written as:

$$V_r = V_m/V_p = (l_m \times b_m \times h_m)/(l_p \times b_p \times h_p) = L_r^3$$

Such that V_r is the volume scale ratio and V_m and V_p are the respective volumes of the model and the prototype.

Kinematic Similarity

The motion in the two systems are kinematically similar, if homogeneous particles lie at homogeneous points at homogeneous times. In a specific sense, the velocities at corresponding points are in the same direction and are related in magnitude by a constant scale factor. This also requires that streamline patterns must be related by a constant scale factor. The kinematically similar flows must be geometric similar because boundaries form the bounding streamlines. The factors like compressibility or cavitations must be taken into account to maintain the kinematic similarity.

Let points 1 and 2 represent the homologous (corresponding) points in the model and prototype. The conditions of kinematic similarity are as follows:

$$V_{1m}/V_{1p} = V_{2m}/V_{2p} = V_r \text{ (Velocity ratio)}$$

Since geometric similarity is a must for kinematic similarity, we can write

$$V_m/V_p = (l_m/T_m)/(l_p/T_p) = L_r/T_r$$

Here T_r is the time scale ratio given as T_m/T_p.

Similarly, acceleration and discharge scale ratios a_r and Q_r can be written as:

$$a_r = a_m/a_p = (l_m/t_m^2)/(l_p/t_p^2) = L_r/t_r^2$$

$$Q_r = Q_m/Q_p = (l_m^3/T_m)/(l_p^3/T_p) = L_r^3/T_r$$

Dynamic Similarity

The two flows are said to be dynamically similar is dynamic similarity means the similarity of forces between model and prototype. two flows have force distributions in such a manner that identical forces are parallel and are related in magnitude by a constant scale factor at all corresponding points. Thus, dynamic similarity exists between the model and the prototype, if the ratios of the forces acting at the corresponding points in the model and prototype are equal. Besides, these forces must act in parallel direction.

$$F_{m1}/F_{p1} = F_{m2}/F_{p2} = F_r$$

Such that F_r is the force scale ratio and F_m and F_p are the respective forces on the model and the prototype. For a model and prototype, the dynamic similarity exists, when both of them have same length-scale ratio, time-scale ratio and force-scale ratio (or mass-scale ratio). It means for dynamic similarity between the model and prototype, the dimensionless numbers should be same for the model and the prototype. It may be quite difficult to satisfy the condition that all the dimensionless numbers are same for the model and prototype. Therefore, models are designed to ensure that the dominating dimensionless numbers are same. Following are the dynamic similarity laws for various kinds of flow.

- Reynolds model law: Pipe flow, sub-marines, and aeroplanes *etc.*
- Froude model law: Free-surface flow, jet from orifice or nozzle *etc.*
- Euler model law: The pressure force is dominant
- Weber model law: When surface tension is dominant
- Mach model law: velocity of flow is comparable to the velocity of sound; compressible flow

Scale ratios for various characteristics are given in Table 10.1 for direct application. As an illustration, the scale ratios of few characteristics for the Froude number are derived, which are also incorporated in Table 10.1.

$$Fr = V/\sqrt{Lg}$$

$$V_r = \sqrt{L_r g_r}$$

$$\text{Time ratio} = L_r/V_r = L_r/\sqrt{L_r g_r} = (L^{1/2} g^{-1/2})_r$$

$$Q_r = V_r A_r = L_r^2 \sqrt{L_r g_r} = L_r^{5/2}\, g_r^{1/2}$$

$$\text{Force} = \text{Density} \times \text{Volume} \times \text{Acceleration} = \rho_r\, L_r^3 g_r$$

Table 10.1. Flow Characteristics and Similitude Scale Ratios

Characteristic	*Dimension*	*Scale Ratios for Laws of*	
		Reynolds	*Froude*
Geometric			
Length	L	L_r	L_r
Area	L^2	L_r^2	L_r^2
Volume	L^3	L_r^3	L_r^3
Kinematic			
Time	T	$(L^2\rho/\mu)_r$	$(L^{1/2} g^{-1/2})_r$
Velocity	LT^{-1}	$(\mu/L\rho)_r$	$(L^{1/2} g^{1/2})_r$
Acceleration	LT^{-2}	$(\mu_2/L^3\rho^2)_r$	g_r
Discharge	L^3T^{-1}	$(L\mu/\rho)_r$	$(L^{5/2} g^{1/2})_r$
Dynamic			
Mass	M	$(L^3\rho)_r$	$(L^3\rho)_r$
Force	MLT^{-2}	$(\mu^2/\rho)_r$	$(L^3g\rho)_r$
Pressure	$ML^{-1}T^{-2}$	$(\mu^2/L^2\rho)_r$	$(Lg\rho)_r$
Impulse and momentum	MLT^{-1}	$(L^2\mu)_r$	$(L^{7/2}g^{1/2}\rho)_r$
Energy and work	ML^2T^{-2}	$(L\mu^2/\rho)_r$	$(L^4g\rho)_r$
Power	ML^2T^{-3}	$(\mu^3/L\rho^2)_r$	$(L^{7/2}g^{3/2}\rho)_r$

Note: The notations in this table have the same meaning as given earlier in this chapter.

Reynolds/Froude Model Laws

$$[Re]_m = [Re]_p = V_m\rho_m d_m/\mu_m = V_p\rho_p d_p/\mu_p$$

Froude's model law

$$[Fr]_m = [Fr]_p = V_m/\sqrt{(g_m L_m)} = V_p/\sqrt{(g_p L_p)}$$

Besides what is stated in the foregoing write-up in this section, following rules may be appropriately applied.

- ☆ For compressible flows, Reynolds number, Mach number and specific heat ratio are correspondingly equal in the model and the prototype
- ☆ For incompressible flows with no free surface, model and prototype Reynolds number are equal.
- ☆ For incompressible flows with free surface, Reynolds number, Froude number, Weber number and cavitation numbers for the model and the prototype must match.

MODELLING AND SCALING LAWS

Scaling laws are essentially required to predict the prototype behaviour using measurements taken from a model study. Consider the problem of drag force that a smooth spherical body experiences in a stream of compressible fluid. The functional relationship between five variables namely, the drag force, F (MLT^{-2})

of a body moving through a fluid is dependent upon density ρ (ML^{-1}), velocity V (LT^{-1}), diameter D (L) and viscosity μ ($ML^{-1}T^{-1}$) is written as:

$$f(F, \rho, V, D, \mu) = 0 \tag{10.40}$$

Using the Bukingham's equation, one can show that

$$R/\rho V^2L^2 = F(\rho VL/\mu) \tag{10.41}$$

This applies whatever the size of the body *i.e.* it is applicable to the prototype as well as its geometrically similar model. For the model and the prototype, one can write:

$$R_m/\rho_m V^2_m L^2_m = F(\rho_m V_m L_m/\mu_m) \tag{10.42}$$

$$R_p/\rho_p V^2 pL^2_p = F(\rho_p V_p L_p/\mu_p) \tag{10.43}$$

Dividing eq. (10.42) by eq. (10.43), we have

$$[R_m/\rho_m V^2_m L^2_m]/[R_p/\rho_p V^2_p L^2_p] = [F(\rho_m V_m L_m/\mu_m)]/[F(\rho_p V_p L_p/\mu_p)]$$

It is difficult to proceed further unless we assume dynamic similarity. For dynamic similarity, Reynolds numbers are the same for both the model and prototype. Therefore,

$$[R_m/\rho_m V^2_m L^2_m]/[R_p/\rho_p V^2_p L^2_p] = 1 \tag{10.44}$$

$$R_m/R_p = \rho_m V^2_m L^2_m/[\rho_p V^2_p L^2_p] \tag{10.45}$$

Thus, scaling law for the resistance force is given by eq. (10.45)

In terms of scaling ratio, the scaling law for resistance of the body is given by the following relation:

$$R_r = \rho_r V_r^2 L_r^2 \tag{10.46}$$

Similar scaling laws are derived for use in case-to-case basis covering various fluid flow problems.

QUESTIONS (THEORY)

1. What do you understand by dimensional homogeneity and consistency? Use an equation to illustrate your answer.

2. Define dimensional variables, dimensional constants and pure constants. You can use an equation to explain your answer.

3. Name the two methods commonly used to carry out dimensional analysis. Explain the stepwise procedure of Rayleigh method. What are its limitations?

4. Explain the stepwise procedure for dimensional analysis using Bukingham's pi theorem.

5. What do you understand by dimensionless numbers? How dimensionless numbers are important in fluid mechanics?

6. Write short notes on the following Numbers: Reynolds number, Froude number and Mach number

7. What do you understand by a mathematical and physical model? Differentiate between a undistorted and distorted physical model.

8. Explain the geometrical, kinematic and dynamic similarity of models and prototypes.

Turbo-Machines

Chapter 11

Introduction to Fluid Machines

Engineers often deal with turbines and pumps in hydraulics and other engineering applications, notably in civil, environmental and industrial engineering. As such, engineers need to understand how the hydraulic machines work, and how to select appropriate machinery for a particular activity. This chapter introduces the subject of fluid machinery. This chapter initiates introduction to the turbine and pumps, while detailed discussions on pumps follow in the next two chapters.

WHAT IS FLUID MACHINE?

The energy stored in a fluid mass is in the form of potential, kinetic and intermolecular energy. A fluid machine is a device, which converts this stored energy into mechanical energy or *vice versa*. The fluid machines use either liquids or gases as the working fluid depending upon the purpose. Machines using liquid (mainly water, for almost all practical purposes) are termed as hydraulic machines. If fluid does work on the rotor of the machine, the machine is known as a turbine. On the other hand, if the rotor of the machine does work on the fluid element then the machine is known as pump. If gas is used as the fluid then the machine is known as compressor, fan or blower. The machine producing mechanical energy is known as reciprocating engine while the machine developing energy of the fluid from the mechanical energy is known as reciprocating pump or reciprocating compressor.

TURBINES

Newton's Law implies that a force is required to change the momentum of the fluid. Conversely, a change in the momentum of fluid generates the force. For every meter of descend, 1 cubic meters of water gives about 9800 Joules of mechanical energy. Similarly, a fall of 1 meter flow at the rate of one cubic meter per second is capable to produce 9800 W of power. A hydraulic turbine uses this waterpower available in the form of potential and kinetic energy and converts it into usable mechanical energy. In a hydraulic turbine, blades or buckets provided on a wheel directed against water alter its momentum. The resulting force turns the shaft of the wheel, which in turns operates the generator to produce power. A turbine producing power using water are known as water turbines or hydraulic turbines. Turbines operating on gases are steam turbine, gas turbine, and air turbine depending upon whether they use steam, gas (the mixture of air and products of burnt fuel in air) or air.

Hydraulic turbines defined as the prime movers that transform the kinetic energy of the falling water into mechanical energy of rotation, whose primary function is to drive an electric generator. Hydropower generation is essentially a controlled method of water descent, which is usefully utilized to generate power. To manage such a wide range of heads, varying from a few meters to as much as 2000 m, many different kinds of turbines differing in their working components are employed (Gasoriek, 2008). The main components of a hydroelectric generation system can be grouped under two groups as follows (Figure 11.1):

- The hydraulic system components, which include the turbine, the associated conduits like penstocks, tunnel and surge tank and their control system
- The electric system components formed by the synchronous generator and its control system.

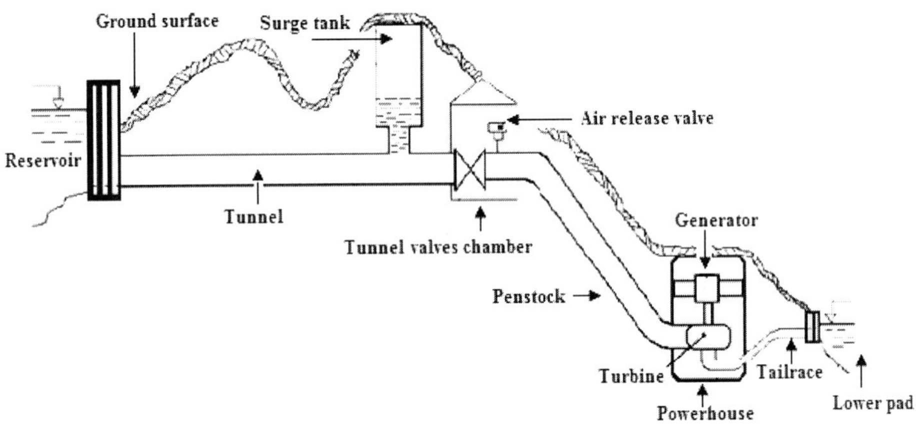

Figure 11.1. Schematic Layout of a Hydroelectric Plant with Surge Tank.

Classification of Turbines

A number of classification criteria are used to classify turbines. Five of these criteria are:

- Type of energy at the inlet
- Direction of flow through the runner
- Head at the inlet of the turbine

☆ Specific speed of the turbine
☆ Disposition of shaft of runner

Type of energy at the inlet

Based on the type of energy at the inlet, turbines are classified as impulse turbines and reaction turbines. In the impulse turbine *e.g.* Pelton turbine, the total head of the incoming fluid is converted into a large velocity head at the exit of the supply nozzle. It means the entire available energy of the water turns into kinetic energy. In reaction turbine *e.g.* Francis and Kaplan turbines, the rotation of the runner or rotor is partly due to impulse action and partly due to change in pressure over the runner blades. The main differences between an impulse and a reaction turbine are listed in Table 11.1.

Table 11.1. Characteristics of Impulse and Reaction Turbines

Impulse Turbine	*Reaction Turbine*
The entire available energy of the water is converted into kinetic energy before use fluid enters the turbine runner	Only a portion of the fluid energy is converted into kinetic energy before the fluid enters the turbine runner
The work is done only by the change in the kinetic energy of the jet	The work is done partly by the change in the velocity head, but almost entirely by the change in pressure head
Flow regulation is possible without loss	It is not possible to regulate the flow without loss
Unit is installed above the tailrace	Unit is entirely submerged in water below the tailrace
Casing does not perform any hydraulic function because the jet is unconfined and is at atmospheric pressure. Thus, casing serves only to prevent splashing of water	Casing is necessary because the pressure at inlet to the turbine is much higher than the pressure at outlet. Unit has to be sealed from atmospheric pressure
It is not necessary that wheel should run full, air has free access to the buckets	Water completely fills the vane passage

Direction of flow through the runner

According to the direction of flow through the runner, turbines are classified as tangential flow turbines, radial flow turbines, axial flow turbines and mixed flow turbines. In a tangential flow turbine, the flow is tangential to the wheel circle. The Pelton turbine is a tangential flow turbine. In a radial flow turbine, the path of the flow of water remains in the radial direction and in a plane normal to the runner shaft. The radial flow turbines are further classified as inward and outward flow turbines. In the former, flow is inward from periphery to center (centripetal type). In the latter type the flow is from center to the periphery (centrifugal type). No pure radial flow turbine is in use these days. In an axial flow turbine, the path of flow remains parallel to the axis of the shaft. For the axial flow reaction turbine, the shaft of the turbine is vertical. The lower end of the shaft, known as hub, is larger. The vanes are fixed on the hub, which also acts as runner for an axial flow reaction turbine. The important axial flow turbines are the propeller and the Kaplan turbines. If the vanes fixed on the hub are non-adjustable, then the turbine is known as propeller turbine. If vanes on hub are adjustable then the turbine is a Kaplan turbine. This type of turbine is quite suitable where large quantities of water are available at low heads. When there is gradual change of flow in the runner from radial to axial, the flow is called mixed flow and the turbine as mixed flow turbine. For example, in a Francis turbine water flows radially but exits axially.

Based on head and quantity of water

The turbines are classified as high, medium and low head turbines based on head and quantity of water available. A high head turbine works under heads exceeding 250 m. The quantity of water required in a high head turbine is usually small. The Pelton turbine is a high head turbine. Medium head turbine works under heads in the range of 45 m to 250 m, requiring medium flows of water. Francis turbine works under the medium head conditions. Low head turbine, such as Kaplan turbine, works under a head of less than 45 m. Large quantity of water is required because of low heads.

Based on specific speed of turbine

The specific speed of a turbine is the speed of a geometrically similar turbine, which would produce unit power (one kilowatt) working under a unit head (one meter). The manufacturer provides information on specific speed along with other ratings. It always refers to the point of maximum efficiency. A turbine for a particular site is often selected in consultation with the manufacturer through a process of trial and error. Criteria for selection of hydro-turbine based on specific speed as per IS: 12837 are given in Table 11.2 (BIS, 2000).

Table 11.2. Criteria for Selection of Turbines (Note the specific speed for various kinds of turbines)

Type of Machine	*Head Variation (Per cent of rated head)*	*Load Variation (Per cent of rated output)*	*Specific Speed (m-mhp)*	*Peak Efficiency (Per cent)*
Pelton	120 to 80	50 to 100	15 to 65	90
Francis	125 to 65	40 to 100	60 to 400	93
Deriaz	125 to 65	50 to 100	200 to 400	92
Kaplan	125 to 65	40 to 100	300 to 800	92
Propeller	110 to 90	90 to 100	300 to 800	92
Bulb	125 to 65	40 to 100	600 to 1200	92

Based on disposition of shaft of runner

Disposition of shaft of the runner determines whether the turbine is horizontal or vertical. Usually, Pelton turbines are set-up with horizontal shafts, whereas other types have vertical shafts. Whereas horizontal shaft turbines are suitable where water availability is less, vertical shaft multi-jet turbines suit to large flow installations.

The Pelton Wheel

The Pelton Wheel or Pelton turbine is named after an American engineer Laster A. Pelton who has largely contributed to its development around the year 1880 (Boyle, 2004). It is the only impulse hydraulic turbine in common use now a days. Major parts of a Pelton wheel turbine are shown in the schematic diagram (Figure 11.2).

1. Nozzle and spear
2. Runner (Rotor) and buckets
3. Casing
4. Breaking jet

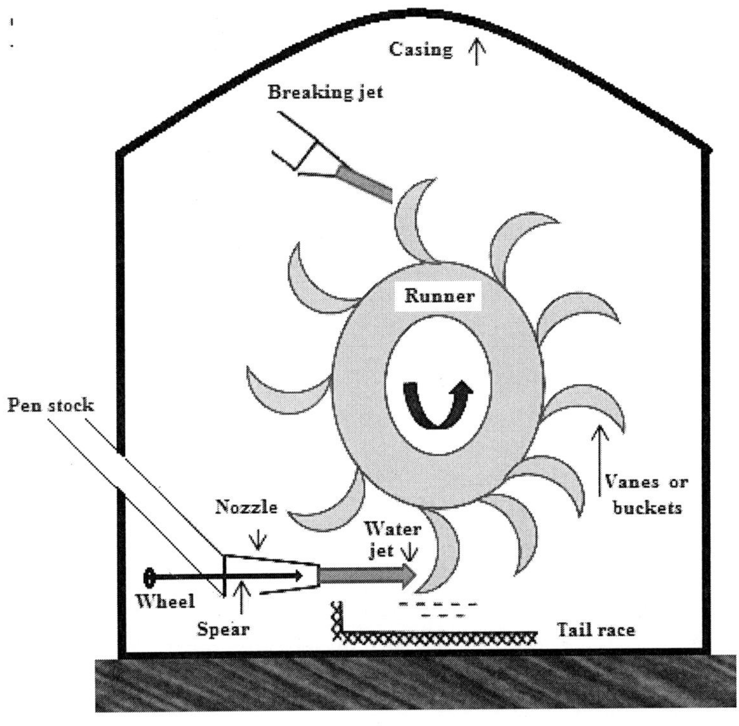

Figure 11.2. Construction Details of a Pelton Wheel.

Nozzle and spear

Nozzles increase the kinetic energy of the water striking the buckets or vanes attached to the runner. A hand wheel or an automatic mechanism operates spear installed inside the nozzle. Movement of the spear in backward and forward directions increases or decreases the rate of water flow respectively. While the numbers of jets in a horizontal shaft turbine are limited to two, for a vertical shaft turbine, it may go up to six.

Runner/rotor and buckets

Runner or rotor, a kind of circular disc, is the rotating part of the turbine. A number of evenly spaced buckets are installed on its periphery. The jets of water discharged by the pressure nozzles drive the wheel. The nozzles are set in such a manner that each nozzle directs a jet along the tangent to the circle passing through the center of the bucket.

The buckets, made of cast iron, cast steel, bronze or stainless steel, are in the form of two hemispherical bowls joined together. Down the center of each bucket is a splitter ridge that divides the jet into two equal streams. The jets flow round the smooth inner surface of the buckets and leave with a relative velocity almost opposite to the direction of the original jet. Although exact 180° deflection of the water jet is desirable to exert maximum driving force on the wheel, but water leaving a bucket in that case may hit the back of the following bucket. It is why; the deflection is always limited to about 160° to 170°. The number of buckets, N on the runner are calculated by the following empirical relation.:

N = 15 +53/Specific speed

Casing

Casing, made of cast iron or fabricated steel plates, is the outer covering of the turbine. Besides acting as a safeguard in the event of any accident, it prevents the splashing of the water. It also guides and discharges the water to the tailrace.

Breaking jet

Spear pushed in the forward direction completely closes the jet of water that in turn stops the water that strikes the runner. Still, the runner keeps on rotating due to its inertia. To stop the runner in the shortest possible time, a small nozzle directs the jet of water at the back of the vanes. The jet of water used to stop the runner is known as the breaking jet.

Working

The water stored into the dam flows through the penstock to the nozzles of the Pelton turbine. Nozzles increase the kinetic energy besides forming jets of water that strike the buckets of the runner. As a result, runner rotates at high speed. A generator attached to the shaft of the runner converts the mechanical energy into electrical energy.

Efficiency

Studies have revealed that conversion of kinetic energy of the jet to mechanical energy of rotation is maximum, when the blade to jet speed ratio is 0.5 (Boyle, 2004). Because of friction in bearings and between the wheel and the atmosphere in which it rotates, the overall efficiency reaches its peak when the blade to jet speed ratio is 0.46, *i.e.* slightly less than the theoretical value of 0.5. Pelton wheel is quite efficient with overall efficiencies in the range of 85-90 per cent. Higher efficiencies can be achieved through:

- ☆ Proper design of buckets, the length, width, and depth of the buckets about 2.5, 4 and 0.8 times the jet diameter
- ☆ Making the bucket surfaces smooth
- ☆ Using notched buckets for smooth entry of the jets

Advantages

- ☆ With an overall high efficiency, it is the most efficient of hydro turbines under the high head and low discharge category
- ☆ It is easy to assemble
- ☆ It operates with a very flat efficiency curve
- ☆ Each bucket splits the water jet in half, thus balancing the side-load forces or thrust on the wheel and the bearings
- ☆ It has a tangential flow, which means that it can have either axial or radial flow
- ☆ No chance of cavitation since water jet strikes only a specific portion of the runner
- ☆ It has fewer parts as compared to Francis turbine having both fixed and guide vanes
- ☆ The whole process of water jet striking the runner and leaving the turbine occurs at atmospheric pressure.

Limitations

- ☆ Only suits high head situations say more than 300 m
- ☆ The efficiency decrease very quickly with time
- ☆ The size of the turbine, runner, generator and powerhouse are large especially under low heads
- ☆ Because of high heads, it is difficult to control the variation in the operating head.

Francis Turbine

Francis turbine, developed by James B. Francis, is a reaction turbine *i.e.* the fluid entering the runner has both the pressure and kinetic energies. The pressure energy is converted to kinetic energy in the runner. It is the most commonly used water turbine today. Since both the reaction and impulse forces are used in the turbine, at the same water head it is better than an impulse or a reaction turbine. Main components of Francis turbine are as follow (Figure 11.3):

1. Spiral casing
2. Stay or stationary vanes
3. Guide vanes
4. Runner blades
5. Draft tube

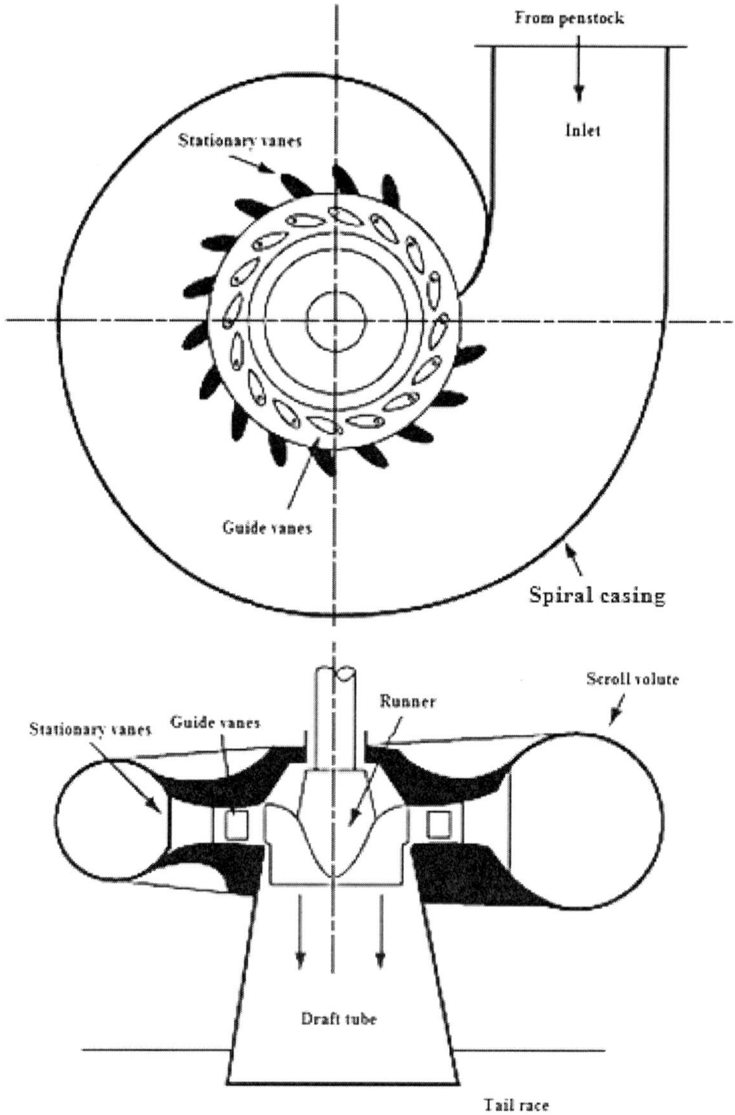

Figure 11.3. Construction Details of a Francis Turbine.

Spiral casing

Like a centrifugal pump (See Chapter 13), the spiral casing has a uniformly decreasing cross- section area, along the circumference. It helps to ensure that the water strikes the runner blades with uniform velocity.

Stationary or stay vanes

The basic purpose of the stationary and guide vanes is to convert a part of pressure energy of the fluid at its entrance to the kinetic energy and then to direct the fluid on to the runner blades at an angle appropriate to the design. Stay vanes reduce the swirling of water due to radial flow, as it enters the runner blades. It helps to increase the efficiency of the turbine.

Guide vanes

Water after passing through stay vanes, enters the runner blades while gliding through guide vanes. Angle of the guide vanes can be changed to control the angle at which water strikes the runner blades. These vanes also regulate the flow rate into the runner blades so that power output is controlled as per the load on the turbine.

Runner blades

Design of runner blades of a mixed flow turbine is such that the upper parts of the blades use the reaction force of water flowing through it. The lower halves of the blades, in the shape of a small bucket, use the impulse action of water. Apparently, the design of the runner blades controls the overall performance of a Francis turbine.

Draft tube

Draft tube, attached to the central axis of the runner, has an increasing cross-section area along its length. It also connects the runner exit to the tailrace. Since the water coming out of runner blades is at considerably low pressure, its expanding cross-section area helps to recover some pressure as it flows towards the tailrace.

Working of Francis turbine

Water, passing through stationary and guide vanes, enters the turbine through spiral casing and the runner blades. In a mixed flow type turbine, water enters the turbine radially and leaves the turbine axially (Gasoriek *et al.*, 2008). Since the runner blades are stationary and cannot pitch or change their angles, guide vanes control the angle at which water strikes the blades. The Francis turbine is quite efficient for heads in the range of 50-400 m and water flow rates in the range of 10-700 m^3/s. The design of the machine such as radius of the runner, angle of vanes and the size of turbine as a whole depends upon the available head, flow rate and the kind of application. The water coming out of runner has lost most of its kinetic and the pressure energies. Although, draft tube recovers some pressure, still the recovered pressure may not be enough to stop the air to enter into the runner housing. Thus, cavitation problem continues to plague the Francis turbine.

Advantages

- Its control is easy even with variable heads
- Runner size is small
- The efficiency does not vary much with time.

Disadvantages

- Muddy water can create problems
- Maintenance is relatively difficult
- Cavitation continues to be a problem

Kaplan Turbine

Kaplan turbine, developed by an Austrian Professor Viktor Kaplan in 1913, is a propeller type (Blades are like a propeller), axial flow water turbine (Figure 11.4). The machine having the combination of automatically adjusted propeller blades and wicket gates is able to achieve high efficiencies over a wide range of water flows and water levels (Admin, 2009, E-Max Group of Institution, 2017). It is capable of efficiently working at low heads in the range of 10 to 70 m and high flow rates. The main parts of this turbine are:

1. Hub
2. Blades
3. Guide vanes or wicket gates
4. Shaft
5. Blade control mechanism
6. Scroll casing
7. Runner blades
8. Draft tube

Hub (Boss): The part of the runner on which blades are mounted.

Blades: Blades of the turbine are like the propeller, resulting in swirling flow of water at the entrance and exit.

Guide Vanes: These vanes guide the water and control the water passage (*i.e.* how much water flow goes in the turbine). The guides are adjustable and their opening and closing depends upon the power demand.

Shaft: The rotating element of the turbine that rotates the generator.

Blade Control Mechanism: The blades having moveable joints, control the angle at which water strikes the blades.

Scroll Casing: The inlet is a scroll-shaped tube around the turbine's wicket gate that guides the water to enter tangentially from the source to the runner in the turbine. It protects the runner, runner blades, guide vanes and other internal parts of the turbine from external damage.

Runner Blades: It is the most important part of the turbine, which is connected to the shaft of the generator and consists of movable vanes and hub (boss). Number of blades fixed at equidistance vary from 3 to 6. The runner blades are twisted along its length to have optimum angle of attack for the whole cross section of blades.

Draft Tube: After passing through the runner, water discharges to the tailrace through a gradually expanding tube. It decelerates the water so that some kinetic energy is recovered. The draft tube is designed to increase water pressure to an extent that it is more than the pressure of water in the tailrace. It prevents the back flow from the tailrace. Moreover, if not enough pressure is recovered then cavitation may occur and damage the turbine.

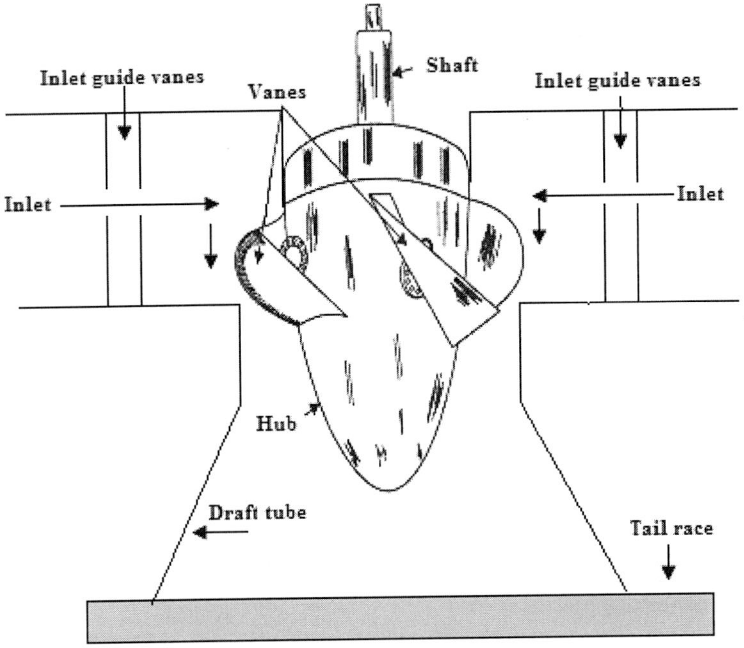

Figure 11.4. Construction Details of a Kaplan Turbine.

Salient features that differentiate a Kaplan and a Francis turbine are listed in Table 11.3.

Table 11.3. Characteristics of Francis and Kaplan Turbines

Character	*Francis*	*Kaplan*
Type of flow	Radial or mixed flow	Axial flow
Vanes	More, 16-24	Less, 4-8
Runner vanes	Fixed on the shaft	Adjustable as per requirement
Type of shaft	Horizontal or vertical	Always vertical
Head and flow rate	Medium head- medium flow rate	Low head- high flow rate
Specific speed	Low, 60-400	High, 300-800
Size	Bulky in comparison to Kaplan	Compact
Efficiency	Less in comparison to Kaplan	Higher than Francis

Working

The water from the penstock flowing over the guide vanes enters the scroll casing. Water at this stage deflects through 90° and enters the adjustable runner vanes, which make some adjustment as per the flow rate. As the water moves over the twisted blades, a lift force is generated in the direction opposite to which water is leaving. This lift force causes the blades to rotate. Power is recovered from both the hydrostatic head and the kinetic energy.

Power and efficiency of Kaplan turbine

Considering that the runner generates a torque, τ with a rotational speed of ω (rev/s), then

Power output, P_o = (Torque)(Angular velocity) = τ ω watts (11.1)

Such that τ = F r and ω = 2πN (Rad/s)

Here τ is the torque in N-m, F is the force in N and r is the radius or distance from the center to the edge, m, and N is the rotational speed of the turbine.

The power input due to water is given as:

Power input, $P_i = \rho g Q H$ (11.2)

P_i is power input, H is the total available head at turbine inlet (m), ρ is the density of water (kg/m^3), Q is volume flow rate of water (m^3/s) and g is gravitational acceleration (m/s^2)

Efficiency, per cent = 100 (P_o/P_i)

Advantages

- ✰ It can work at low head
- ✰ Number of blades is less
- ✰ It requires less space
- ✰ It has adjustable runner vanes for high efficiencies

Disadvantages

- ✰ Cavitation is a problem
- ✰ High flow rate required

POWER TRANSMISSION THROUGH PIPES

A pipeline conveys the water from a reservoir to the turbine. In other words, it transmits the hydraulic power of the water to generate hydropower (Figure 11.5). Power transmitted through pipes depends upon the weight of the liquid flowing through the pipe and the total head available at the end of the pipe. Let us assume that the length of the pipeline from A to B is L, m, and diameter of the pipe is d, m. It results in a flow velocity V in the pipeline (Figure 11.5). Following formulae determine the power transmitted at the outlet of the pipe in kw.

$$P = \frac{\rho g}{1000} \times A \times V (H - H_f)$$

$$P = \frac{\rho g}{1000} \times A \times V[H - \{4fLV^2/(d \times 2g)\}] \quad (11.3)$$

Such that P is the power transmitted, H is the total head available at the inlet of the pipe, H_f is the loss of head due to friction given by Darcy's- Weisbach equation and f is the coefficient of friction.

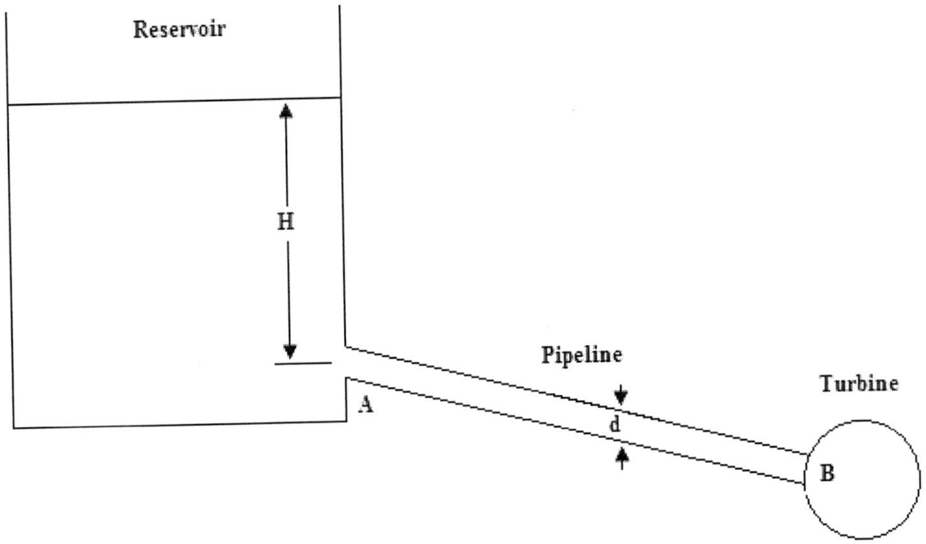

Figure 11.5. Transmission of Hydraulic Power to a Turbine through the Pipeline.

Maximum Transmission of Power

To establish the condition for maximum transmission of power, we differentiate eq. (11.3) with respect to Q or V depending upon the form of eq. (11.3). In the present case, we rewrite equation and differentiate w.r.t. V.

$$P = \frac{\rho g}{1000} \times A[HV - \{4fLV^3/(d \times 2g)\}]$$

$$\frac{dP}{dV} = \frac{\rho g}{1000} \times A[H - \{3 \times (4fLV^2)/(d \times 2g)\}]$$

Equating it to 0.0, the term in the bracket should be equal to 0.0. It means

$$H - 3\,H_f = 0$$

$$H_f = H/3 \qquad (11.4)$$

Since d^2P/dV^2 is negative, P has only a maxima and not a minima.

From eq. (11.4), we can state that maximum power is transmitted when one third of the head available at the reservoir is lost due to friction in the pipe.

Maximum Efficiency of Power Transmission

Following relation gives the efficiency of power transmission.

η = Power available at the inlet, A/Power transmitted at outlet, B

$$\eta = W\,(H-H_f)/WH$$

W is the weight of the water moving from A to B.

$$\eta = (H-H_f)/H \quad (11.5)$$

Assuming that the friction coefficient and other loss coefficients are constant, we can write, $H_f = RQ^2$ such that R is the hydraulic resistance of the pipeline and Q is the volume flow rate.

$$\eta = (H- RQ^2)/H \quad (11.6)$$

One can make the following observations.

- ☆ The efficiency η equals one at Q = 0, a trivial situation
- ☆ As Q increases η decrease from a maximum value of unity to zero.

The zero value corresponds to

$$RQ^2 = H$$

$$Q = \sqrt{(H/R)}$$

The head available at the reservoir is totally lost to overcome friction in the flow through the pipe.

Efficiency at the point of maximum power transmission is

$$\eta = (H- H/3)/H$$

$$\eta = 66.7 \text{ per cent}$$

We can infer that power transmission efficiency through the pipeline is maximum of 66.7 per cent.

PUMPS

The machines, in which mechanical energy from moving parts is transferred to a fluid to increase its stored energy either by increasing its pressure or velocity are known as pumps, compressors, fans or blowers. The machine is termed as a pump when it uses liquid, and is termed as a compressor, a fan, or a blower, when it uses gas. A hydraulic pump is any device that uses input force to create pressure or add kinetic energy to the fluid that in turn creates flow. A compressor is a machine, the main objective of which is to increase the static pressure of a gas. The compressor converts mechanical energy to increase the potential energy of a compressible fluid by increasing its pressure. Fans or blowers, on the other hand, utilize the mechanical energy of the rotor to increase mostly the kinetic energy of the fluid resulting in a high flow of gas. The change in static pressure is quite small. A detailed discussion on pumps is included in the following Chapters 12 and 13.

QUESTIONS (THEORY)

1. Define a fluid machine. Differentiate between a turbine, a pump, a compressor, and a blower.

2. What is a turbine? How will you classify the turbines? Discuss the classification based on flow through the runner.

3. Differentiate between an impulse and a reaction turbine.
4. Briefly discuss the Pelton wheel turbine. List its advantages and disadvantages.
5. Briefly discuss the Francis turbine. How it differs from Kaplan turbine?
6. Briefly discuss the Kaplan turbine. List its advantages and disadvantages.
7. Derive the equation of efficiency of power transmission through pipes. Show that the maximum power transmission efficiency is 66.7 per cent.

Chapter 12

Recriprocating Pumps

Any fluid tends to flow from high to low potential. The pump, a hydraulic machine, imparts energy to the fluid flowing through them to enable them to move from low to high potential (Figure 12.1). Although, the pump as a hydraulic machine was briefly introduced in the previous chapter, this and the next chapter respectively deal with reciprocating and rotodynamic pumps in a detailed manner.

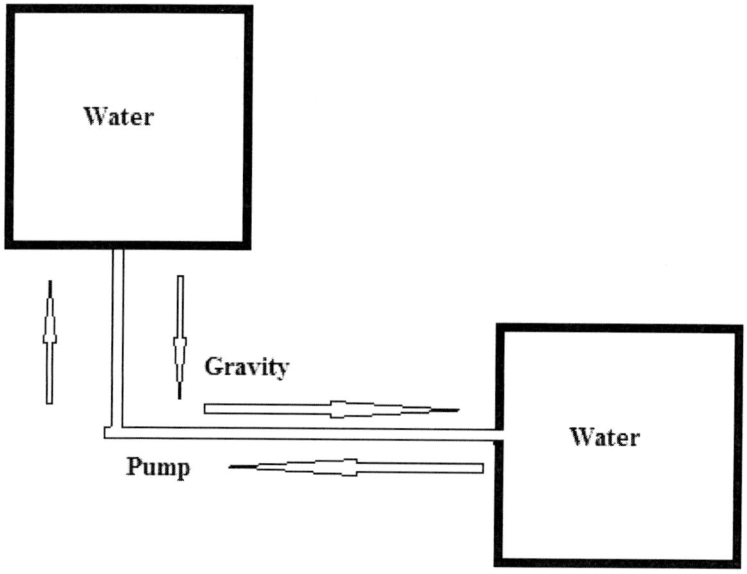

Figure 12.1. Illustration of the Need of a Pump to move Water from Low to High Potential.

WHAT IS A PUMP?

Hydraulic machines, as already stated in the previous chapter, are mechanical devices that either extract energy from a fluid or add energy to a fluid because of dynamic interactions between the device and the fluid, a gas or a liquid. The machines that add energy to the fluids are called pumps. Hydraulic pump is a machine having moving mechanical components that convert the mechanical energy into hydraulic energy appearing as pressure and velocity of the fluid. Thus, a hydraulic pump is also a device that uses input force to create pressure or kinetic energy. The most common applications of pumps include transfer of fluids from one place to another (*e.g.* water from an underground into a water storage tank) and/or distribute/circulate liquids around a system (*e.g.* cooling water or lubricants through machines and equipment).

Need to Study Pumps

An engineer often encounters the problem of fluid transport from one place at a lower elevation to some other point on a higher elevation. Besides, he also deals with several hydraulic machines that use pump as one of the components. In the first instance, an engineer is often called upon to

- ☆ Match pumps to pipelines *i.e.* the most suitable size and type
- ☆ Understand the relationship between head and flow in a pump
- ☆ Determine the power requirement of pumps and plan cost-cutting strategies

He can perform these tasks only if he is conversant with the principles, construction and operation of the pumps.

Experiences with pumps

Hydraulic pumps have a number of applications in automobiles, material handling systems, automatic transmissions, controllers, compressors and household items. Following applications in daily life illustrate the importance of hydraulic pumps.

- ☆ The air pump for filling air in tyres
- ☆ Human heart
- ☆ The hand operated hydraulic pump in a hydraulic jack to lift the car
- ☆ A backhoe using an engine driven hydraulic pump to drive the articulating parts of the mechanical hoe
- ☆ The hydraulic pumps in the automotive vehicles especially in power steering and cooling systems
- ☆ The lift system of a tractor (operated by the hydraulic pumps)
- ☆ The use of pumps in automatic transmissions and material handling systems in industries
- ☆ Commonly used compressors use pumps. Many precise controllers are developed using hydraulic pumps

- ☆ The hydraulic pumps used in routine household systems like power lift and air-conditioners
- ☆ Centrifugal, turbine and submersible pumps used in irrigation tube wells
- ☆ Turbine pumps used in household water supply systems

Basic Working Principle

Let us consider a pump to be a black box (to understand its working). Fluid enters the pump at certain velocity and pressure, which may be even zero, and leaves it with increased energy, in the form of higher velocity and pressure. It is possible because the black box (hydraulic pump) performs two functions.

- ☆ First, its mechanical action creates a vacuum at the pump inlet, which forces the fluid from the reservoir into the inlet line to the pump
- ☆ The black box houses the rotating components, which move the fluid by either confining it in definite volume and then displacing it or imparting energy to the fluid because of dynamic action of the moving parts, which increases velocity and pressure of the fluid.

CLASSIFICATION OF PUMPS

Although pumps are classified in a number of ways, the most common classification categorizes the pumps in the following two groups.

- ☆ Positive displacement pumps, also called hydrostatic pumps, variable delivery pumps or discontinuous or intermittent flow pumps (Miller 1995). On the basis of motion, these pumps are further categorized as:
  - ❒ Reciprocating pumps
  - ❒ Rotary pumps
- ☆ Non-positive displacement pumps, also called hydrodynamic or dynamic pressure pumps, rotodynamic or turbo pumps, constant delivery pumps *etc.* These pumps are all continuous flow pumps (Gülich, 2007).

A hydrostatic pump means that it converts mechanical energy to hydraulic energy with comparatively small quantities and velocities of fluids. The pressure developed and the work done is a result of essentially static forces rather than dynamic effects.

In a hydrodynamic pump, on the other hand, liquid velocities and movements are large; output pressure actually depends on the liquid flow velocity.

Although not very exhaustive, Figure 12.2 presents the classification of the pumps.

DESCRIPTION OF PUMPS

Positive-Displacement Pumps

Positive displacement, as the name suggests, is displacing something positively. Simply stated, a reciprocating pump is any machine that uses reciprocating motion

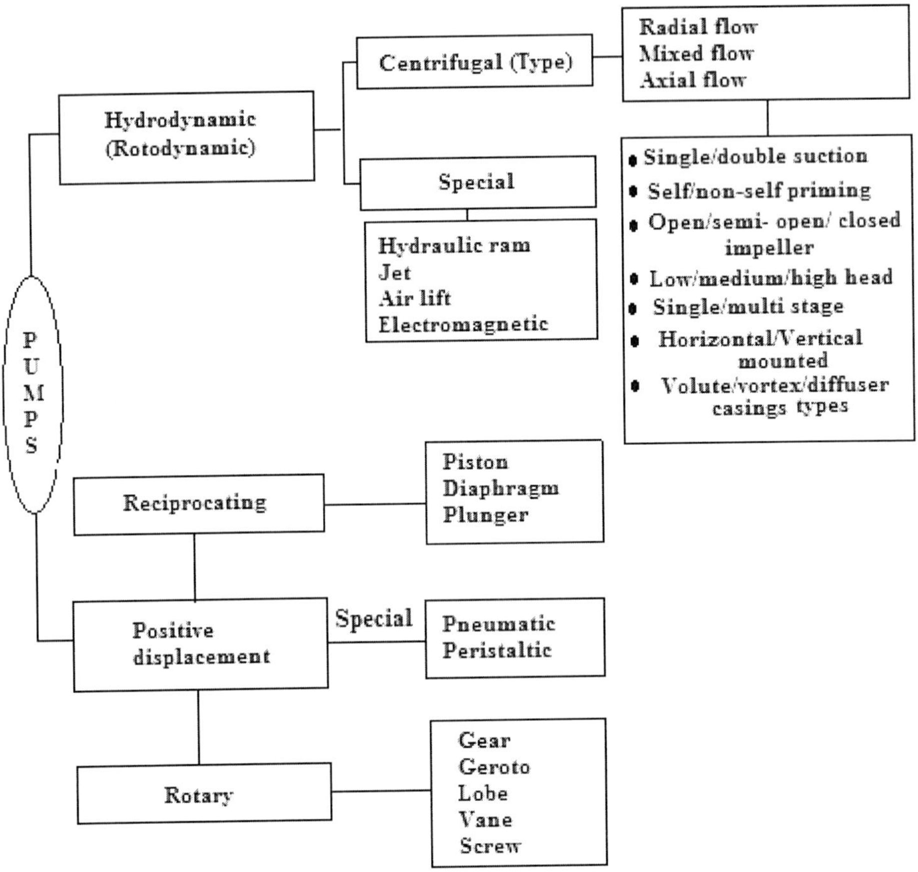

Figure 12.2. Classification of Pumps.

to cause fluid to move from one place to another. It operates on the principle of pushing the liquid by a piston/diaphragm/plunger that executes a reciprocating motion in a closed fitting cylinder. The pumps in this case displace certain fixed quantity regardless of changes in pressure against which the pump is operating. The general principle of operation is that fluid is drawn into the chamber at a low pressure. Once the liquid fills the cylinder, the inlet closes and the outlet port opens. The fluid is then forced out of the chamber by reducing its volume. Positive-displacement pumps are of either fixed or variable displacement. The output of a fixed displacement pump remains constant during each pumping cycle and at a given pump speed because of the close-tolerance fit between the pumping element and the pump case. It is possible to change output in a variable displacement pump by altering the geometry of the displacement chamber. Slippage in a positive-displacement pump is negligible compared to the pump's volumetric output. If the output port were plugged, pressure increases instantaneously to the point that the pump's pumping element or its casing would fail (may even explode, if the drive shaft does not break first), or the pump's prime mover would stall. An important advantage associated with these pumps is that the high-pressure and low-pressure areas (means input and output region) are separated and hence the fluid cannot leak back due to higher pressure at the outlets. This and other features make the positive displacement pump most suited and universally accepted for hydraulic systems.

The simplest example of piston pump is a hand pump used to lift water for domestic use (Figure 12.3). Upon lowering the pump handle, the inlet check valve opens allowing the groundwater to enter the pump chamber. When the handle is raised, the outlet check valve opens simultaneously closing the inlet check valve. It brings the water on the upper side of the chamber, which is then released through the water outlet during lowering the pump handle.

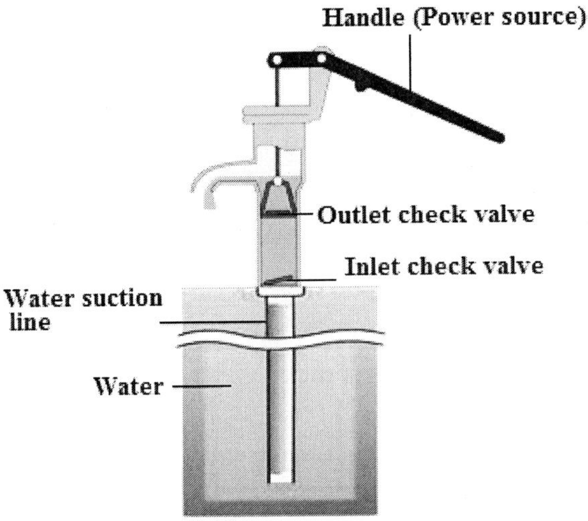

Figure 12.3. Schematic View of a Hand Pump.

The important advantages of positive displacement pumps over non-positive displacement pumps are (Miller, 1995):

- Capability to generate high pressures
- High volumetric efficiency
- High power to weight ratio
- Efficiency variation throughout the pressure range is small
- It can operate over a wide range of pressure and speed. The fluid flow rate of these pumps ranges from 0.1 to 3500 m^3/hr, the pressure head ranges between 0.5 to 7000 kg/cm^2 and specific speed is less than 500.

The major disadvantages of reciprocating pumps are:

- High initial cost
- High maintenance cost

- ☆ Pulsating flow
- ☆ Low discharge capacity
- ☆ Non-uniform torque

Non-Positive Displacement Pumps

Non-positive displacement pumps work on the principle of relative motion of the fluid and the moving part of the machine. These pumps impart velocity (kinetic energy) to the fluid, which is converted to pressure energy upon exiting the pump casing. The rotating element of the machine, rotor or impeller, usually consists of a number of vanes or blades. It is housed in the fixed part, known as stator or casing. Rotodynamic machines provide smooth and continuous action resulting in a pulsation free flow. If the output port of a non-positive-displacement pump is blocked for one reason or the other, the pump continues to operate yet the output at the outlet port is zero. It is because of slippage within the pump. Thus, dynamic pumps are able to operate safely even under closed valve conditions for short periods with no high shock loads. Notwithstanding this statement, it is always better to avoid such situations (Michael *et al.*, 2008). Rotodynamic pumps provide compact reliable solutions and are ideal for high flow low discharge heads. Centrifugal and propeller pumps are two examples of non-positive-displacement pumps.

Positive versus non-positive displacement pumps

Table 12.1 lists some major differences in positive and non-positive displacement pumps. Figure 12.4 also illustrates some of these differences.

Table 12.1. Differentiating Characters of Positive and Non-positive Displacement Pumps

Positive Displacement Pumps	*Non-positive Displacement Pumps*
The flow rate (capacity) does not change with head (Figure 12.4 left)	The flow rate decreases with head (Figure 12.4 left)
The flow rate is not much affected by the viscosity (reported in units of Saybolt Seconds Universal) of fluid (Figure 12.4 middle)	The flow rate decreases with the viscosity (Figure 12.4 middle)
Efficiency is almost constant with head (Figure 12.4 right)	Efficiency first increases with head and then decreases (Figure 12.4 right)
Suitable for low rates at high pressures	Suitable for large volumes at moderate pressures in a single stage
Fluctuating flow	Delivery is smooth and continuous
Can be used to compress air	Cannot be used for air
Low head pumps have low efficiencies	Low head pumps have high efficiencies
Priming is not required	Priming is required

RECIPROCATING PUMPS

Reciprocating pumps are available in various designs and models. These pumps are categorized based on source of power, kind of rotating part, number of cylinders and presence of air vessel as follows.

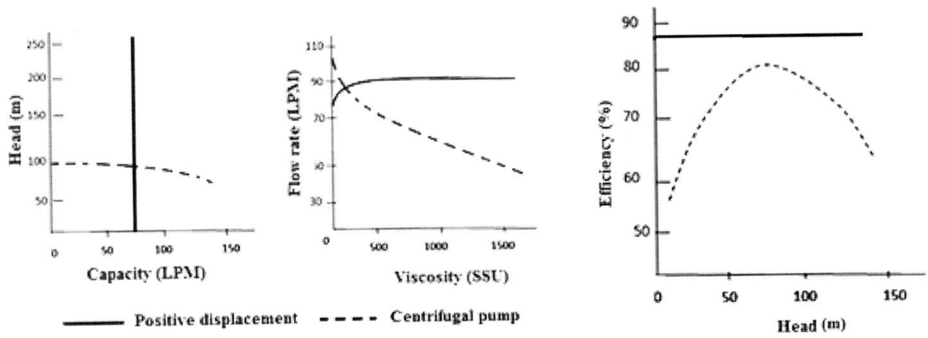

Figure 12.4. Performance Curves for Positive and Non-positive Displacement (Centrifugal) Pumps.

- ✰ Based on the source of power: manual, animal or auxiliary power sources
- ✰ Based on the kind of rotating part: piston pump, plunger pump, diaphragm pump, pneumatic and peristaltic pumps.
- ✰ Based on number of cylinders: single cylinder, double cylinder, triple cylinder and multi-cylinder pumps
- ✰ Based on the use of air vessels: pumps with or without air vessels.

Air vessels is a cast iron closed chamber having an opening at its base. The presence of vessel(s) in a reciprocating pump helps:

- ✰ To ensure continuous flow of water at a uniform rate
- ✰ To reduce the amount of work in overcoming the frictional resistance in the suction and delivery pipes to save the power required to drive the pump.
- ✰ To run the pump at high speed without flow separation

Piston Pumps

Major components of a reciprocating piston pump are (Figure 12.5):

Crank and Connecting Rod: These components translate the rotary motion of the power source to reciprocating motion to move the piston in the cylinder.

Piston or Plunger: The component that reciprocates in a closely fitted cylinder.

Cylinder: It is a hollow cylinder in which piston moves. Suction and delivery takes place in the cylinder.

Suction Pipe: While one end of the suction pipe dips in the fluid source, the other end is attached to the inlet of the cylinder. It transports the fluid from the source to the cylinder.

Delivery Pipe: One end of delivery pipe is attached to delivery port while the other end to the discharge point. It takes the fluid from the cylinder to the delivery tank.

Suction and Delivery Valves: Suction and delivery valves are non-return valves provided at the suction and delivery ends respectively.

Strainer: Provided at the beginning of the suction pipe, it prevents solid particles from entering the suction pipe.

Air Vessels: Air vessels provided on suction or delivery or on both sides aids in reducing flow fluctuations, acceleration head and friction head.

Reciprocating piston pumps are of two types namely the single acting and the double acting pumps.

Single Acting Piston Pump: These pumps have liquid chamber only on one side of the piston and accordingly the valve arrangement is only on that side (Figure 12.5). We refer Figure 12.5 to understand the working of the pump of this type. As the crank rotates (moves from A to C, from 0° to 180°), the piston moves from the farthest point on left side of the cylinder to the top of the cylinder on right side. It is the suction stroke. Because of the vacuum created in the cylinder, the suction valve opens and fluid enters into the cylinder. In the next stroke called the delivery stroke, the crank rotates from C to A (from 180° to 360°). As the piston moves to the left side of the cylinder, pressure inside the cylinder increases. In this stroke, suction valve closes while the delivery valve opens. The fluid is forced into the delivery pipe, which is delivered to the required height. It may be noted that fluid is discharged only once in a crankshaft cycle *i.e.* only in the forward stroke of the piston (Figure 12.5). Apparently, the flow is intermittent (Figure 12.6 top). The air vessel on the discharge side smoothens the flow (Figure 12.5).

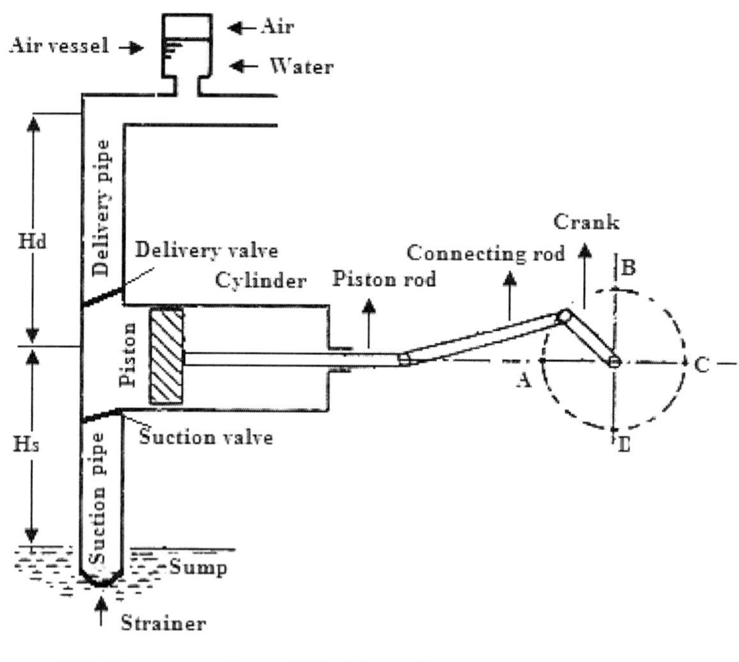

Figure 12.5. Components and Working of a Reciprocating Piston Pump.

Double Acting Piston Pump: In this type of pump, liquid chamber and valves arrangement exist on both the sides of the piston (Figure 12.7). Thus, it has two suction and two delivery pipes (appropriately joined at some point). When there is a suction stroke on one side of the piston, there is a delivery stroke at the same time on the other side of the piston. Therefore, for one complete revolution of the crank, there are two suction and two delivery strokes. The pump delivers water during both the delivery strokes from one side in the first half of the cycle and from the other side in the second half. As a result, the discharge per cycle is more and smoother compared to the single acting pump (Figure 12.6 middle).

Based on number of cylinders, we categorize the pumps as follows.

Simplex Pump: It is the simplest of all pumps having only one piston-cylinder assembly.

Duplex Pump: This kind of pump has two piston-cylinders assemblies. Both the pistons are coupled to the single crankshaft through separate connecting rod. The connecting rods are coupled to the crankshaft at an angular distance of 180° from each other so that when one piston is in upstroke, the other is in down stroke. Each piston-cylinder can be single acting or double acting.

Triplex Pump: This pump has three piston-cylinder assemblies. The three pistons are coupled to the single crankshaft through separate connecting rods. There is an angular separation of 120° between any two adjacent connecting rod and crankshaft couplings. The discharge per cycle is more and smoother compared to the single or double acting pumps (Figure 12.6 bottom).

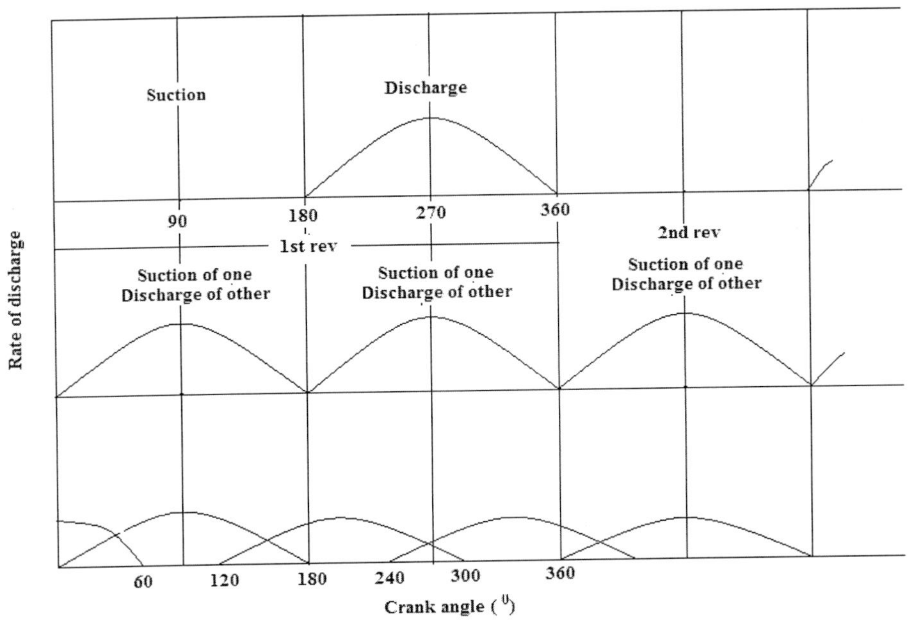

Figure 12.6. Variation of Discharge in Single Cylinder Single Acting (Top), Single Cylinder Double Acting (Middle) and Triplex Single Acting Reciprocating Pump.

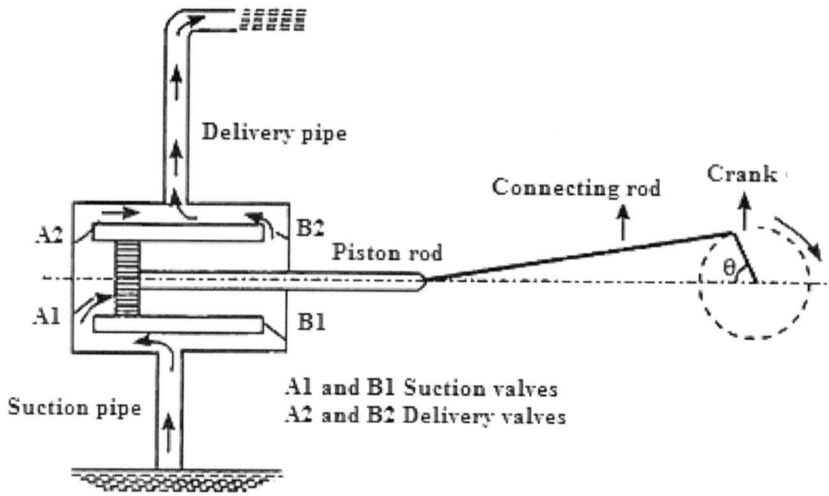

Figure 12.7. Parts of a Reciprocating Double Acting Piston Pump: Note the Arrangement of Valves on both Sides.

AXIAL AND RADIAL PISTON PUMPS

Multi-cylinder piston pumps are of two types namely the axial and radial types. Both are available as either constant displacement pumps or variable displacement pumps. An axial piston pump has several pistons in a circular array within the cylindrical block arranged parallel to the drive shaft. Fixed displacement axial piston pumps have a constant geometric displacement and convey a constant flow rate depending on the rotation speed. A variable displacement axial piston pump on the other hand adjusts the geometric output volume from maximum to zero, as a result of which flow rate may vary. In radial piston pumps, the pistons are arranged radially in a cylinder block; they move perpendicularly to the shaft center line. Displacement can be varied by moving the reaction ring that increases or decreases piston travel by varying eccentricity. Several controls are available for this purpose.

Axial Piston Pump

As stated earlier, axial piston pumps contain a number of pistons attached to a cylindrical block, which move in the same direction as the block's center line (axially). Much of the pressure and flow control circuitry is included internally, allowing for reliable operation and simple design of the associated hydraulic system. These pumps are available in two designs:

- ☆ Bent-axis-type piston pump
- ☆ Swash-plate-type piston pump

Bent-axis-type piston pump

Schematic cut section of a bent axis piston pump shown in Figure 12.8, contains a cylinder block rotating with a drive shaft. All components namely drive shaft, thrust plate, piston block shaft and piston block revolve. Connecting rods connected to the thrust plate also revolve with it. However, the center line of the cylinder block is set at an offset angle relative to the center line of the drive shaft. The cylinder block contains a number of pistons arranged along a circle. The piston rods are connected to the drive shaft flange by a ball and socket joint. The pistons are forced in and out of their bores as the distance between the drive shaft flange and cylinder block changes. A universal link connects the cylinder block to the drive shaft to provide alignment and positive drive. The volumetric displacement of the pump depends on the offset angle θ, which can vary from 0° to a maximum of about 30°. For a fixed displacement, units have offset angles ranging between 23° or 30°. There will be no flow when the cylinder block is center line. As the pump revolves, half the chambers suck the fluid as they pass over the intake port. The other half discharges the fluid through the output port.

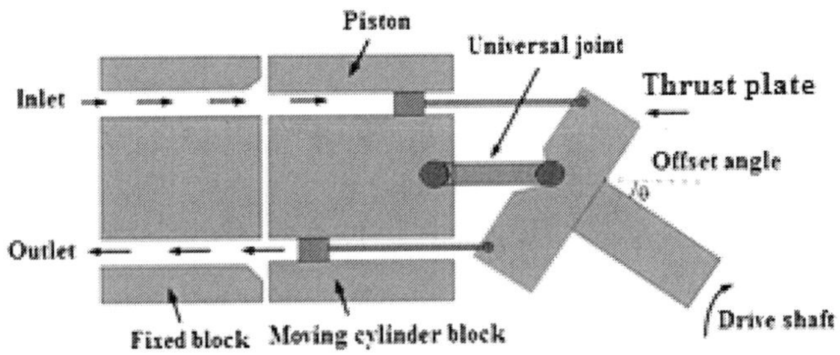

Figure 12.8. Operation of a Bent-axis-type Piston Pump.

Swash-plate-type piston pump

Schematic diagram of swash-plate-type piston pump is Figure 12.9. In this type, the cylinder block and drive shaft are located on the same center line. The pistons connected to a shoe plate bear against an angled swash plate. As the cylinder rotates, the pistons reciprocate because the piston shoes follow the angled surface of the swash plate. The outlet and inlet ports are located in the valve plate so that the pistons pass the inlet while pulling out and pass the outlet as they are being forced back in. This type of pump can be designed to have a variable displacement capability by changing the angle of the swash plate, which can be remotely controlled. The discharge will be zero for a vertical swash plate. The maximum swash plate angle is limited to 17.5°.

Radial Piston Pump

Radial piston pumps contain pistons arranged as wheel spokes around a cylindrical block. A drive shaft rotates this cylindrical block, which pushes or slings the pistons, causing compression and expansion. The eccentricity between the piston housing and cylinder block center lines determines the piston stroke. These pumps have a low noise level, very high loads at the lower speeds, and high efficiency. These are further categorized as radial piston and radial plunger pumps.

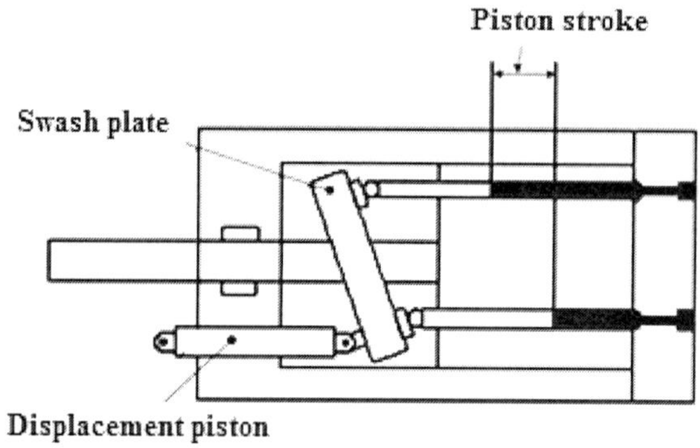

Figure 12.9. Operation of a Swash-plate-type Piston Pump.

Radial piston pump

The typical construction of radial piston pump comprises several pistons aligned radially in a cylindrical block (Figure 12.10 left). It consists of a pintle, a cylinder barrel with pistons and a rotor containing a reaction ring. Most have odd number of pistons to reduce the flow pulsation. The piston shoes ride on an eccentric ring, which causes them to reciprocate as they rotate. The eccentricity determines the stroke of the pumping pistons. Each piston is connected to inlet port when it starts extending while it is connected to the outlet port when start retracting. The connection to the inlet and outlet port is controlle by the timed porting arrangement in the pintle. Eccentric movement of the reaction ring with respect to the pintle or shaft axis initiates the pumping action. As the cylinder barrel rotates, the pistons on one side travel outward. This draws the fluid in as the cylinder passes the suction port of the pintle. It continues until the maximum eccentricity reaches. When the piston passes the maximum eccentricity, reaction ring forces the pintle inwards. This forces the fluid to flow out of the cylinder and enter in the discharge (outlet) port of the pintle. The radial piston pumps are either inside or outside impinged depending upon whether the filling of workspace of the pumping pistons is from inside (over a hollow shaft) or from outside. The radial piston pumps are capable of working on high pressure (up to 1000 bar). It is possible to use the pump with various hydraulic fluids like mineral oil, biodegradable oil, HFA (oil in water), HFC (water-glycol), HFD (synthetic ester) or cutting emulsion.

Radial plunger pump

The pumping action occurs because of the rotating rotor and reciprocating plunger. In a rotary plunger rotary pump, the axes of the plungers are perpendicular to the rotational axis of the rotor or at an angle of not less than 45° to the axis. The rotor is located eccentrically with respect to the axis of the case (Figure 12.10 right). Suction and forced delivery of the liquid occur with the reciprocating motion of the plungers because of centrifugal forces and spring action. Rotary pumps of this type may have as many as 72 plungers arranged in multiple rows.

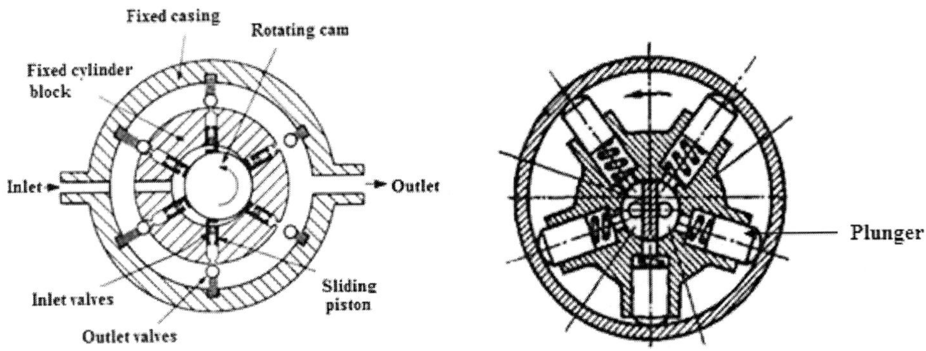

Figure 12.10. Radial Piston Pump (Left) and Plunger Pump (Right).

Plunger Pump

In a plunger pump, a plunger is attached to the rotor via the connecting rod. The plunger, made of hard and durable ceramic material, is quite resistant to wear. As the ceramic plunger moves out of the cylinder, it creates suction to draw in the fluid through the intake. As the rotor turns, the plunger is pushed back. It creates pressure that sends the fluid out through the outflow valve. High-pressure seal in this case is stationary, and only the plunger slides through the seal (Figure 12.11). It is unlike the piston pump, where seal slides along with the piston. As such, plunger pumps are capable of generating much higher pressures than piston pumps. A plunger pump is a single acting pump and therefore, two plungers are required for a double acting plunger pump (Tackett *et al.*, 2008). These are commonly used to transfer municipal and industrial sewage, besides being ideal for offshore tasks.

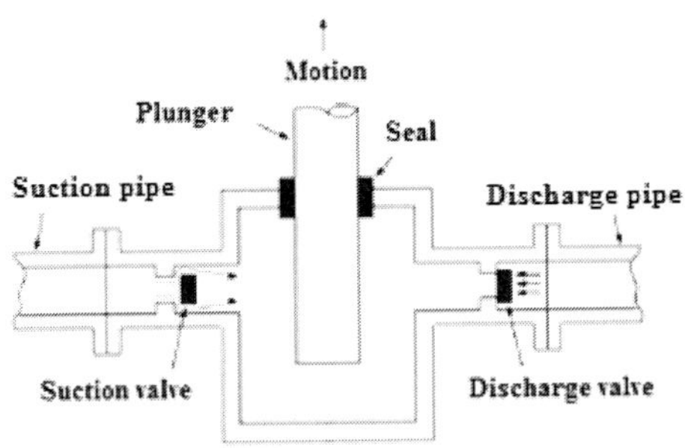

Figure 12.11. A Schematic View of a Plunger Pump.

Diaphragm Pump

Diaphragm pumps have a reciprocating diaphragm in a liquid chamber that produces the pumping action. The diaphragm is made of a rubber, thermoplastic or Teflon having suitable non-return check valves to pump the fluid. It is why; this type of pump is also named as membrane pump. The arrangement of valves and their functioning is similar to the piston pumps. The basic operation of the pump is illustrated in Figure 12.12.The liquid chamber is the fixed body of the pump with diaphragm fitted to one side of the chamber. The actuators make the diaphragm bulge in and out of the liquid chamber. When the diaphragm bulges out of the liquid chamber, the inlet valve opens and allows the liquid to enter the chamber. When the diaphragm bulges in the chamber, the pressure increase opens the outlet valve to pump the liquid out of the chamber.

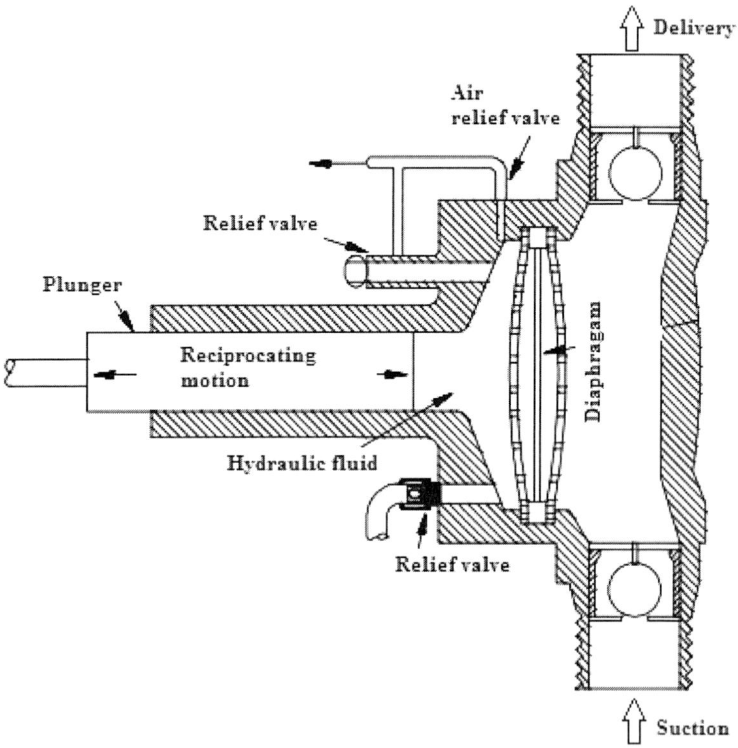

Figure 12.12. Illustration of the Basic Principle of Diaphragm Pump.

Actuation of the diaphragm is achieved either by hydraulic plunger or by air pressure. Accordingly, the diaphragm pumps are either hydraulically operated or air operated pumps. In the former, driving hydraulic fluid is on one side of the diaphragm, while the liquid to be pumped is on the other side. The driving fluid actuates the diaphragm, which in turn pumps the liquid contained on the other side. As against this, air operated diaphragm pumps have two chambers, each divided into two parts by flexible diaphragms. The liquid side of both the chambers has inlet and outlet non-return valves. A rigid rod interconnects the centers of the two diaphragms. The pressurized air enters one of the air chambers and exits from the other chamber at a time. The chamber receiving the air pushes the fluid out through the outlet valve while the other chamber takes in the fluid through inlet valve. After the end of the stroke, the air control valve reverses the direction of airflow in the air chambers. It also reverses the flow of the liquid in the liquid chambers. This configuration makes the diaphragm pump to act like a double acting pump.

In a diaphragm pump, there is no direct contact between the power source and the liquid to be pumped. Moreover, there is no frictional motion between the pump chamber elements. Thus, diaphragm pump does not require seals and liners. There is least possibility of leakage and contamination of the liquid. Moreover, there is no heat generation due to friction. Thus, the diaphragm pumps are ideal, safe and reliable to pump creams, gels and oils, corrosive or volatile solvents, abrasive, viscous, toxic and inflammable liquids.

Pneumatic Pump

Pneumatics, a branch of technology, uses the force of compressed gases to generate mechanical effects. Pneumatic pumps essentially use air in the same way that hydraulic pumps use fluids to move fluids through a piping system. These pumps comprise a double piston system, with one piston having a substantially larger diameter than the other piston. Between the two pistons are airtight chambers filled with either liquid or another compressed gas. The compressed gas from the outside pushes the larger-diameter piston, which in turn pushes against the gas or liquid in the intermediate chamber. Because there is no pressure loss between the larger piston and the smaller one, the smaller piston receives a highly amplified force, which translates into powerful mechanical action. Specifically designed for oil and gas production applications, these pumps are used in many other industries, including pulp and paper, water treatment, fertilizer dosing, and food processing. These pumps are quite durable than other pumps.

Peristaltic Pump

The peristaltic pump is a type of positive displacement pump. It comprises a pump segment (and tubing); a manifold and a rotary pump head. The pump head is a rotor containing two or more rollers. These rollers press the tubing that creates vacuum. It helps to draw the fluid through the tube (Figure 12.13). The fluid flow is generated in the equipment by periodically pressing a tube segment to the pump housing (the manifold). The fluid is expelled when the next roller passes over the tube. The complete closure of the tube when squeezed between the roller and the casing (track) gives the pump its positive displacement action. It also prevents backflow and eliminates the need of having check-valves when the pump is not running. A particular advantage of peristaltic pumps is that the fluid remains inside the tubing. As a result, the risk of contamination of either the product or the pump itself is minimal. The tubing can be easily cleaned and even sterilized after every use.

ROTARY POSITIVE DISPLACEMENT PUMPS

The pumping action in rotary pumps is due to the relative movement between the rotating elements and the stationary elements of the pump. The operation of these pumps is quite different compared to reciprocating or centrifugal pumps. As per the design of rotary pumps, a continuous seal is maintained between inlet and outlet ports by the action and position of the pumping elements and close running clearances of the pump. Therefore, rotary pumps do not require valve arrangements as in reciprocating pumps.

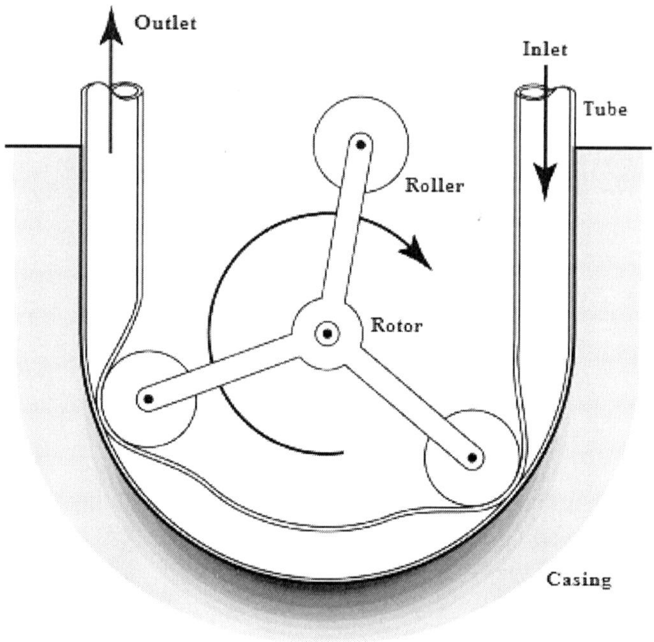

Figure 12.13. Construction and Working of a Peristaltic Pump.

In a rotary-type positive displacement pump, rotating parts form a moving cavity that carry the fluid from one side and discharges it to the pump outlet. According to the type of element that transmits the liquid, rotary pumps are gear-, lobe-, vane-, or screw-type pumps. Although the construction, shape and design of gear, vane and lobe pumps are bit different, all these pumps are categorized as rotary pumps (Sahu, 2018). Based on the number of rotors, rotary pumps are mainly of two types.

- Single rotor
- Multiple rotor

Each of these types of pumps has different design features. Rotary pumps, being highly sensitive, require special care so that no abrasive material enters the pump, lest it damages the helices and clearances in the pump.

Gear Pump

Gear pumps are most robust and simple positive displacement pumps. Depending upon the setting of the gears, gear pumps are classified as external gear and internal gear pumps.

External gear pump

External gear pump has a rotary housing containing two identical intermeshing gears. A typical housing has an inlet and an outlet. While motor drives one of the gears, it in turn drives the other idler gear. Gear pumps come with a straight spur, helical, or herringbone gears. Straight spur gears are widely used, being easiest to cut (Figure 12.14). Helical and herringbone gears work quietly, but are bit costly. A shaft supports each gear with bearings on both sides of the gear. Typically, all four bearings operate in the pumped liquid. Fluid in a gear pump is carried in between the chambers formed between teeth of adjacent gear teeth enclosed by the pump housing and side plates (also called wear or pressure plates). Apparently 3 factors namely volume of cavities between the teeth, speed of gears, and amount of fluid that slips back to the inlet regulate the flow. As the gear teeth unmesh, a partial vacuum is created at the pump inlet. Fluid flows in to fill the space and moves further between the chambers formed. As the teeth mesh again, fluid is forced out at the outlet end. Running clearances between gear faces, gear tooth crests and the housing create an almost constant loss for any pumped volume at a fixed pressure. This means the volumetric efficiency at low speeds and flows is poor. It is why, gear pumps are operated close to their maximum rated speeds. Moreover, external gear pumps perform well under high-pressure applications because the gears are supported on both sides. An important drawback of external gear pump is the unbalanced side load on its bearings, caused by high pressure at the outlet and low pressure at the inlet. Besides reduced life of the bearings, it results in slower speeds and low-pressure ratings.

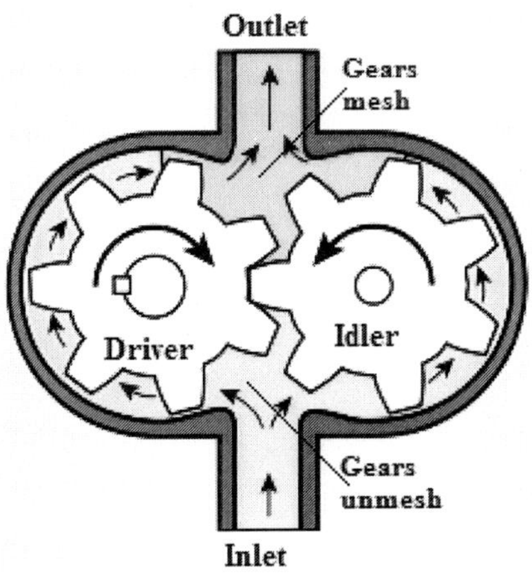

Figure 12.14. Construction Details of an External Gear Pump.

Internal-gear pump

These pumps are small, compact units, having an internal and an external gear, separated by a crescent-shaped seal (Figure 12.15). Outer gear has teeth on the inside. The two gears rotate in the same direction, inner gear rotating faster than the outer; having one or two teeth less than the outer gear. The inner idler gear rotates on a stationary pin and operates inside the rotor gear. The fluid is drawn into the pump at the point where the gear teeth begin to separate and is carried to the outlet in the space between the crescent and the teeth of both gears. The contact point of the gear teeth forms a seal, as does the small tip clearance at the crescent. The crescent is integral with the pump head. It prevents liquids flowing from the discharge to the suction port. Since only one bearing is used in the pumps, the life of the bearing gets reduced because of overheating load. These pumps are not suitable for high speed and high-pressure applications.

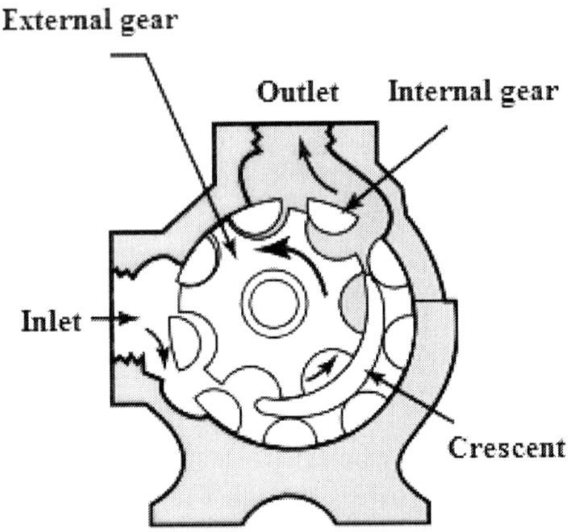

Figure 12.15. Important Components of an Internal Gear Pump.

Common applications of gear pumps are:

Petrochemicals: Pure or filled bitumen emulsions, pitch, diesel oil, crude oil, lube oil *etc.*

Chemicals: Sodium silicate, acids, plastics, mixed chemicals *etc.*

Pulp and paper: Acid, soap, dye, black liquor, kaolin, lime, latex, sludge *etc.*

Food: Chocolate, cocoa butter, fillers, sugar, vegetable fats and oils, molasses, animal food *etc.*

Paint and ink

Resins and adhesives

Gerotor Pump

Gerotor pump, the name derived from Generated Rotor, consists of two rotors, one inner with N teeth, and the outer with N+1 teeth (Figure 12.16). The inner rotor is located off-center and is driven while the outer is idler. The geometry of the two rotors partitions the volume between them into N different dynamically changing volumes. The volume of each partition changes continuously. Therefore, any given volume first increases, and then decreases. An increase in volume creates vacuum that helps to suck to the fluid. As the volume decreases, compression occurs. During this part of cycle, fluid is pumped at the outlet. Gaseous fluids are compressed at this time of the cycle. Gerotors are widely used throughout industry, and are available in a variety of shapes and sizes. These pumps primarily suit low-pressure applications such as lubrication systems or hot oil filtration systems. These are also common in low to moderate pressure hydraulic applications.

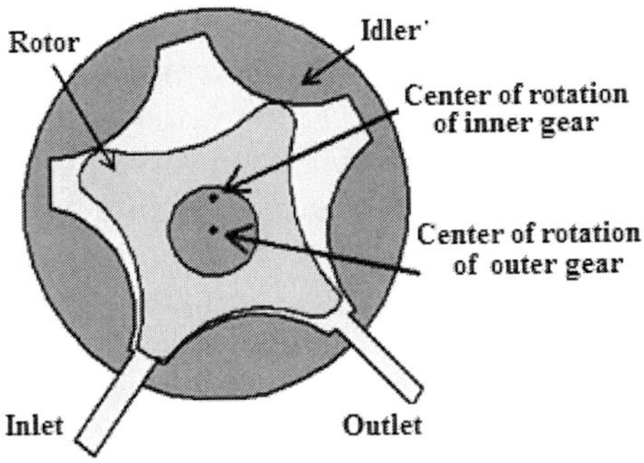

Figure 12.16. Construction Details of a Gerotor Pump.

Lobe Pump

Lobe pump derives its name from the rounded shape of its rotors' radial surfaces. Single or multiple lobe pumps, are similar to external gear pumps in operation, except that the pumping elements or lobes do not make contact with each other (Figure 12.17). External timing gears prevent the contact between the lobes. All lobes get drive through suitable drive gears outside of the pump casing chamber. As the lobes come out of mesh, they create expanding volume on the inlet side of the pump. Liquid flows into the cavity and the lobes trap the incoming fluid. Fluid travels around the interior of the casing in the pockets between the lobes and the casing. Finally, the meshing of the lobes forces liquid through the outlet port under pressure (Figure 12.17). Since there is no lobe -to -lobe contact, the wear in abrasive applications is minimal. Because of superb sanitary qualities, high efficiency, reliability, corrosion resistance and good clean in place and steam-in-place (CIP/SIP) characteristics, lobe pumps are widely used in industries such as pulp and paper, chemical, food, beverage, pharmaceutical and biotechnology *etc.* These pumps can handle solids, slurries, pastes, and a variety of liquids.

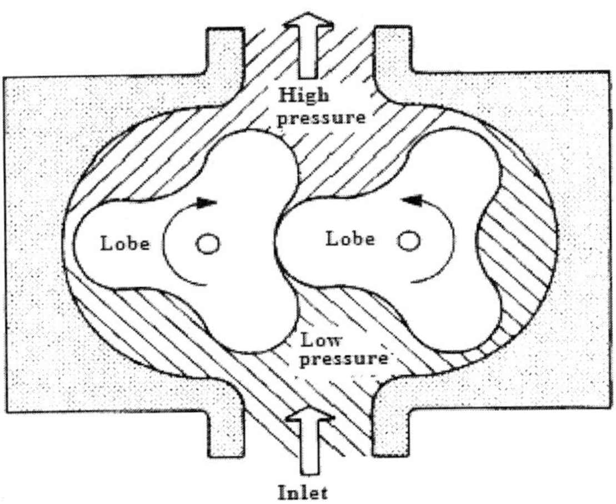

Figure 12.17. A View of a Lobe Pump.

Vane Pump

A main disadvantage of gear pumps is due to some leakage between gear teeth and the pump housing. Vane pumps help to overcome this limitation. Since vane pumps comprise spring or hydraulically loaded vanes placed in the slots of driven rotor (Figure 12.18), the leakage is minimal. A vane pump consists of a rotor, vanes, ring and a port plate with inlet and outlet ports. The rotor eccentrically placed inside a cam ring is connected to the prime mover through a shaft. The rotor is sealed into the cam by two side plates. When the prime mover rotates the rotor, the vanes are thrown outward due to centrifugal force. The vanes track along the ring. It provides a tight hydraulic seal to the fluid. As the rotor rotates, it produces a suction cavity in the ring, creating vacuum at the inlet pushing the fluid into the pump. Vanes carry the fluid around to the outlet, and their retraction expels the fluid at the outlet. The eccentricity, expansion of vanes, width of vanes and speed of the rotor govern the capacity of the pump. There would be no flow for zero eccentricity.

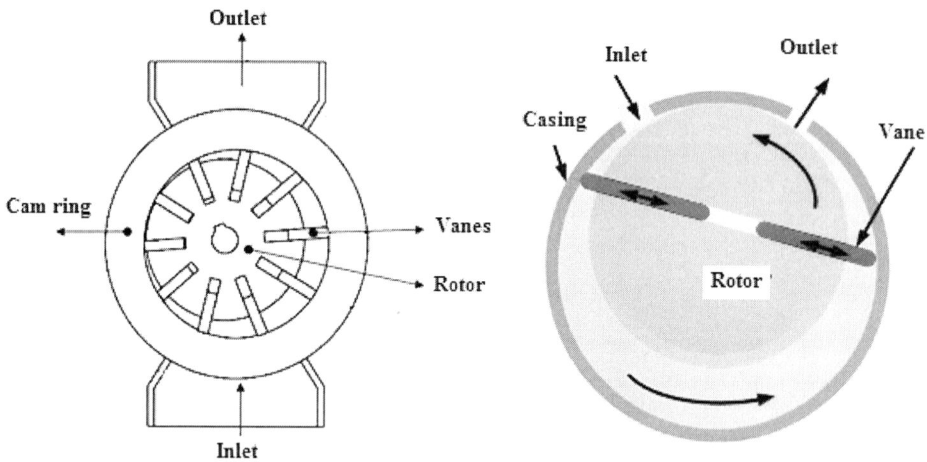

Figure 12.18. A Simple Vane Pump (Left) and its Working Principle (Right).

Vane pumps are available in a number of vane configurations including sliding vane, flexible vane, swinging vane, rolling vane, and external vane *etc.* Each type of vane pump has its own advantages. Capacity and pressure ratings of a vane pump are generally lower than the gear pumps, but reduced leakage gives an improved volumetric efficiency. These pumps can handle low viscosity liquids at relatively higher pressure. These pumps are capable of developing good vacuum because of negligible leakage. These pumps are capable of running dry for some time without any failure. However, these pumps are not suitable for high-speed applications and for the high viscosity fluids or fluids carrying abrasive particles. The maintenance cost is often high due to many moving parts. The operating range of these pumps is from -32°C to 260°C.

Vane pumps are categorized as balanced, unbalanced and variable vane pumps. In a balanced pump, the rotor and vanes are contained within a double eccentric cam ring and there are two inlet segments and two outlet segments for each revolution. The double pumping action not only gives a compact design, but also ensures that the forces at the two outlets are equal and opposite, completely cancelling each other. As a result, there are no net loads on shaft bearings. Consequently, the life of this type of pump in many applications is exceptionally good. In an unbalanced design, all pump action takes place in the chambers located on one side of the rotor and shaft. Variable displacement feature is brought into vane pumps by varying eccentricity between the rotor and the cam ring. In such a pump, a spring-loaded piston holds the stator ring. The system pressure acts directly through a hydraulic piston on the right side. This forces the cam ring against a spring-loaded piston on the left side. If the discharge pressure is large enough, it overcomes the compensated spring force and shifts the cam ring to the left. This reduces the eccentricity and decreases the flow. If the pressure continues to increase, the eccentricity may become zero such that there will be no flow.

Screw Pump

Different types of screw pumps are the single- Archimedes screw, two-screw, and three-screw pumps. In the single-screw pump, a spiralled rotor rotates eccentrically in an internal stator (Figure 12.19). The two-screw pump consists of two parallel intermeshing rotors rotating in a housing machined to close tolerances. The three-screw pump consists of a central-drive rotor with two meshing idler rotors; the rotors turn inside of a housing machined to close tolerances. As the rotors rotate, the hydraulic fluid surrounding the rotors at the inlet is trapped. As the rotor rotates, the fluid uniformly moves along the axis and is forced out at the other end. The inter-meshing of the screw threads on the rotors and the fine clearances of the surrounding housing creates a seal between the pumps inlet and outlet. Due to the axial flow pattern and low internal velocities, the screw pumps have certain advantages in applications where churning and agitation of fluid is undesirable.

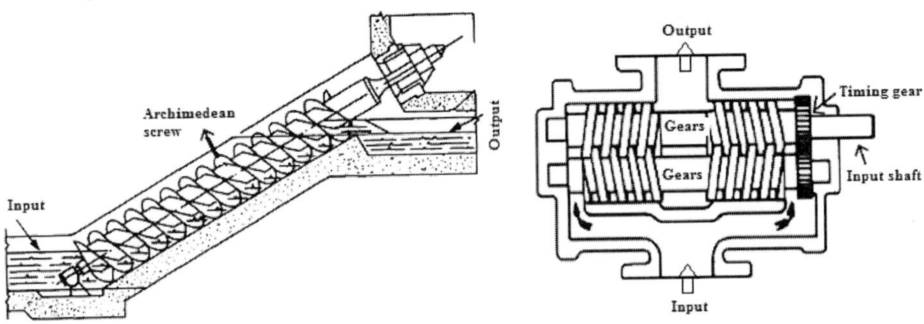

Figure 12.19. Screw Pumps: Archimedean (Left) and Two Screws (Right).

Advantages of screw pump

- ☆ Little noise making these pumps as quite pumps
- ☆ These can work under a wide range of flows and pressures

- ☆ They are capable of accommodating a wide range of liquids and fluid viscosities
- ☆ Screw pumps have high speed capability
- ☆ They have low internal velocities
- ☆ All the screw pumps are self-priming and thus, have good suction characteristics
- ☆ Due to the close arrangement of rotating parts, screw pumps have high tolerance for entrained air and gases.

Disadvantages of screw pump

- ☆ Because of close tolerances and running clearances, the manufacturing cost is high
- ☆ Any changes in the viscosity of the fluid causes high variations in the pump performance
- ☆ A screw pump for high-pressure capability will require high pumping elements, which will increase the overall size of the pump.

THEORY OF RECIPROCATING PUMPS

Capacity of Single Acting Pump

The theoretical capacity or discharge (Q_t) of a single acting piston pump (Figure 12.5) is proportional to its displacement per unit time. Equation for the calculated capacity of the pump in terms of displacement volume is:

$$Q_t = ALN/60 \qquad (12.1)$$

Such that Q_t is the theoretical discharge in m^3/s, A is the cross sectional area of the cylinder, m^2, L is the stroke length, m and N is the speed of rotation, revolution per minute (RPM). The constant 60 is the conversion factor that converts revolution per minute (RPM) into revolution per second (RPS). The above equation assumes 100 per cent hydraulic efficiency. Considering the number of cylinders, n and number of acting, C the general discharge equation is (Minz, 2018):

$$Q_t = ALNnC/60 \qquad (12.2)$$

Simply stated the discharge of a double acting pump is double the discharge of a single acting pump. Note that the discharge of a double acting pump in practice will not be exactly double (as will be shown in the following section), yet for all practical purposes it can be taken as double the discharge of a single acting pump.

Capacity of Double Acting Pump

A double acting pump is used to reduce the pulsation in discharge (Figure 12.7). Theoretical discharge of a double acting reciprocating pump is the sum of the discharge during forward and backward strokes as:

Discharge = Discharge in forward stroke + Discharge in reverse stroke

$$Q_t = ALN/60 + (A-A_p)\,LN/60 \qquad (12.3)$$

Here A_p is the area of the piston rod. All other parameters have the same meaning as explained previously. If D is the diameter of the piston and d that of the piston rod, then

$$A\text{-}A_p = \pi D^2/4 - \pi d^2/4$$

$$= \pi\,(D^2 - d^2)/4 \qquad (12.4)$$

If d is quite small in comparison to D, then it can be neglected such that $(A\text{-}A_p)$ is given as:

$$A\text{-}A_p = \pi D^2/4 \sim A \qquad (12.5)$$

Incorporating eq. (12.5) in eq. (12.3), we have

$$Q_t = 2ALN/60 \qquad (12.6)$$

Comparing with eq. (12.1), we can say that the discharge of a double acting pump is nearly double the discharge of a single acting pump.

Power Requirement

Theoretical work done or power, P of a single acting pump is given as:

$$P = \gamma Q_t\,(H_s + H_d) \qquad (12.7)$$

Such that γ is the specific weight and is given as ρg, ρ is the density of the fluid and g acceleration due to gravity, m/s^2, H_s is the suction head and H_d the delivery head. Incorporating eq. (12.1) for Q_t in eq. (12.7), for a single cylinder single acting pump, we have:

$$\text{Power, } P = \gamma\,(ALN)\,(H_s + H_d)/60 \text{ Watt}$$

Incorporating eq. (12.2) for Q_t in eq. (12.7) for multi-cylinder multi-acting pump, we have:

$$\text{Power, } P = \gamma\,(ALNnC)\,(H_s + H_d)/60 \qquad (12.8)$$

Due to leakage and frictional losses, actual power input will be more than the theoretical power. Dividing eq. (12.8) with efficiency, we get

$$\text{Actual power } P_a = \gamma\,(ALNnC)\,(H_s + H_d)/(60\,\eta) \qquad (12.9)$$

Here η is the pump efficiency.

Performance of the Pump

The performance of a pump depends upon the precision with which it is manufactured. An ideal pump has zero clearances between all mating parts. Since it is not possible to have a pump with zero clearances, working clearances need to be as small as possible but enough to maintain proper oil films for lubrication between rubbing parts. Following efficiencies are commonly used to describe the performance of a pump.

Volumetric efficiency

Volumetric efficiency, η_v indicates the amount of leakage that takes place within the pump, due to manufacturing tolerances and flexing of the pump casing under designed pressure and other operating conditions. It can happen from one or all of the following factors.

- Delay in closing the suction and delivery valves as a result of which some of the fluid goes back while valves are closing/seating
- Leakage through valves, gland, and piston
- Liberation of air from the liquid being handled

Volumetric efficiency, defined as the ratio of actual flow rate of the pump to the theoretical flow rate of the pump, is always less than 1.0. The smaller is this ratio, the better the volumetric efficiency. Mathematically, η_v can be expressed as:

$$\eta_v = Q_a/Q_t \tag{12.10}$$

Such that Q_a is the actual flow rate of the pump and Q_t is the theoretical flow rate of the pump. It is also possible to calculate theoretical flow rate by multiplying the pump's displacement per revolution by its driven speed.

$$Q_t = V_d \times N_1 \ (m^3/rev \times rev/s = m^3/s) \tag{12.11}$$

Such that V_d is the theoretical volume displaced per revolution and N_1 the revolution per second (N/60). For example, if the pump has a displacement of 100 cc/rev and is being driven at 1000 RPM, its theoretical flow is 100 L/min.

The volumetric efficiency for piston pumps is around 90–98 per cent while for gear pumps it is in the range of 80–90 per cent.

Mechanical efficiency (η_m)

Mechanical efficiency indicates the amount of energy loss that occurs for reasons other than leakage. This includes friction in bearings and between mating parts. This also includes the energy losses due to fluid turbulence. Mechanical efficiency is defined as the ratio of the pump output power assuming no leakage to actual power delivered to the pump.

Mechanical efficiency = Pump output power assuming no leakage/actual power delivered to the pump

We write the following relationship to describe the mechanical efficiency.

$$\eta_m = p\, Q_t/(T_a\, \omega) \tag{12.12}$$

Where p is the pump discharge pressure in Pa or N/m^2, Q_t is the theoretical flow rate of the pump in m^3/s, T_a is the actual torque delivered to the pump in N-m and ω is the speed of the pump in rad/s such that ω is $2\pi N/60$.

$$T_a = \text{Actual wattage delivered by the motor, } P/\omega\ [(Nm/s)/rad/s = N\text{-}m] \tag{12.13}$$

Mechanical efficiency, η_m in terms of torque is as follows:

η_m = Theoretical torque required to operate the pump (T_t)/Actual torque delivered to the pump (T_a)

$$\eta_m = T_t/T_a \tag{12.14}$$

The theoretical torque T_t required to operate the pump is the torque required if there were no leakage. Following relation is used to calculate the theoretical torque (https://nptel.ac.in/courses/112106175/Module per cent 201/Lecture per cent 2011.pdf):

$$T_t = V_d \times p/2\pi \quad (m^3\,N/m^2 = N\text{-}m) \tag{12.15}$$

Mechanical efficiencies are about 90–95 per cent.

Overall efficiency

Overall efficiency is defined as the ratio of actual power delivered by the pump to actual power delivered to the pump.

η_o = Actual power delivered by the pump/Actual power delivered to the pump

Overall efficiency, η_o considers all energy losses and mathematically described as follows:

$$\eta_o = \eta_v \times \eta_m \tag{12.16}$$

$$\eta_o = (Q_a/Q_t) \times p\,Q_t/(T_a\,\omega) \tag{12.17}$$

Slip

Another way to express volumetric efficiency is in terms of slip. The slip of the pump is defined as:

$$\text{Slip} = Q_t - Q_a \tag{12.18}$$

The major component of slip is the leakage of fluid back through the discharge or suction valve as it is closing (or seated). It can reduce calculated displacement from 2 to 10 per cent depending upon valve design and condition, normal value being 2 per cent. The slip of the pump in percentage is given as:

$$\text{Slip (per cent)} = [(Q_t\text{-}Q_a)/Q_t]100 = (1 - C_d) \times 100 \tag{12.19}$$

Here C_d is the coefficient of discharge given as ratio of the actual discharge to the theoretical discharge (Q_a/Q_t). Since the actual discharge is always less than the theoretical discharge, the value of coefficient of discharge is around 0.95 to 0.98 for pumps in good condition. The slip of the pump is usually positive although under special conditions it may be negative as well. It is possible when the length of suction pipe is considerably long, delivery head low and pump is running at high speed.

Example 12.1

Calculate the theoretical discharge of single acting single cylinder reciprocating pump if the area of cylinder is 0.30 m^2, length of stroke is 0.15 m, and speed of pump is 50 rpm.

$A = 0.30$ m^2, $L = 0.15$ m, $N = 50$ RPM, $N = 1$ and $C = 1$

$Q_t = ALNnC/60$ (eq. 12.2)

Incorporating the values, we have

$Q_t = 0.30 \times 0.15 \times 50 \times 1 \times 1/60 = 0.0375$ m^3/s

Example 12.2

What is the theoretical flow rate from a fixed-displacement single acting piston pump with a nine-bore cylinder operating at 1800 RPM? Each bore has a diameter of 20 mm and stroke is 25 cm. What is the actual flow rate, if the volumetric efficiency is 91 per cent ?

The theoretical flow rate is calculated using the following parameters

$A = (\pi/4)\ d^2 = 0.00031\ m^2$ (Since diameter =0.02 m), L =0.25 m, N =1800 RPM, N =9 and C =1

Incorporating in eq. (12.2), we have

$= 0.00031\times 0.25\times (1800/60) \times 9\times1 = 0.021\ m^3/s = 21\ L/s.$

Since the volumetric efficiency is 91 per cent, the actual flow rate is:

Actual flow rate $= 21\times91/100 = 19.11\ L/s$

Example 12.3

Calculate the power required to drive the single acting reciprocating pump for water of following specification.

Area of cylinder = 0.85 m^2, Length of stroke = 0.35 m, Number of cylinder = 1, Speed of pump = 25 rpm, Suction head = 0.6 m, Discharge head = 0.9 m. Assume the density of water being pumped as 1000 kg/m^3.

Using eq. (12.8), we have

Power $= 1000\times9.81\times0.85\times0.35\times25\times1\ (0.6 +0.9)/60 = 1824$ watt or 1.82 kW

Power in terms of HP is calculated as

Power $= 1000\times0.85\times0.35\times25\times1\ (0.6 +0.9)/60\times76 = 2.44$ HP

The factor 76 is to convert kilogram- meter/second in horsepower.

Example 12.4

A pump having a volumetric efficiency of 94 per cent delivers 30 LPM of oil at 1000 RPM. What is the volumetric displacement of the pump?

Volumetric efficiency of the pump, $\eta_v = 94$ per cent

Discharge of the pump = 30 LPM

Speed of pump, N = 1000 RPM

Since $\eta_v = Q_a/Q_t$, we have

$Q_t = 30/0.94 = 31.91$ LPM

Volumetric displacement is obtained using eq. (12.11) as follows.

$V_d = Q_t/N = 31.91/1000 = 0.0319$ L/rev

Example 12.5

Determine the overall efficiency of a pump driven by a 12.5 HP prime mover if the pump delivers fluid at 40 LPM at a pressure of 12.6 MPa.

Output power = $12.6 \times 10^6 \times 40/(1000 \times 60) = 8399$ watt ~8500 watt

Input power = $12.5 \times 746 = 9325$ watt

$\eta_o = 8500/9325 = 0.912$ or 91.2 per cent.

Example 12.6

How much hydraulic power would a pump produce when operating at 120 bar and delivering 0.0012 m^3/s of oil? Determine the power rating of the electric motor to drive this pump if its overall efficiency is 86 per cent.

Operating pressure of the pump = 120 bar

Flow rate, $Q = 0.0012\ m^3/s$.

1 bar = $10^5\ N/m^2$ or 10.19 m of water

Using the first conversion, we have

Power of pump = Pressure × flow rate = $120 \times 10^5 \times 0.0012 = 14400$ watts or 14.4 kW

Using the second conversion, we have

Power of the pump = $0.0012 \times 10^3 \times 120 \times 10.19/76 = 19.3$ HP or 14.4 kW.

Since pump efficiency is 86 per cent

Power required = $14.4/0.85 = 16.74$ kW

Example 12.7

A single acting reciprocating pump discharges 0.025 m^3/s of water when running at 65 RPM. Stroke length is 50 cm and the diameter of the piston as 25 cm. If the total lift is 15 m, calculate the theoretical discharge, slip and percentage slip, coefficient of discharge and power required to run the pump. Assume any appropriate value required to solve the problem.

$Q_t = 0.5 \times 65 \times \pi/4 \times (0.25)^2/60 = 0.5 \times \pi/4\ (0.25)^2 = 0.0265\ m^3/s$

Slip = $0.0265 - 0.025 = 0.0015$

per cent slip = $(.0015/0.0265) \times 100 = 5.66$ per cent

Coefficient of discharge = $0.025/0.0265 = 0.943$

Power input = $\rho \times g \times H_s \times Q_t = 1000 \times 9.81 \times 15 \times 0.0265 = 3899$ W = 3.9 kW

The above calculation assumes that there is no leakage and frictional losses. In practice, actual power input will be more than the theoretical power. To calculate the actual power required, divide the theoretical power required by the pump efficiency, η. Assuming pump efficiency as 0.93, we have

Power required to run the pump = $3.9/0.93 = 4.2$ kW.

Example 12.8

A pump has a displacement volume of 98.4 cm^3. It delivers 0.00152 m^3/s of oil at 1000 RPM and 70 bar. If the prime mover input torque is 124.3 Nm. What is the overall efficiency of pump? What is the theoretical torque required to operate the pump?

Volumetric discharge = V_d = 98.4 cm^3

Theoretical discharge is

$Q_t = V_d \times N/60 = 98.4 \times 1000/60 = 1640$ cc/s $=1.64\text{x}10^{-3}$ m^3/s

Volumetric efficiency = $(1.52\times10^{-3}/1.64\times10^{-3})$ 100 = 92.68 per cent

Actual power = $0.00152\times70\times10^5$ = 10640 W (1 bar =100000 Pa)

Power given to the pump = Torque $\times\omega = 124.3 \times 2\pi N = 124.3 \times2\times1000\times3.142/60$ =781101.2/60=13018.4

Overall efficiency = (13018.5/10640)100 = 81.7 per cent

Overall efficiency = volumetric efficiency × Mechanical efficiency

Mechanical efficiency = (81.7/92.68) 100 = 88.2 per cent

Mechanical efficiency can also be calculated using eq. (12.12)

Mechanical efficiency = $p\times Q_t$/(Torque × w) = $(70\times10^5\times0.00164)/13018.4$= (11480/13018.4)100 = 88.2 per cent

Theoretical torque = Actual torque × mechanical eff. = 124.3 × 0.882 = 109.6 Nm

Example 12.9

A hydraulic motor drives a load at 600 rpm with 800 N-m of torque. What is the output power?

N = 600 rpm such that w = $600\times2\pi/60$ = 62.84 rad/s

T_a = 800 N m

Power = $T_a \times w = 800 \times 62.84$ = 50272 Watts = 52.360 kW.

NET POSITIVE SUCTION HEAD

Net positive suction head (NPSH) is the head of water at the center line of the pump suction required to push the suction valve from its seat, to overcome friction losses and acceleration head within the fluid cylinder. Normally expressed in pressure units, it describes the fluid pressure above the vapour pressure of the fluid. Two terms commonly used to describe NPSH are the net positive suction head required (NPSHR) and the net positive suction head available (NPSHA). NPSHA should always be more than NPSHR+ Friction losses on the suction side + acceleration head. The pump manufacturers specify NPSHR values for any plunger size, rotating speed, pump capacity and pressure. Several thumb rules help to decide the NPSHA. One of the rules states that it should be 1.03 to 1.25 times the NPSHR. Lower the specific gravity of the fluid, higher should be the safety margin.

AIR VESSEL

The flow from a single cylinder single acting reciprocating pump is pulsating and intermittent in nature. On the contrary, for smooth running of the pump, continuous flow is desirable during both the suction and the delivery strokes. An air vessel either on delivery (Figure 12.20) or suction or on both sides is used to achieve the following objectives:

☆ Air vessel in the discharge pipe dampens out the pressure variations during discharge so as to obtain a uniform discharge

☆ Air vessel reduces the work to be done by overcoming friction in suction and delivery pipes

☆ Air vessel facilitates the pump to run at high speeds without flow separation.

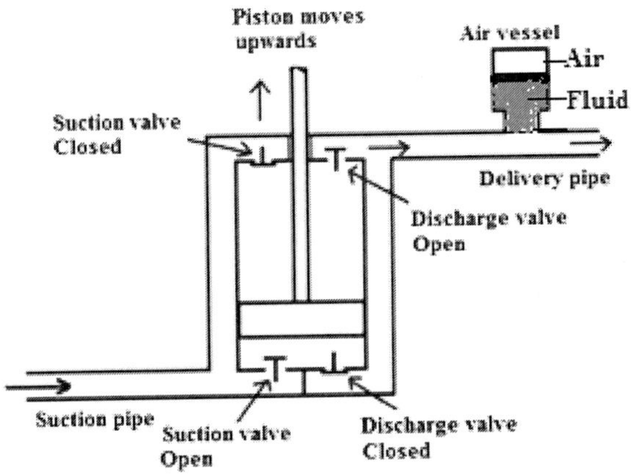

Figure 12.20. A Double Acting Reciprocating Pump with Air Vessel on the Discharge Side.

An air vessel is a cast iron closed chamber having an opening at its base. While the lower half contains the fluid being pumped, the upper half has compressed air, air and water being separated by a flexible diaphragm. The diaphragm in the air vessel is movable as per difference of pressure between the two fluids. As the discharge pressure rises, the air in the vessel is compressed. On the other hand, air expands when there is fall in the pressure. The peak pressure energy is thus stored in the air vessel and returned to the system when pressure falls. Apparently, the friction head remains constant throughout the stroke. To get best results, air vessels are fitted on the suction and delivery pipes close to the cylinder of the pump. Saving of work and consequently of the power is about 84.8 per cent in a single acting reciprocating pump and 39.2 per cent in a double acting reciprocating pump. One should not use air vessels on reciprocating boiler feed pumps as they may introduce air into the de-aerated water.

AFFINITY LAWS

Although affinity laws are often associated with centrifugal pumps, but positive displacement pumps also exhibit these relationships. The affinity laws of positive displacement pumps are quite straightforward.

Flow: Flow varies directly with change in speed. Doubling the rotational speed doubles the flow.

Pressure: Pressure is independent of the change in speed. If one ignores efficiency losses, the pressure generated at any given rotational speed will be that required to support flow.

Horsepower: Horsepower varies directly with the change in speed. Doubling the rotational speed will require twice the original power.

SELECTION OF HYDRAULIC PUMPS

From the foregoing discussions, it emerges that different types of pumps work best for different applications and operating conditions. Accordingly, pump designs with wide ranging operating characteristics are available in the market. While some of these have been included in this chapter, others are discussed in the next chapter. A user must select the pump carefully to achieve the functional objective as well as to minimize the cost of pumping/circulation. Users must choose an optimum pump after carefully evaluating the various options available with him (Figure 12.21). The main factor is the system's overall requirement. For example, it may be wrong to use a pump with high delivery when the requirement is that of a low delivery rate. Similarly, using a pump that must work continuously at its peak just to meet the minimum requirements of the system will be equally wrong. While selecting a pump, one must balance various requirements of the whole system. Besides balancing the various requirements, keep the following points in view.

- Flow rate requirement
- Operating speed of pump
- Pressure rating
- Performance/application
- Reliability
- Initial investments
- Operation and maintenance cost
- Noise level of the pump
- Fluid compatibility
- Type of pump control
- Pump contamination tolerance
- Availability of pump and spares

Capacity-pressure range of reciprocating, rotary and centrifugal pumps illustrated in Figure 12.21 can be used to make a preliminary decision on the kind of pump required to meet specific requirements.

PIPING SYSTEM

For a pump to operate efficiently, piping system design besides the selection of pumps is equally important. A poorly designed system can cause several problems that may reduce the pump life and/or may prove to be a potential hazard to personnel and associated equipment.

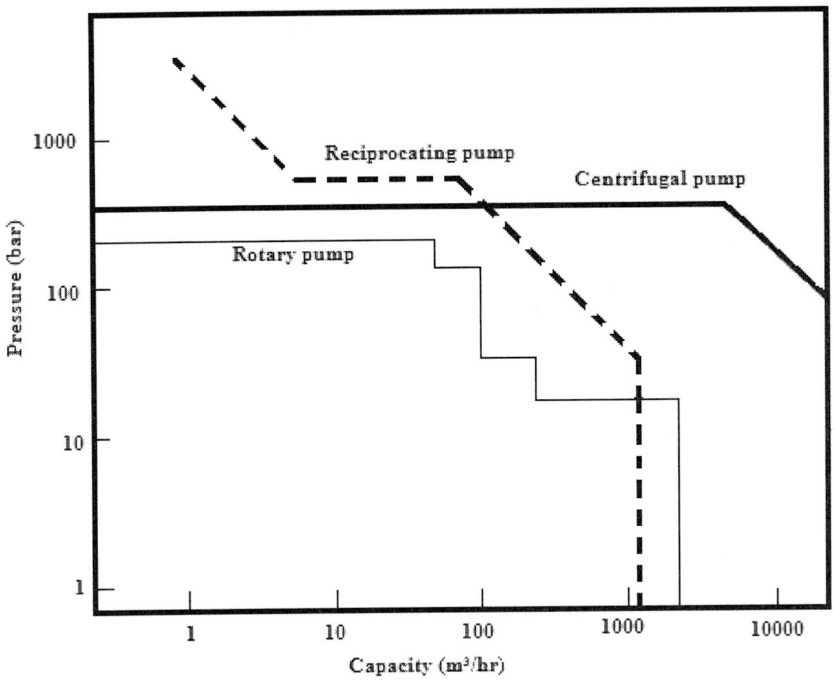

Figure 12.21. Approximate Upper Limits of Capacity and Pressure Range for Centrifugal, Reciprocating and Rotary Pumps.

Fluid Flashing: Entrained gases in the fluid coming out when pressure in the piping or pump falls below fluid vapour pressure.

Cavitation: Free gases in a fluid are forced back into the fluid. The implosions of bubbles cause severe pressure spikes that cause pitting to damage the pump internal parts.

Piping Vibrations: It can result from improper piping support, cavitation or normal reciprocating motion of the pump.

Noisy Operation: It normally happens when pump is cavitating.

Reduced Capacity: It is mainly due to fluid flashing as gases are likely to fill the pump chamber.

QUESTIONS (THEORY)

1. Name the main classification criterion of pumps. List the various kinds of pumps under each group.

2. Discuss the non-positive and positive displacement pumps. Differentiate between the two kinds of pumps.

3. Draw a neat sketch showing the single acting piston pump. Label the sketch showing various components. Explain the working of single acting piston pump.
4. Draw a neat sketch showing the double acting piston pump. Explain the working of double acting piston pump.
5. What are simplex, duplex and triplex pumps? Show the discharge characteristics of these pumps with the help of a neat sketch.
6. Differentiate between a piston and a plunger pump.
7. What do you understand by rotary positive displacement pumps? Differentiate between an internal and an external gear pump.
8. Derive the equation to assess the capacity of single acting piston pump. Show that capacity is nearly double if the pump is a double acting pump.
9. Derive an equation to determine the power of the prime mover required to drive a pump.
10. What do you understand by volumetric, mechanical and overall efficiency of a reciprocating pump?
11. Define slip. Under what conditions it could be negative.
12. Write a short note on bent-axis-type piston pump, swash-plate-type piston pump, radial piston pumps, radial plunger pump, peristaltic pump, net positive suction head, and air vessel.

Chapter 13

Hydrodynamic Pumps

Hydrodynamic or non-positive displacement pump is a type of velocity pump that adds kinetic energy to the fluid by increasing its flow velocity. On the other hand, for the fluid to flow in a pipeline, we need high pressure and not the high velocity, since the head losses in the pipes are proportional to square of the flow velocity. Moreover, high velocity might otherwise be damaging to the system because of significant water hammer effects. To overcome these effects, the volute of a hydrodynamic pump converts a part of the kinetic energy to potential energy as per the Bernoulli's principle discussed in a previous chapter of this book. The hydrodynamic pumps are used under low-pressure high-volume applications being unable to withstand high pressures. These pumps are primarily used to transport fluids and have little application in the hydraulic or fluid power industry. This chapter discusses various kinds of hydrodynamic pumps, their selection, operation and power requirement with emphasis on centrifugal pumps.

KINDS OF PUMPS

Depending upon the nature of the flow through the impeller, pumps are categorized as radial flow (centrifugal), axial flow (propeller) or mixed-flow types.

Radial Flow Pumps

Fluid enters a radial flow pump along the axial plane. It is accelerated by the impeller and exits at right angle to the shaft (radially, Figure 13.1). A spiral casing, known as volute, surrounds the impeller of the radial flow pump. It is why these pumps are also known as volute type pumps. The water leaving the vanes is directed to circumferentially flow in the volute chamber. As the area of the volute chamber gradually increases in the direction of flow, some of the kinetic energy is converted to pressure energy. As the water approaches the delivery pipe, a considerable part of kinetic energy is already converted to pressure energy. Some loss of energy does take place because it is not possible to completely avoid the formation of eddies and due to increasing quantity of water through the volute chamber. Radial flow pumps compared to axial and mixed flow pumps operate at higher pressures and lower flow rates.

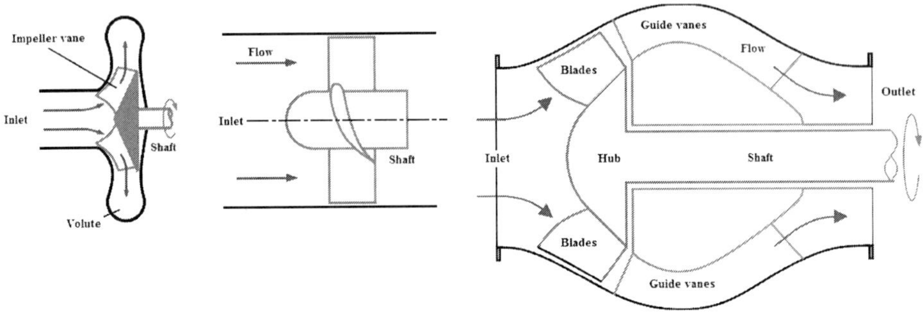

Figure 13.1. Categorization of Pumps on the Basis of Flow: Radial Flow (Left), Axial Flow (Middle) and Mixed Flow (Right).

The name centrifugal pump most often applies to describe all impeller type hydrodynamic pumps including the radial, axial and mixed flow types. Strictly speaking, the term centrifugal pump applies only to the radial flow type pumps. Centrifugal pumps are most appropriate for ordinary (*i.e.* low to moderate viscosity) liquids under a wide variety of flow conditions. These are commonly used in systems where low viscosity fluids are used such as handling of water or milk in dairy plants.

Axial Flow or Propeller Pumps

In many applications such as those associated with drainage and irrigation, one often requires high discharges but at low heads. Axial-flow pumps are commonly used under such conditions. In an axial flow pump fluid enters and exits along the same direction parallel to the rotating shaft (Figure 13.1 middle) in contrast to radially in a radial flow pump. The impeller rotates within a cylindrical casing with fine clearances between the blade tips and the casing walls. Fluid particles, in course of their flow through the pump, do not change their radial locations. The inlet guide vanes properly direct the fluid to the rotor. The outlet guide vanes eliminate the whirling component of velocity at discharge port. The number of impeller blades usually varies from 2 to 8, and the hub diameter to impeller diameter ratio ranges from 0.3 to 0.6. The impeller action lifts the fluid instead of accelerating it. The propeller pump can be oriented in vertical, horizontal, or angled position. Except at design capacity, the efficiency of an axial flow pump is less than that of a centrifugal pump. The axial flow pump can overload the drive motor if the flow rate is significantly less than the design capacity. The head curve for the axial flow pump is much steeper than that of the centrifugal pump.

Mixed Flow Pumps

Mixed flow pumps, as the name suggests, function as a compromise between radial and axial flow pumps. In this pump, fluid experiencing both radial acceleration and lift exits the impeller somewhere between 0-90° from the axial direction (Figure 13.1 right). Therefore, mixed flow pumps operate at higher pressures than axial flow pumps while delivering higher discharges than radial flow pumps. The exit angle of the flow dictates the pressure head-discharge characteristic of this pump viz-a-viz radial and mixed flow.

Advantages and Disadvantages

The advantages of hydrodynamic pumps are:

- ✰ The flow through these pumps is smooth and continuous
- ✰ Have fewer moving parts
- ✰ Initial and maintenance cost is low
- ✰ They are capable of handling almost all kinds of fluids including slurries and sludges
- ✰ The operation is simple and reliable
- ✰ No damage to pump may occur in the event of closure of discharge pipe during pump operation.

The disadvantages are:

- ✰ These pumps are not self-priming and hence must be positioned below the fluid level or some arrangement for priming has to be made
- ✰ Discharge is a function of output resistance
- ✰ Volumetric efficiency is low.

Other Classifications

The other factors used to classify the pumps are:

- ✰ Type of casing
- ✰ Number of stages
- ✰ Number of entrances for fluid
- ✰ Disposition of the shaft
- ✰ Direction of flow of water
- ✰ Working head
- ✰ Specific speed

Following paragraphs discuss some of these classifications.

Type of casing

Volute: The name is derived from the spiral shape of the casing, which is so constructed so as to collect the liquid as it leaves the outer edge of the impeller vanes. Offsetting the impeller effectively creates a curved funnel such that cross-sectional area increases towards the pump outlet. The increasing area of flow decreases the exit velocity simultaneously increasing pressure in the casing towards the outlet (Figure 13.2). Volute flow pump is a radial flow pump discussed in the previous section.

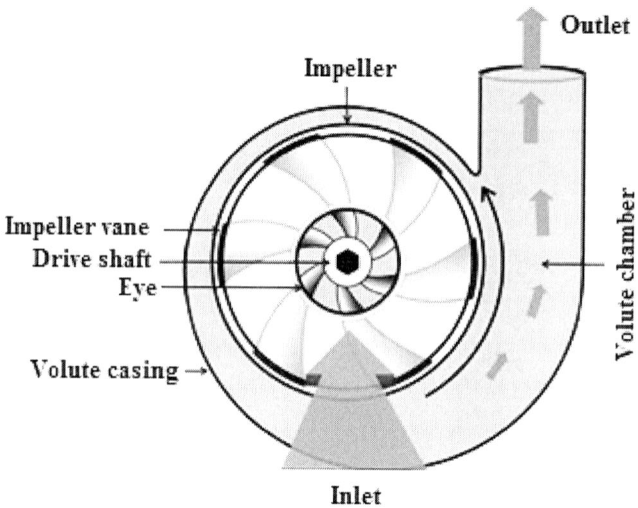

Figure 13.2. A Schematic View of Volute Casing.

Vortex: In a vortex casing, a circular chamber is used in between the casing and the impeller (Figure 13.3 left). Vortex casing reduces eddies formation to a considerable extent thereby increasing pump efficiency.

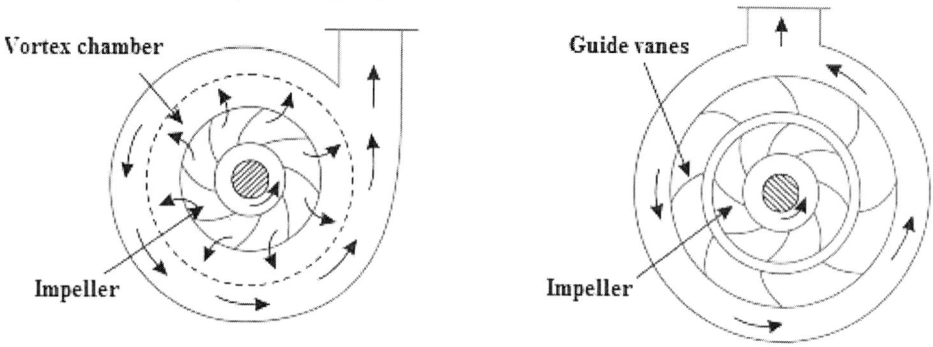

Figure 13.3. Casings: Vortex (Left) and Diffuser (Right).

Diffuser: In diffuser casing, a diffuser is provided as an additional component. The diffuser surrounds the impeller periphery has many vanes with gradually enlarging passages. The fluid leaving the impeller passes through these guide vanes or diffuser (Figure 13.3 right). The fluid pressure increases as the fluid is expelled between a set of stationary vanes that diffuse the liquid. This provides a more controlled flow and a more efficient conversion of velocity head into pressure head. Provision of fixed diffuser increases the efficiency of the pump up to 90 per cent. Because of their resemblance to a reaction turbine, diffuser pumps are also named as turbine pumps.

Classification based on stages

Single-Stage: Single-stage centrifugal pumps have only one impeller fitted to the pump shaft. It is usually a low-lift (low-head) pump.

Multi-Stage: When the pressure required is more than what is practicably or economically furnished by a single stage, additional stages are used. Each impeller and matching casing is a stage. A pump with more than one stage is a multi-stage pump. The impellers surrounded by guide vanes led the water from the outlet of one stage to the entrance of the next stage through a by-pass channel. The water finally discharges into a wide chamber from where it is pushed to the delivery pipe. Multi-stage pumps commonly used are the vertical turbine and submersible type pumps. The distance between the water source to the outlet determines the number of stages required for a particular installation. Vertical turbine pumps can have as many as 25 stages.

Classification based on pump suction

Single Suction: The impeller of a single suction pump, also known as end-suction pump, has only one water inlet (Figure 13.4 left). When the fluid in the impeller is pushed towards the peripheral direction, low-pressure area is formed at the center of impeller. The total static pressure difference helps to suck the fluid into the impeller center.

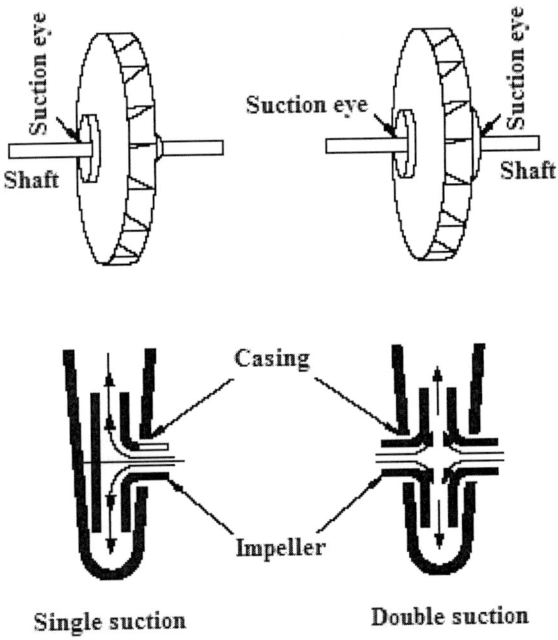

Figure 13.4. Single and Double Suction Impellers.

Double Suction: A double suction pump has two ends; having the benefit of decreasing the axial force and increasing the flow quantity. Practically, a double suction pump impeller consists of two back-to-back impellers, from which the water flow into the volute (Figure 13.4 right). Thus, one double suction pump works like two same diameter end-suction pumps. As such, its capacity is doubled for the same outer diameter impeller. The entry and exit ends of a double suction pump are in the same direction, being perpendicular to the pump shaft. It helps in the layout and installation of pumps and water pipes. Double suction pumps are widely used in drainage and irrigation of farmland, waterworks, pump stations, power stations, fire fighting and shipping, *etc.*

Classification based on disposition of pump shaft

The primary difference between these pump types is the position of the shaft. In a horizontal centrifugal pump, the vertical impeller is mounted on a shaft in a horizontal position. Sometimes, it is in an overhung position or placed between bearings. In case of vertical pump, the horizontal impeller is mounted on a vertical shaft. It is always an overhang and of radial-split case type design.

Classification based on head

Low Head: A low-head pump is capable to produce heads up to 15 m. A volute having no guide vanes usually surrounds the impeller. The shaft is generally horizontal and water may enter the impeller from one or both sides depending on the discharge requirement.

Medium Head: Normally provided with guide vanes, these pumps are capable of generating heads in the range of 15 to 40 m. Water may enter from one or both sides depending on the quantum of water pumped.

High Head: They can generate heads exceeding 40 m. They are generally multi-stage pumps because an ordinary single impeller cannot build up such a high pressure. High-head centrifugal pumps may be horizontal or vertical; the latter mostly used in deep wells.

Classification based on specific speed of the pump

Specific speed, a variable obtained from the service data, has great practical significance in the design and choice of pumps. It expresses the relation between speed, discharge and head. The specific speed (N_s) is a characteristic coefficient derived from the similarity conditions, which allows a comparison of impellers of various pump sizes even when their operating data differ. Specific speed is defined as "the speed of an ideal pump geometrically similar to the actual pump, which when running at this speed will raise a unit of volume, in a unit of time through a unit of head". In metric units, the specific speed of a centrifugal pump is the required speed of a geometrically similar pump, which when running at this speed will deliver a flow rate of $1 m^3/s$ at a head of 1 m. For the working parameters of the model pump, we generally use a subscript s such that N_s is the specific speed given as:

$$N_s = N\, Q^{1/2}/H^{3/4} \tag{13.1}$$

Where N is the pump speed in RPM, Q is the capacity in m^3/s at the best efficiency point and H is the total head in m per stage at the best efficiency point.

From this equation, it is clear that the specific speed increases with N and *Q* and decreases with *H*. Clearly, high values come from high-flow and low-head pumps, and low values result from high-head and low flow pumps. Thus, depending on the specific speed, it is possible to select an appropriate pump. Table 13.1 reports the range of specific speeds; N_s for different types of pumps.

Table 13.1. Types of Centrifugal Pumps and their Specific Speed

Sl.No.	*Type of Pump*	*Specific Speed (N_s)*
1	Low-speed radial flow	10-30
2	Medium-speed radial flow	30-50
3	High-speed radial flow	50-80
4	Medium speed mixed flow or screw type	80-160
5	High speed axial flow or propeller type	160 to 500

The values of specific speed in the above equation depend upon the units used. Following equation establishes a relationship that determines a dimensionless characteristic coefficient known as dimensionless specific speed.

$$N_s^* = N\,Q^{1/2}/(g\,H^{3/4}) \qquad (13.2)$$

Here g is acceleration due to gravity and is equal to 9.81 m/s^2 and N_s^* is the dimensionless specific speed. The following relationship exists between the numerical values of the dimensional and dimensionless coefficients:

$$N_s = 333\,N_s^* \qquad (13.4)$$

Example 13.1

A pump is to be operated at 3000 rpm to deliver a discharge of 0.2 m^3/s to a head of 6 m. Calculate the specific speed and comment upon what sort of pump will be required.

Using eq. (13.1)

$$N_s = 3000\,(0.2)^{1/2}/(6)^{3/4} = 3000\times 0.447\times 0.26 = 348.7$$

Therefore, a high speed axial or propeller pump will be required (Refer Table 13.1).

COMPONENTS OF A PUMP

Shaft and Impeller

A pump comprises rotating and stationary components (Figure 13.5 left). The rotating components are mainly the shaft and the impeller. The shaft couples the pump to the motor that drives the pump (Figure 13.5 left). The shaft made of steel or stainless steel is carefully sized to support the impeller. An undersized shaft can cause vibrations, shorten the bearing life while an oversize shaft will increase the cost of the pump. In most pumps, a shaft sleeve covers the portion of the shaft under the sealing arrangement. The shaft sleeve positions the impeller correctly on the shaft, and protects the shaft from wear and tear. The shaft runs through a stuffing box provided at the outside of the back plate of the pump. A stuffing box is an assembly used to house a gland seal, packing or other type of sealing element such as a mechanical seal to prevent leakage of fluid between rotating and stationary parts of the pump. Bearings support and hold the shaft in place. The bearings handle the loads resulting from the rotation of the impeller.

The impeller made of iron, steel, bronze, brass or aluminium, driven by the motor imparts energy to the fluid (Figure 13.5 right). As previously discussed, based on the direction of flow with reference to the axis of rotation, impellers or pumps are radial, axial and mixed flow types. According to another classification, impellers are open, semi-open and closed types.

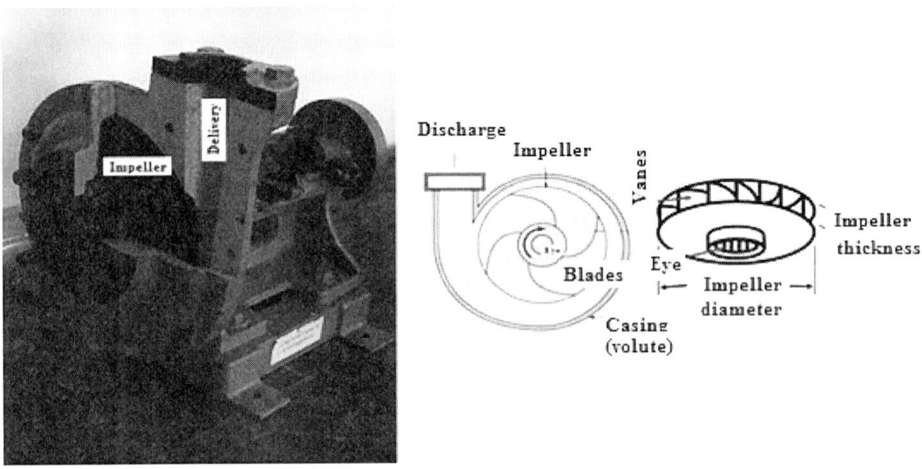

Figure 13.5. Components of a Centrifugal Pump.

Open impeller

An open impeller has vanes attached to a center hub mounted directly on to a shaft. There is no wall surrounding the vanes (Figure 13.6 left). It makes open impellers weaker than closed or semi-closed impellers. On the other hand, open impellers are easier to clean and repair. Open impellers are mostly used in smaller pumps or pumps used to handle suspended solids. The number of vanes normally vary from 2 to 8 but can be more.

Semi-open impeller

Semi-open or semi-closed impellers have a back wall that adds strength to the impeller (Figure 13.6 middle). A common problem with such impellers is their reduced efficiency, but the ability to pass solids is an important trade-off.

Closed impeller

Closed impellers have a back and front wall (plate) around the vanes (Figure 13.6 right). It adds to strength. The liquid moves in between the cavity between the vanes and plates. Closed impellers are primarily used in large pumps, where liquid fluids are involved. They do not perform well with fluids having solids. These are difficult to clean when clogged.

Figure 13.6. Open Impeller (Left), Semi-open Impeller (Middle) and a Closed Impeller (Right).

Casing

The most important stationary part of a pump is the casing, the cover housing the impeller. The main function of casing is to enclose the impeller at suction and delivery ends and thereby form a pressure vessel. Purpose of the casing is to collect the liquid discharged from the periphery of the impeller. While the rotating impeller increases the kinetic energy of the liquid, the casing converts this kinetic energy into pressure energy. The pressure at suction end of a single-stage pump may be as little as one-tenth of atmospheric pressure and at delivery end may be twenty times the atmospheric pressure. The fluid from the delivery end is discharged through the discharge pipe.

Suction and Delivery Pipes

A suction pipe has its one end connected to the pump inlet, the opening (eye) of the impeller, and the other end dips into liquid source from where liquid is to be pumped. Discharge pipe connects the delivery end of the pump casing to the delivery site.

Foot Valve with a Screen

A foot value (optional) is a non- return value fitted at the lower end of the suction pipe, which retains the water after the pumping to avoid frequent priming (Figure 13.7). It also prevents trash from entering the suction side.

Reflux Valve

The main purpose of reflux valve is to retain water in the suction pipe section between the pump and the reflux valve. It helps in priming. When water is pumped to overhead tanks or is conveyed long distances, a reflux valve provided on the delivery side permits flow only in the forward direction. Besides checking the return flow, it also prevents water hammer if pump suddenly stops to operate.

Figure 13.7. A Photographic View of the Foot Valve.

Self-priming Device

This device is an optional attachment to achieve self-priming of the pump. This device eliminates the need for external priming. The pump fitted with a self-priming device is a self-priming centrifugal pump.

Working Principle of Centrifugal Pumps

The centrifugal pump operates on the principle of transfer of energy (or angular momentum) to the fluid from a rotating impeller housed inside a casing. The rotating impeller creates partial vacuum that helps to suck the liquid from the water reservoir. The fluid entering the axis or eye of the impeller is discharged from the impeller periphery. The fluid gains kinetic energy and momentum because of the angular momentum imparted by the high-speed impeller. A part of the kinetic energy of the fluid changes to pressure energy (head) as the fluid passes through a diverging area between the impeller periphery and the casing. The fluid flows radially outward or axially into a diffuser or volute chamber before exiting the pump. The maximum possible suction lift of a centrifugal pump at the mean sea level is 10.33 m of water (corresponding to 1 atmosphere). The suction lift in practice is much less than this theoretical value. Maximum suction at any location depends upon pressure at the liquid surface, which decreases with altitude as the pressure at any location is a function of its altitude above the mean sea level. It also depends upon the vapour pressure of liquid at pumping temperature. The suction lift decreases with increasing vapour pressure of the fluid. Maximum suction also varies inversely with pump capacity.

The head developed by these pumps depends upon the pump design and size, shape, and speed of the impeller. The rotational speed of the impeller and the flow resistance of the system govern the flow capacity. The flow decreases with increasing system resistance (load), as some of the fluid slips back at higher resistance. These pumps, within limits, operate at approximately constant head and variable flow rate. Centrifugal pumps can be operated in a closed off condition (*i.e.* closed discharge line), because the liquid can re-circulate within the pump without causing damage. However, such a condition is better to avoid, as the operation of the pump in this way is unstable. Moreover, energy dissipation within the pump may overheat the fluid and the pump. It will have adverse consequences on the system health.

TERMINOLOGY

Capacity: It is the water handling capability of a pump commonly expressed as water pumped per unit time. In metric units it is expressed in cubic meter per second (m^3/s). For small capacities, unit of L/s is also used.

Cavitation: It is a phenomenon caused by implosion of vapour bubbles (See section on cavitation in this chapter).

Datum: The reference in the horizontal plane used to measure all the elevations and heads.

Discharge Port: The point where the discharge hose or pipe is connected to the pump.

Dynamic Discharge Head: It is the sum of static discharge head and the friction head in the discharge line referred as total discharge head.

Dynamic Suction Head: It is the sum of static suction lift and the friction in the suction line referred as total suction head.

Friction Head: It is the head required to overcome the resistance to flow in the pipe and fittings. It is dependent upon the size, condition and type of pipe, number and type of pipe fittings, flow rate, and nature of the liquid.

Friction Head Losses: It refers to the loss in pressure or head due to friction and turbulence as the liquid flows through pipes, and pipe fittings. In a pumping system, frictional resistance offered to the liquid flowing through suction and delivery pipes causes friction head losses, which in several instances constitute significant components of the total pumping head. Friction head losses are classified as major friction head losses, and minor friction head losses. Major losses are the friction head losses in the pipe network that depend upon the pipe diameter, length of the pipe, material and the roughness of the inner surface of pipe-network, besides the viscosity and density of the flowing fluid. Minor losses are the head losses due to pipe fittings, bends, entry and exit, sudden expansion and contraction, valves, screen, foot valve and so on. The sum of major and minor friction head losses constitutes the 'total friction head losses'. The major friction head losses (H_f) in the suction and delivery pipes of a pumping system are calculated by the Darcy-Weisbach equation or the Hazen-William equation discussed in Chapter 9. Although, it is possible to calculate the minor head losses but more often depending upon the number and kind of fittings, these losses are taken as some fraction of major losses.

Manometric head: Manometric head is the head against which a centrifugal pump has to work *(Minz, 2018)*. It is the sum of the actual lift + the friction losses in the pipes + the discharge velocity head. For special pumps, allowance is also made for the flow velocity at the suction intake and any pressure differences at the water surfaces in the supply and receiving tanks. Following equation is used to calculate the manometric head.

$$h_m = h_s + h_d + h_{fs} + h_{fd} + (V_d^2 - V_s^2)/2g \quad (13.5)$$

Here, h_m is the manometric head, h_s is the suction head, h_d the delivery head, h_{fs} is the frictional head loss in suction pipe, h_{fd} is the frictional head loss in delivery pipe and V_d is the velocity of pipe in delivery pipe or at the outlet and V_s is the velocity in the suction pipe, being small is often neglected.

Pressure Head: It is prudent to take into account the pressure head, if the pumping system either begins or terminates in a tank, which is under pressure different from the atmospheric pressure. For such cases, first convert the pressure in such a tank to head of fluid. For example, if P is the pressure, then head in m of liquid is P/γ. Eq. (13.5) in such cases modifies to

$$h_m = h_s + h_d + h_{fs} + h_{fd} + P/\gamma + (V_d^2 - V_s^2)/2g \quad (13.6)$$

Note that vacuum in the suction tank or a positive pressure in the discharge tank adds to the system head (Figure 13.8). A positive pressure in the suction tank or vacuum in the discharge tank reduces the system head.

Static Suction Head (h_s): It is the vertical distance between the center of the impeller to the free surface of the water from where water is pumped. Suction head exists when the source of supply is above the center line of the pump (Figure 13.8 top). Suction lift exists when the source of supply is below the center line of the pump (Figure 13.8 bottom). If the liquid level is above pump center line, h_s is positive. If the liquid level is below pump center line, h_s is negative. The condition where h_s is negative is denoted as a suction lift. While calculating the manometric head, suction lift is added while suction head is subtracted.

Static Discharge Head (h_d): Static discharge head is the vertical distance between the pump center line and the point of free discharge or the surface of the liquid in the discharge tank (Figure 13.8).

Total Discharge Head (h_{dt}): It is the sum total of discharge reservoir pressure head (h_{pd}), static discharge head (h_d), the velocity head at the pump discharge flange (h_{vd}) and the total friction head in the discharge line (h_{fd}). If the water is discharging at the atmosphere pressure, h_{pd} is zero. However, if the water is discharging in open air, there may be some velocity head at the outlet.

$$h_{dt} = h_{pd} + h_d + h_{vd} + h_{fd} \quad (13.6)$$

Total Dynamic Suction Lift (h_s): It is the static suction lift minus the velocity head at the pump suction flange plus the total friction head in the suction line. The total dynamic suction lift, as determined during pump tests, is the reading of a gauge on the suction flange, converted to m of liquid and corrected to the pump center line, minus the velocity head at the point of gauge attachment.

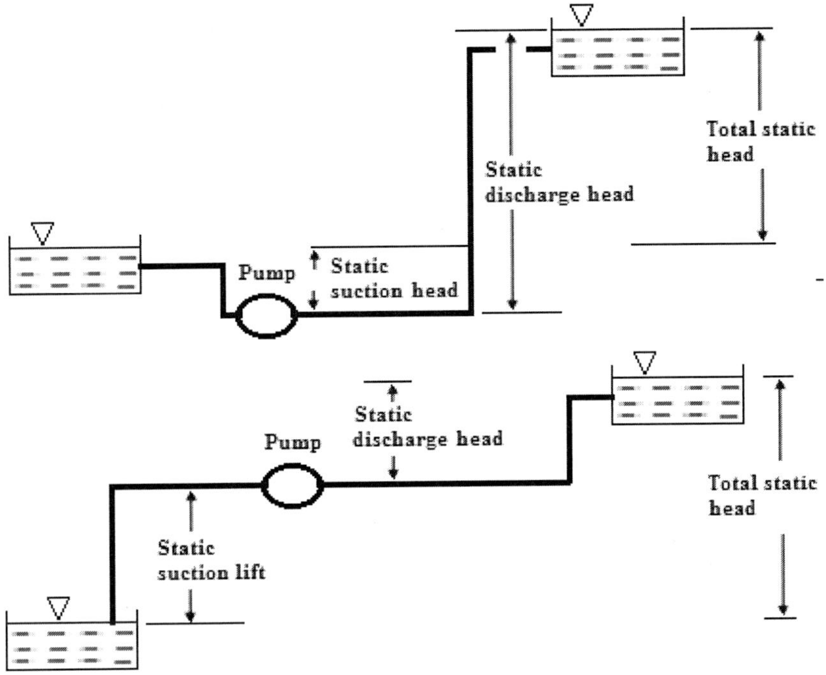

Figure 13.8. Illustration of Static Suction Head (Top) and Static Suction Lift (Bottom).

Total Head (H) Or Total Dynamic Head (TDH): It is the total dynamic discharge head minus the total dynamic suction head or plus the total dynamic suction lift. Total head generated by a pump is called 'pumping head' which is also known as 'manometric head' or 'total dynamic head' (TDH). Whereas eq. (13.5) gives the total dynamic head, Figure 13.9 illustrates the procedure to calculate the various parameters of eq. (13.5).

Total Static Head: It is the vertical distance between the free level of the source of supply and the point of free discharge or the free surface of the discharge tank.

Total Suction Head (h_s): It is the suction reservoir pressure head (h_{ps}) plus the static suction head (h_s) plus the velocity head at the pump suction flange (h_{vs}) minus or plus the friction head in the suction line (h_{fs}).

$$h_s = h_{ps} + h_s + h_{vs} \pm h_{fs} \tag{13.7}$$

Velocity Head (h_v): It is the energy of a liquid because of its motion at velocity, V. It is the equivalent head in m through which the water would have to fall to acquire the same velocity, calculated from the formula: $h_v = V^2/2g$. It is possible to ignore the velocity head in most high head systems. It may not be possible to ignore it in low head systems, where it may be significant relative to the total head.

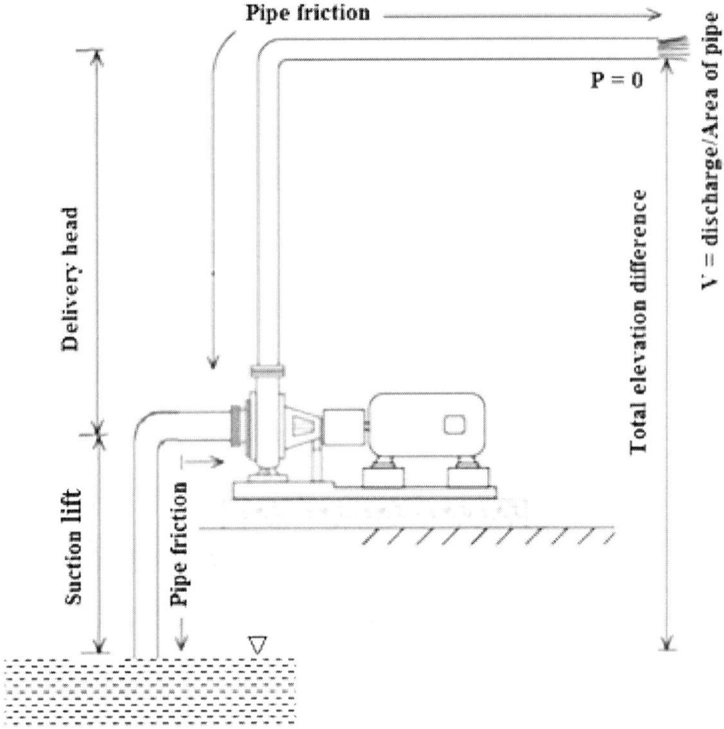

Figure 13.9. Calculation of Total Dynamic Head.

Example 13.2

Calculate the friction head losses in the following set-up shown in Figure 13.10. P_1 and P_2 are the gauge pressures.

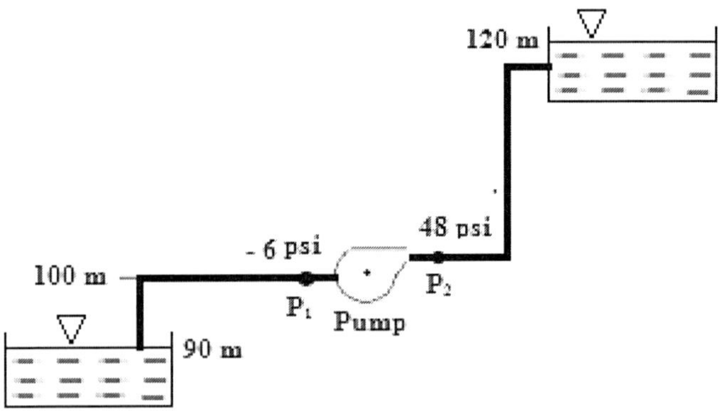

Figure 13.10. Schematics of a Water Pumping System.

Using the reading of the pressure gauges before and after the pump, we have

Total dynamic head generated by the pump = 48 - (-6) = 54 PSI

In term of head of water = 54 PSI × 0.703 = 37.96 m

Total static head = 120 -90 = 30 m

Friction head losses = 37.96 – 30 =7.96 m

PUMP PERFORMANCE OR CHARACTERISTIC CURVES

The pump performance curves are of great significance to the engineer responsible for selecting pumps for a particular flow system (Figure 13.11). These curves comprise (Sahu, 2017).

- Head capacity curve which shows how much water a given pump will deliver at a given head
- Overall efficiency curve representing the relationship between the pump efficiency and the discharge
- Brake horse power curve, giving the relation between the discharge and horse power
- Net positive suction head required over the capacity range

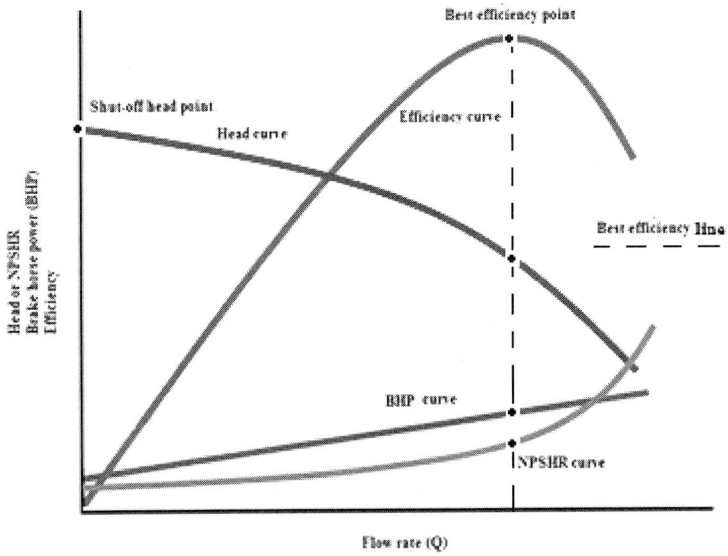

Figure 13.11. Characteristic Curves of a Centrifugal Pump.

The typical characteristic curves show the total dynamic head, brake horsepower, efficiency, and net positive suction head all plotted over the capacity range of the pump at a constant impeller speed (Figure 13.12). Sometimes, the curve relating the net positive suction head required against the capacity range may not be included in the performance curves. The curves shown in Figure 13.12 indicate the variation of head (H), power (P) and discharge (Q) with respect to speed of the impeller. The former is a more common way to represent the performance curves. Some important terms and few characteristics related to performance curves are described as follows.

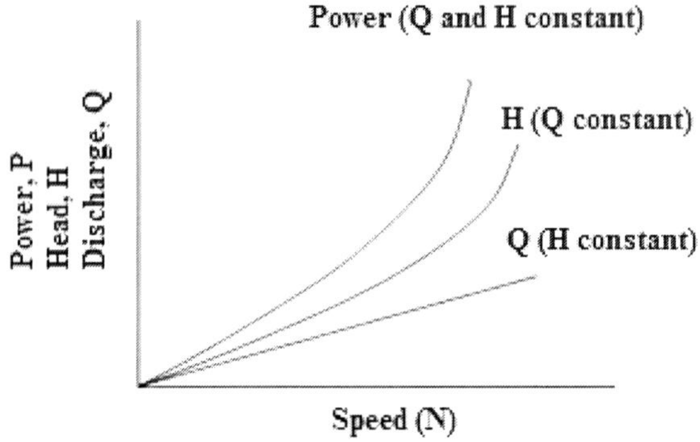

Figure 13.12. Plots of the Characteristic Curves of a Centrifugal Pump.

Head Curve: The head curve rises with decreasing flow rate.

Shut-off Head: It indicates the head developed by the pump at zero discharge. It represents the rise in pressure head across the pump with the discharge valve closed.

Brake Horse Power: The brake horsepower initially increases with increasing discharge, subsequently falling as maximum discharge is approached.

Efficiency: The efficiency, a function of the flow rate, initially increases reaching a maximum value at some flow rate, the normal or design flow rate or capacity of the pump. There is fall in its value thereafter.

Best Efficiency Points (BEP): The points on the various curves corresponding to the maximum efficiency.

Following is the list of some general features of characteristic curves of a centrifugal pump.

- ☆ Head, H is almost constant at low flow rates
- ☆ Maximum 'H'(shut off head) is at zero flow rate
- ☆ Head drops to zero as discharge, Q approaches Q_{max}
- ☆ Q is not greater than Q_{max}
- ☆ Efficiency is always zero at $Q = 0$ and $Q = Q_{max}$
- ☆ Efficiency, η is not an independent parameter
- ☆ $\eta = \eta_{max}$ at roughly $Q = 0.6\ Q_{max}$ to $Q = 0.93\ Q_{max}$
- ☆ $\eta = \eta_{max}$ is the best efficiency point (BEP)
- ☆ All the parameters corresponding to η_{max} are called the design points, Q, H, and BHP
- ☆ Pumps design should be such that the efficiency curve should be as flat as possible around η_{max}
- ☆ BHP rises almost linearly with flow rate

MATCHING A PUMP TO A PIPELINE

Pump characteristics curves alone are not sufficient to select a pump. The pump must match with the system *i.e.* pipeline, which is a mixture of hydraulics, site engineering and economics. By considering the Darcy-Weisbach expression for the head loss in pipes, it is apparent that a small increase in diameter causes a significant reduction in head loss. A further restriction in the pump selection may be because of the retention time of the fluid in the pipe, which may be limited. Considering all these factors, selected pipe diameter enables one to calculate the friction head as a function of discharge using a system curve. A system curve is a graphical representation of a piping system's energy requirement response to a range of flows. The selected pump needs to supply total head in any piping system comprising of the total system head, the sum of static head and friction head.

$$\text{Static head due to elevation} = (z_2 - z_1) \tag{13.8}$$

$$\text{The frictional head loss} = H_f$$

Therefore, piping system's energy requirement, H_{sys} is given as:

$$H_{sys} = (z_2 - z_1) + H_f \tag{13.9}$$

For laminar flow

$$H_{sys} = (z_2 - z_1) + bQ \tag{13.10}$$

H_{sys} is the total system curve head, m, and $(z_2 - z_1)$ is the static head, m, simply the difference in height of the supply and destination reservoirs (Figure 13.13), b is a constant and can be determined from Darcy-Weisbach equation. According to this equation $H_f = 128\mu LQ/(\pi\rho g D^4)$ such that $b = 128\mu L/(\pi\rho g D^4)$.

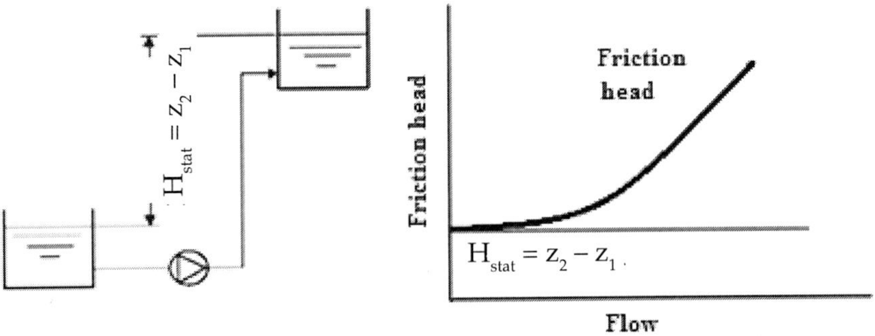

Figure 13.13. Illustration of the Preparation of Pipe System Curve.

For turbulent flow including the minor friction head losses (See eq. 9.40), H_{sys} is given as:

$$H_{sys} = (z_2 - z_1) + cQ^2 \tag{13.11}$$

H_{sys} is the total system curve head, m, c is a constant and can be determined from Darcy-Weisbach equation. Eq. (13.11) is a generic expression of major and minor head losses and is commonly applied in pump selection. Reduction in the system head wherever possible, may help to reduce both the cost of the installation and the cost of pumping the fluid (Gülich, 2007). For a simple pipe system, the system head curve given by eq. (13.11) is parabolic in nature beginning from the static head (Figure 13.13). Proper selection of pump requires matching the system and the pump curves in such a way that the two intersect the pump curve near the high efficiency point (Figure 13.14).

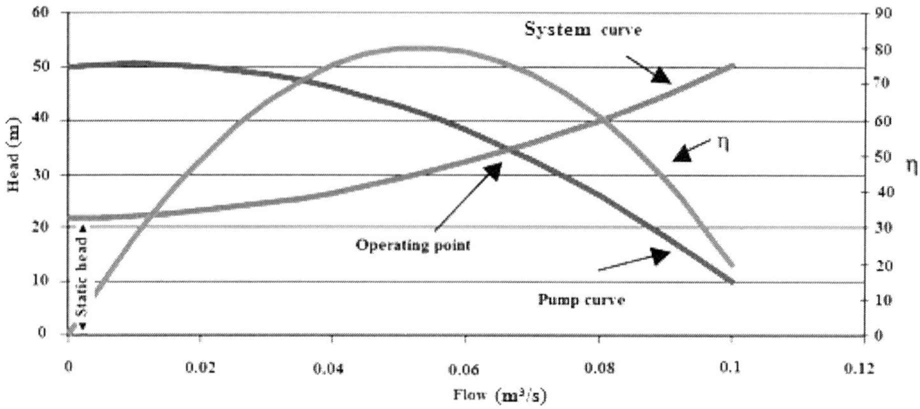

Figure 13.14. Matching the Pump Curve with Water Supply System Curve.

Example 13.2

A rotodynamic pump delivers water from a river at an elevation of 102 m to a reservoir with a water level of 135 m, through a 350 mm diameter cast-iron pipe. The frictional head loss in the pipeline is given by $H_f = 550\ Q^2$, where H_f is the head loss in m and Q is the discharge in m^3/s. Minor head losses from valves and fittings amount to $50\ Q^2$ in the same units. Given the pump characteristics (Table 13.2), develop the system head curve. Determine the discharge of the pump and head to which water can be raised.

Use the following equation to calculate H_f reported in Table 13.2.

$$H_f = Z_2 - Z_1 + 550\ Q^2 + 50\ Q^2$$

$$H_f = 135 - 102 + 600\ Q^2$$

$$H_f = 33 + 600\ Q^2$$

Plot Q versus H_f to obtain the match point at X (Figure 13.15)

The value of discharge at the match point is 0.137 m^3/s and the head is 44 m.

Table 13.2. Pump Discharge, Head, Efficiency and Total System Head (Calculated)

Q (m^3/s)	0.0	0.05	0.10	0.15	0.20
H (m)	60	58	52	41	25
η (per cent)	—	44	65	64	48
H_f		34.5	39	46.5	57

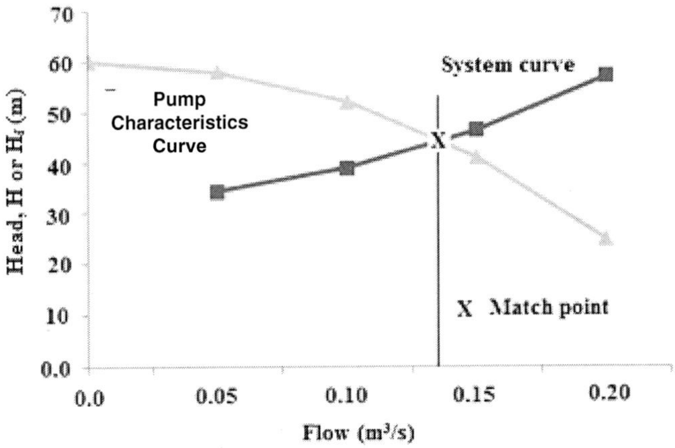

Figure 13.15. Matching the Pump Characteristics Curve with System Head Curve.

AFFINITY LAWS

Effect of Impeller Speed

A centrifugal pump is a dynamic device where a rotating impeller generates the required head. There is a distinct relationship between the impeller peripheral velocity and the head generated. For a fixed impeller diameter, peripheral velocity directly relates to shaft rotational speed (Michael *et al.*, 2008). Thus, varying the rotational speed has a direct effect on the performance of the pump. Therefore, it is important to know as to how the performance parameters vary so that one can control the pump speed accordingly. The equations relating pump performance parameters of flow, head generated and power required to speed of the pump are known as affinity laws. Mathematically, these are expressed as:

Flow is proportional to the speed

$$Q_1/Q_2 = N_1/N_2 \tag{13.12}$$

Head is proportional to the square of the speed

$$H_1/H_2 = N_1^2/N_2^2 \tag{13.13}$$

Power is proportional to the cube of the speed

$$P_1/P_2 = N_1^3/N_2^3 \tag{13.14}$$

Here Q is the flow rate, H is the head, P is the power required and N is the rotational speed. The subscripts 1 and 2 indicate these parameters before and after the change in speed. It can be seen from the above laws that doubling the speed of the centrifugal pump increases the power consumption by 8 times. Conversely, a small reduction in speed will result in significant reduction in power consumption. This forms the basis for energy conservation in centrifugal pumps with varying flow requirements.

NPSHR also varies approximately with the square of speed in the same way as pump head. Thus, following equation helps to assess NPSHR at one speed to some other speed.

$$NPSHR_1/NPSHR_2 = N_1^2/N_2^2 \tag{13.15}$$

There is a minimum NPSHR plateau at very low speeds and as such, NPSHR does not tend to zero at zero speed. Therefore, it is necessary to consider NPSHR carefully in variable speed pumping.

Effect of Impeller Diameter

Changing the impeller diameter gives a proportional change in peripheral velocity, so it follows that equations similar to the affinity laws exist for variation in the pump performance as a function of the impeller diameter D.

Flow is proportional to the impeller diameter

$$Q_1/Q_2 = D_1/D_2 \tag{13.16}$$

Head is proportional to the square of the impeller diameter

$$H_1/H_2 = D_1^2/D_2^2 \tag{13.17}$$

Power is proportional to the cube of the impeller diameter

$$P_1/P_2 = D_1^3/D_2^3 \tag{13.18}$$

Here parameters have the same meaning as described previously. The subscripts 1 and 2 indicate these parameters before and after the change in impeller diameter.

Example 13.3

The impeller diameter of a pump was reduced from 20 cm to 15 cm while its speed remained constant. What are its flow capacity, head and power requirement, if the initial values of these parameters are 20 L/s, 20 m and 5 BHP?

To calculate the flow capacity, we use eq. (13.16):

$$Q_2 = Q_1\,(D_2/D_1) = 20\,(15)/(20) = 15 \text{ L/s}$$

To calculate the head, we use eq. (13.17):

$$H_2 = H_1\,(D_2/D_1)^2 = 20(15)^2/(20)^2 = 11.25 \text{ m}$$

Use of eq. (13.18) calculates the power consumption:

$$P_2 = P_1\,(D_2/D_1)^3 = 5\,(15/20)^3 = 2.1$$

Pump Efficiency

The overall efficiency (η_o) of a pump is defined as the ratio of the power output from the pump to the power input to the pump *i.e.* from the prime mover.

$$\eta_o = \gamma Q H/\text{Shaft power in Watt} \tag{13.19}$$

$$\eta_o = \gamma Q H/(1000 \times \text{Shaft power in kW}) \tag{13.20}$$

Here Q is the discharge of the pump (volume of liquid actually delivered per second by the pump), and γ is the specific weight of the liquid and H is the total head delivered by the pump.

The overall efficiency described by eq. (13.20), η_o comprises manometric efficiency, volumetric efficiency and mechanical efficiency such that overall efficiency is the product of these three efficiencies (Fox and McDonald, 1994).

$$\eta_o = \eta_m \times \eta_v \times \eta_{me} \tag{13.21}$$

Here η_m is the manometric efficiency, η_v is the volumetric efficiency and η_{me} is the mechanical efficiency.

Manometric efficiency

Also known as hydraulic efficiency, η_m is the ratio of the manometric head developed by the pump to the head imparted by the impeller to the liquid. That is,

$$\eta_m = H/(H+\Delta H) = H/H_{th} \tag{13.22}$$

Here H is the actual head develop by the pump and H_{th} is the theoretical head developed by the impeller given as H+ΔH.

Volumetric efficiency

Practically, there is some water loss due to leakage of the fluid between the back surface of the impeller hub plate and the casing, or through other pump components. The ratio of the quantity of liquid discharged per second from the pump to the quantity of liquid passing per second through the impeller is the volumetric efficiency. Volumetric efficiency is expressed as:

$$\eta_v = Q/(Q+\Delta Q) \tag{13.23}$$

Where, Q is the actual discharge of the pump and ΔQ, the leakage loss in the centrifugal pump.

Mechanical efficiency

Mechanical efficiency (η_{me}) is defined as the ratio of the power actually imparted by the impeller to the power supplied to the shaft by the prime mover. That is

$$\eta_{me} = P_h/P_s \tag{13.24}$$

$$P_h = (\gamma\,(Q+\Delta Q)\,H_{th}/1000) \tag{13.25}$$

$$P_s = \text{Shaft horse power}$$

Overall efficiency

Combining all the efficiencies as per eq. (13.21)

$$\eta_o = H/H_{th} \times Q/(Q+\Delta Q) \times (\gamma\,(Q+\Delta Q)\,H_{th}/1000)/\text{Shaft horse power} \tag{13.26}$$

Simplifying

$$\eta_o = (H \times Q \times \gamma)/(1000 \times \text{Shaft horse power}) \tag{13.27}$$

It is the same equation as the overall efficiency (eq. 13.20) given in the beginning of this section.

COST OF PUMPING

Water power is the theoretical power required to pump any fluid. It is the power required by a pump in lifting a given quantity of fluid in a unit time assuming no loss of power in the pump (Miller, 1995). If Q is the discharge of pump and H the total head against which the pump is working, then the water power generated by the pump or minimum power required to operate the pump is given as:

$$WP = \gamma Q H \text{ Newton-m/s} \tag{13.28}$$

Where, WP is the water power, γ is the specific weight of the water (N/m^3), Q is the discharge of the pump (m^3/s), and H is the total dynamic head (m). Since 1 Newton-m/s is equal to 1 W, the water power in eq. (13.28) is in Watts. Since γ (N/m^2) is also written as ρg where ρ is the density of the liquid (Water ~ 1000 kg/m^3) and g is acceleration due to gravity (9.81 m/s^2), we can write

$$WP = 1000 \times 9.81\, Q\, H = 9810\, QH \tag{13.29}$$

Since 1 watt is also equal to 0.00134 horsepower, we have

$$WHP = 9810\, QH \times 0.00134 = QH/0.076 \tag{13.30}$$

Writing Q in L/s, we have:

$$WHP = QH/76 \sim QH/75 \tag{13.31}$$

Eq. (13.28) is valid for fluids including water but eq. (13.29) onwards are specific to water.

Shaft Horse Power: It is the power required at the pump shaft and can be derived as:

$$\text{Shaft Horse Power} = \text{Water Horse Power/Pump efficiency} \tag{13.32}$$

Brake Horse Power (BHP): The actual horsepower a engine or an electric motor must supply to drive the pump.

$$\text{For direct driven pump, BHP} = \text{Shaft Horse Power} \tag{13.33}$$

With belt or indirect drives, BHP = Water Horse Power/(Pump efficiency × Drive efficiency)

Horse Power Input to Electric Motor (IHP):= Water Horse Power/ (Pump Efficiency ×Drive Efficiency × Motor Efficiency) (13.34)

Kilowatt input to electric motor or energy consumption in kilowatt hours is given as:

$$\text{Energy in Kilowatts} = \text{IHP} \times 0.746 \text{ or BHP} \times 0.746/\text{Motor Efficiency} \tag{13.35}$$

Cost of operation of electric motor = Energy in kilowatts× Hours of pumping×Cost per Kilowatt Hour (13.36)

Pump selection criteria

We are by now quite familiar with reciprocating and hydrodynamic pumps discussed in the previous and this chapter. It is clear that numbers of pumps are available within and across the groups to suit a specific condition. It is also clear that not all the pumps suit a given setting. Therefore, one has to be very careful while choosing a pump. The selection criteria for pumps discussed in previous chapter remains valid. Readers are advised to refer chapter 12 for detailed information on this issue.

Example 13.4

A centrifugal pump delivers 950 L/min to fill a square tank having an area of 1.0 ha. If the pump operates for 7 hours a day, what is the depth of water in the tank at the end of the day?

Water delivered per day (m^3) = $(950 \times 60 \times 7)/1000 = 399\ m^3$

Depth of water in the tank (mm) = Water delivered/Area

Depth of water in the tank (mm) = $399 \times 1000/(1.0 \times 10000) = 39.90$ mm

Example 13.5

A pump operates for 15 hours a day. What should be its capacity to fill a tank having an area of 0.25 ha to 50 mm?

If Q is the capacity of the pump, then

The quantity of water pumped in a day = $Q \times 15 = 15\ Q\ m^3$

Water required = $50 \times 0.25 \times 10000/1000 = 125\ m^3$

Therefore, $Q \times 15 = 125$

$Q = 8.33\ m^3/hr$ or 2.33 L/s

Example 13.6

A pump delivers 21000 liters of water to a height of 45 m in every five minutes. If the efficiency of the pump is 75 per cent, calculate the horsepower required to drive the pump.

Water delivered per second = $21000/(5 \times 60) = 70$ L/s

WHP = $70 \times 45/75 = 42$ (eq. 13.31)

Assuming that the pump is direct driven, we have

SHP = BHP = $42/0.75 = 56$ HP

Example 13.7

A pump lifts 100 m^3/hr of water against a total head of 20 m. Calculate the water power of the pump. If the pump efficiency is 75 per cent, what size of prime mover is required to operate the pump? If a direct driven electric motor with an efficiency of 90 per cent drives the pump, compute the cost of electric energy in the month of April. Assume that the pump operates 10 hours a day for the whole month. Assume the cost of electricity charges as Rs 6.00 per unit.

Discharge of the pump (Q) = $100/(60 \times 60) = 0.028\ m^3/s$, Total head (H) = 20 m

Water power, WP = $\gamma \times Q \times H = 9810 \times 0.028 \times 20 = 5493.6$ W or 5.5 kW (Eq. 13.28)

Check: Water power, WP = $Q \times H/75 = (28 \times 20/75) \times 0.746 = 5.6$ kW (Nearly same)

Since the efficiency of the pump is 75 per cent, the shaft power required is

Shaft power = 5.5/0.75 = 7.33 kW

For a direct driven pump, the shaft power determines the required size of the prime mover. Since motor efficiency is 90 per cent, Input Horse Power (IHP) of the motor is given as:

Input horse power = 7.33/0.90 = 8.14 kW

Total energy consumption per month = Input power × Hours of pump operation × Number of days in a month

= 8.14 × 10 × 30= 2442 kW-h

Since 1 kW-h is equal to 1 unit, therefore, the cost of electric energy = Rs. (2242 × 6) = Rs. 14652 or 14650.

Example 13.8

A centrifugal pump is required to lift water at the rate of 100 L/s. Calculate the brake horsepower of the engine from the following data if pump discharges water at the same level as the pump center.

- ☆ Suction head = 6 m
- ☆ Efficiency of pump = 65 per cent
- ☆ Delivery head may be taken as zero
- ☆ Diameter of pipe = 15 cm
- ☆ f of the Darcy's-Weisbach equation is= 0.01

$Q = 100$ L/s or $0.1\ m^3/s$

Delivery head, $H_d = 0$ (As pump discharges water at the same level as the pump center)

$H = 6$ m, $f = 0.01$, $d = 15$ cm $= 0.15$ m

The length of the pipe where frictional effect may occur is taken equal to the suction head, so

$H_f = f\,LV^2/(2gd) = 8\,f\times L\times Q^2/(\pi^2 \times g\times d^5)$ (Note: $Q = V/A$)

$H_f = 8 \times 0.01\times 6 \times (0.1)^2/\{(3.142^2 \times 9.81 \times (0.15)^5\} = 0.0048/(96.846\times 0.000076)$

$H_f = 0.65$ m

Neglecting the minor head losses, we have

Total head, $H = H_s + H_d + H_f = 6 + 0 + 0.65 = 6.65$ m

Efficiency, $\eta = 65$ per cent $= 0.65$

Horse Power (HP) = $QH/(75\,\eta) = 100\times 6.65/(75\times 0.65) = 13.64$ or nearest 15 HP

Example 13.9

Calculate the cost of pumping 4 million liters of water from a well with a centrifugal pump from the following data.

Suction head = 3 m; Delivery head = 7 m; Friction head = 1.5 m

Output of the pump = 120000 L/hr; Pump efficiency = 70 per cent

Motor efficiency = 85 per cent ; Cost of electricity = Rs. 2 per unit.

Total head = 3+7+1.5 +0.1×1.5 = 11.65 m (Assuming minor friction losses as 10 per cent of major friction losses)

From eq. (13.31)

Water Horse Power = (1200000/3600) ×11.65/75 = 5.17

Brake horse power = WHP/pump efficiency = 5.17/0.7 = 7.4 ~ 7.5

Time required to pump 4 million liters of water = $4\times10^6/4\times10^4$ = 100 hr

Therefore units consumed = 7.5×100/0.85 = 882.4 kwH

The cost of pumping = 882.4×2 = Rs. 1764.8 ~ Rs. 1765

Example 13.10

Calculate the water horsepower developed by the centrifugal pump if flow rate is 4500 L/m and head added to the flow is 10 m. What is the brake horse power required, if the direct driven pump has efficiency of 85 per cent ?

Q = 4500/(1000×60) =0.075 m^3/s

From eq. (13.28)

Water power = ρ×g×H ×Q = 1000×9.81×0.075×10 = 7357.5 Watts

HP = 7357.5× 0.00134 = 9.85 HP

BHP = 9.85/0.85 = 11.6 HP

PUMPS IN SERIES AND PARALLEL

If a single pump fails to provide adequate head or flow capacity, connecting two or more than two pumps either in series or in parallel can resolve the problem (Figure 13.16). For pumps in series, the discharge side of the first pump connects to the suction side of the second pump (Figure 13.16 top). Two similar pumps in series, operate in the same manner as a two-stage centrifugal pump. Thus, in series arrangement, each pump handles same flow rate, but the total head produced by the combination of pumps is additive. For example, if pumps generate head H_1 and H_2 corresponding to a flow Q, the total head developed is $H_t = H_1 + H_2$, where H_1, H_2 are the heads developed by the two pumps in series at the common flow rate Q. Following points are important while connecting two or more pumps in series.

- ☆ Both pumps must have the same impeller diameters as the difference in capacities will cause cavitation if the first pump is unable to supply enough liquid to the second pump
- ☆ For the same reason, both pumps must run at the same speed
- ☆ Be sure the casing of the second pump is strong enough to resist the high pressure. High strength material, ribbing, or extra bolting may be required

- ✰ The stuffing box of the second pump likely to experience discharge pressure of the first pump, must have a high-pressure mechanical seal
- ✰ Be sure both pumps are filled with liquid during start-up and operation
- ✰ The second pump should be operated while the first pump is in operation.

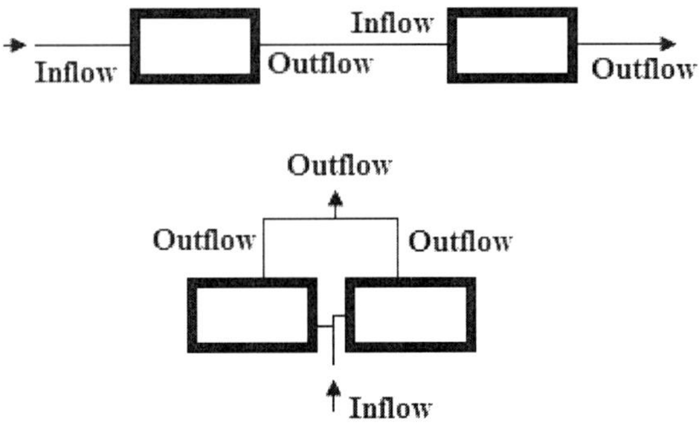

Figure 13.16. Installation of Pumps in Series (Top) and Parallel (Bottom).

Pumps are said to be in parallel when two or more pumps are connected to a common discharge line, and share the same suction conditions (Figure 13.16 bottom). The total flow rate Q_t splits between the inlet into Q_1 and Q_2 ($Q_t = Q_1 + Q_2$). Each pump develops the same head H at the corresponding capacity. Thus, the first pump at capacity Q_1 develops the same head H as the second pump at capacity Q_2. Some important points that must be kept in view while putting pumps in parallel are as follows.

- ✰ Both pumps must produce the same head. It usually means that the two pumps must have the same impeller diameter and run at the same speed
- ✰ Two pumps in parallel will deliver less than twice the flow rate of a single pump in the system because of the increased friction in the piping.
- ✰ The shape of the system curve determines the actual increase in capacity. If there is additional friction in the system from throttling, two pumps in parallel may deliver only slightly more than a single pump operating by itself

The efficiency curve of two or more identical pumps in series or parallel will be the same as that of individual pump.

PRIMING OF CENTRIFUGAL PUMPS

Most centrifugal pumps lose prime due to ingestion of air or vapour into the pump. When vapours or gases fill the pump casing, the pump impeller becomes gas-bound and is incapable of pumping. It is because the clearances and displacement ratios in pumps used for liquids and other more viscous fluids cannot displace the air due to its lower density. Under such a situation, pump requires priming. The operation in which suction pipe, casing of the pump and a portion of delivery pipe is filled by the fluid from outside is known as priming. This process helps to remove air from the impeller and casing necessary to pump the liquid. To ensure that a centrifugal pump remains primed and does not become gas-bound, it is necessary to attach a reflux or foot valve at the end of the suction pipeline or use a self-priming gadget.

NET POSITIVE SUCTION HEAD (NPSH)

Normally two terms are used in the context of NPSH. These are NPSH available (NPSHA) and NPSH required (NPSHR). The difference between the head at the suction flange of the pump and the saturation pressure head of the pumped fluid is NPSHA. NPSHA is normally obtained from the following equation to judge how close the suction head is to the saturated pressure head.

$$NPSHA = H_s - H_{sat}$$

Here H_s is the suction head and H_{sat} the saturation pressure head. A more elaborate equation (eq. 13.37) is normally used to calculate NPSHA or the distance from the water surface to the center line of the suction of the pump (Z_1) so that pre-decided NPSHA is available at the suction flange.

$$NPSHA = P_s/\rho g - Z_1 - H_f - P_{sat}/\rho g - V^2/2g \tag{13.37}$$

Here Z_1 is the distance between the water surface and at the center line of the pump, P_s is the pressure at the water surface on the suction side, usually atmospheric at that location. Pump manufacturers, on the other hand, prescribe NPSHR. To avoid cavitation, NPSHA should be greater than NPSHR. Eq. (13.37) is commonly used to find out the value of Z_1 so that the NPSHA >NPSHR. Following example illustrates its application.

Example 13.11

A pump requires 2.5 m of NPSHR at 30 m^3/hr. The pump installed at an altitude of 1500 m above sea level pumps water at 25°C. What is the maximum suction lift it can attain? Ignore friction and velocity heads.

To calculate air pressure above sea level, we use the approximate equation

$$P_s\,(Pa) = 101325\,(1 - 2.25577\;10^{-5}\,h)^{5.25588}$$

Here h is the altitude above mean sea level. Incorporating, h =1500 m, we have

$$P_s = 82493 \text{ Pa}$$

Similarly, to find P_{sat}, we use the following approximate relation

$$P_{sat} = Exp\,(20.386 - 5132/T)$$

Here P_{sat} is the vapour pressure in mm Hg and T is the temperature in Kelvin (°C +273.15).

P_{sat} for 25°C is given as:

$$P_{sat} = Exp\,(20.386 - 5132/298.15) = 23.88 \text{ mm of Hg}$$

$$P_{sat} = P_{sat}\,(\text{mm of Hg}) \times 133.322 = 3183.73 \text{ Pa}$$

Incorporating the calculated values in eq. (13.37), we get

$$NPSHA = 82493/9810 - Z_1 - 0 - 3183.73/9810 - 0$$

Since NPSHR is 2.5 m, minimum NPSHA should be about 0.6 m (2 ft) more than the NPSHR, we have

$$2.5 + 0.6 = 8.41 - Z_1 - 0.32$$

$$Z_1 = 4.99 \sim 5.0 \text{ m}$$

For more accurate value of Z_1, we should also use the actual values of H_f and velocity head in eq. (13.37).

WATER HAMMER

Water Hammer is a pressure surge or wave that occurs when there is a sudden change in momentum of a fluid (the motion of the fluid abruptly forced to stop or change direction) within an enclosed space. The most recognizable sign of water hammer is a loud banging in the pipes, as if being hit with a hammer. Vibrations of the piping system are also an indicator of water hammer. Water hammer commonly occurs in pipelines if a valve at the end of the pipeline where the velocity of the fluid is high, is suddenly closed (Figure 13.17). It results in reversal of fluid momentum. The pressure wave so created propagates within the pipeline, resulting in a loud noise (hammer) and vibrations in the system. Another cause of water hammer is sudden stoppage of the pump. The sudden stoppage also produces the momentum change causing the water hammer effect. In addition, severe water hammer can also occur when the flow reverses because of instantaneous closure of the foot valve. Water hammer causes significant damage to pipe systems. The high pressure in the system can result in cracking of pipes. Pipes may even burst in worst cases. Water hammer may also cause cavitation in the pipelines. Plastic pipe systems are particularly susceptible to damage due to water hammer. While it is impossible to eliminate water hammer, several strategies can be adopted to minimize its adverse effects.

- ☆ Use slightly oversized valves, especially valves designed to minimize pressure surging

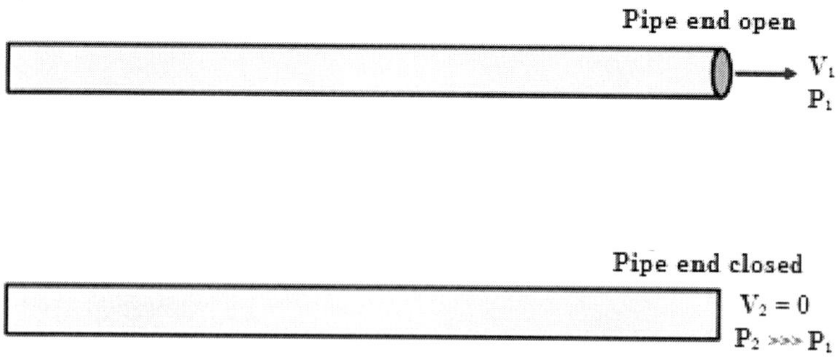

Figure 13.17. Schematic Illustration of the Pressure Surge Due to Sudden Stoppage of Flow.

- For residential or small commercial systems use pipes that are rated at least twice the design static pressure, and limit the velocities to 3 m/s or less
- Install automatic air release valves at high points in the pipelines
- The most common step is to use a check valve on the discharge line (at the pump outlet). It protects the pump from any back surge down the pipeline. Use silent instead of conventional check valves. Silent check valves close while the water is still flowing forward, thus preventing the reverse flow
- Use water hammer arresters. These devices use the energy from the pressure wave to compress air. The most inexpensive water hammer arrester is a vertical pipe, slightly larger than the supply line, with a pressure release valve at the upper end. The pressure release valve is set at a pressure less than the bursting strength of the supply line, but more than the operating pressure of the pipe system.

CAVITATION

Cavitation is defined as the phenomenon of formation and collapsing of vapour bubbles of a flowing liquid in the pumping system. Bubbles form in a region where the pressure in the system falls below liquid's vapour pressure. The bubbles collapse upon reaching in a region of higher pressure. Collapsing of vapour bubbles creates very high pressure in the system. Cavitation causes pitting action on the surface as it is subjected to high pressure. It results in the formation of cavities. Besides, cavitation results in considerable noise and vibrations (Similar effect as that of a water hammer) and reduces the efficiency of the pumps. Common causes of cavitation are:

Drop in Pressure at the Suction Nozzle due to Low NPSHA: Each fluid at the given operating conditions has a specific vapour pressure. If the pressure at the pump suction drops below the vapour pressure of the liquid, fluid vaporizes to form gas bubbles. The explosion of these gas bubbles as per the process discussed in the previous paragraph is the major cause of cavitation.

Increase in Temperature of the Pumped Liquid: The vapour pressure of the liquids increases with increasing liquid temperature. Thus, there are chances for the operating pressure to fall below the vapour pressure with rising temperature. It will lead to formation of bubbles and cavitation.

Increase in the Fluid Velocity at Pump Suction: An increase in fluid velocity at the pump suction can typically occur when liquid flow rates are higher than the design flow rates. As per Bernoulli's principle, higher the liquid velocity, lower will be the pressure head. Moreover, frictional pressure in the piping system increases with increasing flow rates. Chances of low pressure developing resulting in cavitation under this situation is expected.

Reduced Flow at Pump Suction: Certain minimum flow is required to keep the centrifugal pumps from running dry. If flow rate falls below this limit, there is every possibility of vapours forming in the pump and consequent cavitation.

Undesirable Flow Conditions: Sharp elbows, valves, other fittings and/or any obstruction in the piping system cause higher frictional losses in the pump suction. For the same suction lift, pump suction head will decrease because of increased friction loss. It may lead to cavitation.

Inappropriate Pump Selection: Every centrifugal pump has a certain requirement of positive suction head (NPSHR). With improper selection of the pump, NPSHA might fall below the NPSHR, causing cavitation. The minimum value of NPSHA needed to prevent cavitation is slightly higher than the recommended NPSHR by the manufacturer.

From the foregoing discussion, it emerges that impeller inlet region is more prone to cavitation since the lowest pressure in the system is experienced here. Besides, cavitation is most likely to occur near the point of discharge in radial and mixed flow impellers where the velocities are high. In axial flow pump, the blade tip is most vulnerable to cavitation. It is quite difficult to predict the exact location where cavitation may occur. In a similar manner, it is quite difficult to judge the exact extent of damage. On the other side, several options are available to prevent cavitation. For an existing pump, pump operators can adopt the following strategies.

- ✰ Reduce suction lift or increase suction head so as to always operate in NPSHA > NPSHR range
- ✰ Wherever possible, avoid using excessive number of valves, bends, and curves in the piping system. If necessary use the higher radii fixtures
- ✰ Use short suction line with largest possible pipe diameter
- ✰ Control the rise in temperature of the fluid
- ✰ Inject high pressure air into the flowing liquid in low pressure zones
- ✰ Reduce the pump speed

A designer at the design stage of the pump can consider the following options. It will minimize chances of cavitation during its use.

- ✰ Design the system so that there is no separation of flow. Avoid sharp corners or curvatures of flow in the system that may produce vertices or eddies
- ✰ Use cavitation resistant materials such as starlite, bronze, nickel *etc.* (Note: No material is cavitation proof but can only resist cavitation)
- ✰ Ensure high degree of finish of inside surfaces as it helps to reduce the occurrence of cavitation as compared to a low surface finish
- ✰ Use impellers with double curvature blades drawn well forward into the suction orifice
- ✰ Reduce thickness of the impeller blades

TURBINE PUMP

Vertical turbine pump or deep well turbine pump is vertical axis centrifugal or mixed flow type pump comprising stages, which accommodates rotating impellers and stationary bowls with guide vanes. Turbine pumps are used whenever water level in the tube wells or in the open well or the sump is below the practical suction lift limit of the centrifugal pumps. Although a turbine pump also operates on the principle of centrifugal pump, it differs from volute type as stationary guide vanes guide upward the water thrown by the impeller to the periphery. The gradually enlarging vanes guide the water to the casing, thus converting kinetic energy into potential energy. Therefore, a turbine pump generates heads several times that of a volute pump. The vertical turbine pump consists of the five basic elements, namely pump bowl assembly, column pipe assembly, line shafting, surface discharge head and the driver. It has a shaft on the bottom end of which is the turbine and on the top a driving motor, placed above the ground level (Figure 13.18 left).

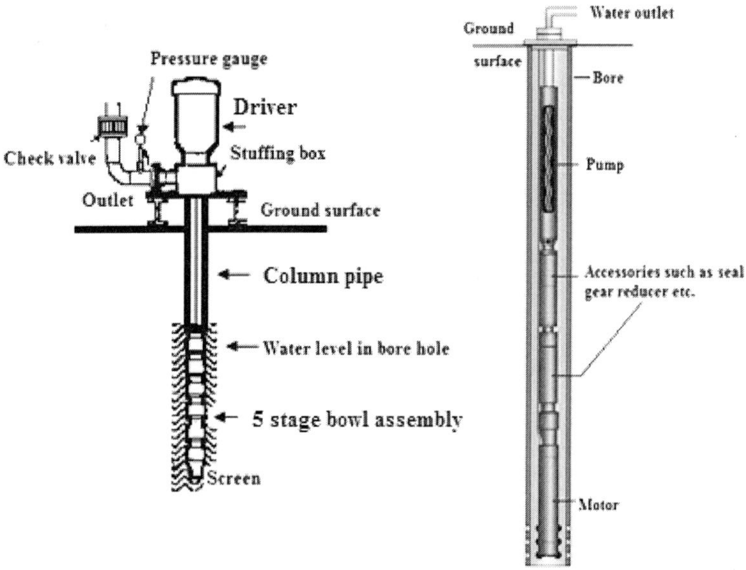

Figure 13.18. A Schematic Diagram of a Vertical Turbine (Left) and a Submersible Pump (Right).

Pump Bowl Assembly: Bowl assembly is the main pumping element of a vertical turbine pump. The number of bowls or stages in an assembly depends upon the head requirement of an individual installation. A typical single stage pump bowl assembly is made up of a suction case or bell, an impeller (enclosed or semi open type), a top intermediate bowl and a discharge case. Impellers are connected to the pump shaft by split cone shaped impeller lock collets or by key. The suction case/bell serve as the intake of pump bowl assembly. Liquid moved by the impeller flows through the pump bowls. The top inter bowl acts as a diffuser and guides the flow into the discharge case, which in turn delivers it into the column pipe. For units of two or more stages, an intermediate bowl and an impeller are added for each additional stage. Every inter bowl acts as an inter stage diffuser and directs the fluid vertically to the next stage impeller. Since the pump unit remains submerged in water, the pump does not require priming. To ensue some minimum submergence at the beginning of suction, pumping unit must be located below the drawdown level of the water source. Pump shaft bearings are located in the suction bell, top inter bowl/discharge case, and in each inter bowl (for multi stage pumps) between stage impellers. Basic design features of a bowl assembly differ in different types of lubrication arrangement made to lubricate the shaft *i.e.*, self water lubrication or oil lubrication. From a sanitary point of view, lubrication of pump bearings by water is always preferable, since lubricating oil may leak and contaminate the water.

Column Pipe Assembly: The column pipe assembly conveys water from bowl assembly, located deep in the well or sump, to discharge head at ground level. Column pipe is either threaded or flanged, depending upon the pump size or customer's requirements. A threaded column features straight threads on both the column and column couplings. Flanged column pipes have male and female spigots to ensure proper alignment of the two adjacent lengths of column pipes. The flanged column pipes bolted together with a gasket between the flanges help to eliminate leakage.

Line Shafting: A vertical turbine pump's line shaft assembly rotates inside the column pipe and transmits torque from the pump driver to the bowl pump assembly. A lead washer inserted between the two shaft ends facilitates dismantling the joint when necessary. Bearings at specific intervals support the shafting.

Surface Discharge Head: Surface discharge heads perform multiple functions. Firstly, it provides a base from which the pump is suspended. Secondly, it directs the fluid flow from the pump column to piping system. Thirdly, it also provides a way to seal the line shaft.

Driver: It is generally a three-phase induction motor. A well-maintained pump provides trouble free service for several years.

The vertical turbine pumps require high initial investments, being costly. Moreover, vertical turbine pumps are difficult to install and repair.

SUBMERSIBLE PUMP

These are special kind of centrifugal pumps where long vertical shaft connecting the motor and pump unit (as in the case of turbine pump) is replaced by a short shaft. The prime mover and pump become closely coupled and both are submerged in water (Figure 13.18 right). Thus, a submersible pump is a centrifugal pump driven by a closely coupled electric motor constructed for submerged operation as a single unit. All submersibles are multi-stage pumps mounted on a single shaft, and all rotating at the same speed. Each impeller passes the water to the eye of the next impeller through a diffuser. The shape of the diffuser is such that it helps to slow down the flow of water and convert velocity head to pressure head. Each impeller and matching diffuser is called a stage. The water inlet into the submersible pump is between the pump and the motor. Submersible pumps are suitable for tube wells having a bore of 100 mm or more. The electrical wiring to the submersible motor must be waterproof. Use of properly grounded electrical controls minimizes the possibility of shorting, which may otherwise damage the entire unit. The discharge pipe supports the pump and motor assembly. For the same output, submersible pumps consume less power besides requiring less space. These pumps are ideal for lifting water from deep tube wells for irrigation, domestic and industrial applications.

QUESTIONS (THEORY)

1. Explain the working principle of a centrifugal pump with appropriate sketch(es).

2. Discuss the pump characterization based on the nature of the flow through the impeller. Explain each of these kinds using neat diagrams showing the kind of flow.

3. Name and describe the various types of casings used in hydrodynamic pumps.

4. Define specific speed. Discuss the pump characterization based on specific speed.

5. Name various kinds of impellers highlighting their use under particular conditions.

6. Define the term: suction head, delivery head, and manometric head.

7. What do you understand by pump's characteristics curves? Show various curves using a neat diagram briefly explaining the various characteristics.

8. What do you understand by piping system curve? What is its role in pump selection? Draw a neat diagram to explain your answer.

9. Discuss the various affinity laws related to impeller speed.

10. Discuss the effect of impeller diameter on the pump capacity, head developed and power requirement.

11. Write brief notes on pump efficiencies, cavitation, water hammer, turbine pump and submersible pump.

12. How will you proceed to calculate the monthly expenditure on power consumed to pump a given quantity of water to a given height?

13. What do you understand by priming of a centrifugal pump? Why you need to prime a centrifugal pump before its start?

14. How will you calculate the maximum static suction lift from known value of net positive suction head required? Clearly mention the assumptions while explaining the process.

15. Differentiate between the following: single and double suction centrifugal pumps, pumps in series and parallel

Objective Type Questions and Answers

Mark the correct or most appropriate answer out of the given choice (A, B, C and D)

CHAPTER 1

Q 1. Characteristic(s) of gases

(a) Is compressible (b) No fixed volume

(c) Have no free surface (d) All of these

Q 2. Characteristic(s) of liquids

(a) Hard to compress (b) Has fixed volume

(c) Has a free surface (d) All of these

Q 3. Which of the three is most viscous

(a) Water (b) Milk

(c) Honey (d) All are equally viscous

Q 4. Viscosity can be expressed in

(a) Poise (b) Centi-poise

(c) N s/m^2 (d) Any of A, B and C

Q 5. Viscosity of liquids ______ with increasing temperature

(a) Increases (b) Decreases

(c) Remains constant (d) None of these

Q 6. Viscosity of gases ______ with increasing temperature

(a) Increases
(b) Decreases
(c) Remains constant
(d) None of these

Q 7. Kinematic viscosity, η is given as

(a) μ/ρ (μ is the dynamic viscosity and ρ is the density)
(b) $\mu \rho$
(c) ρ/μ
(d) None of these

Q 8. An ideal fluid is assumed to be

(a) Incompressible (so that its density does not change)
(b) Flows at a steady rate
(c) Non-viscous
(d) All of these

Q 9. Isothermal compressibility of water in the temperature range of 0-100° C

(a) Increases with temperature
(b) Decreases with temperature
(c) Initially increases with temperature than decreases
(d) Initially decreases with temperature than increases

Q 10. Specific volume, V and density, ρ are related as

(a) $V = \rho$
(b) $V = 1/\rho$
(c) $V = \rho^{1/2}$
(d) None of these

Q 11. Mercury has

(a) A very high vapour pressure
(b) A very low vapour pressure
(c) None of A and B
(d) None of these

Q 12. Boiling point of water

(a) Increases with increasing altitude above the mean sea level
(b) Does not change with altitude above the mean sea level
(c) Decreases with increasing altitude above the mean sea leve
(d) None of these

Q 13. Why a small liquid drop has spherical shape?

(a) Surface tension property of the fluid
(b) Low vapour presure of the fluid
(c) High vapour pressure of the fluid
(d) None of these

Q 14. Surface tension force in a soap bubble is given as (σ, p and d are the surface tension, pressure inside the droplet and diameter of the drop)

(a) $F = 4\sigma (\pi d)$ (b) $F = 2\sigma (\pi d)$

(c) $F = 2\sigma (\pi d^2)$ (d) None of these

Q 15. Water has

(a) Only cohesive properties (b) Only adhesive properties

(c) Both cohesive and adhesive properties (d) None of these

Q 16. Dew drops on the leaves in the morning hours illustrates the

(a) Surface tension phenomenon (b) Cohesion

(c) Adhesion (d) All of these

Q 17. Height of liquid level in the capillary is

(a) Directly proportional to the radius of bore of the capillary

(b) Inversely proportional to the radius of bore of the capillary

(c) Not dependent upon the radius of bore of the capillary

(d) None of these

Q 18. Jurin's law states that capillary rise in two capillaries of radius r_1 and r_2 is related as (h_1 and h_2 are the height of liquid levels in capillaries of radius r_1 and r_2 respectively)

(a) $r_1h_1 = r_2h_2$ (b) $r_2h_1 = r_1h_2$

(c) $r_2 - r_1 = h_1 - h_2$ (d) None of these

Q 19. Compressibility is

(a) Reciprocal of the bulk modulus of elasticity

(b) Same as bulk modulus of elasticity

(c) Not related to bulk modulus of elasticity

(d) None of these

Q 20. Bulk modulus of elasticity, k

(a) Is lower for water than air (b) Is higher for water than air

(c) Is similar for water and air (d) None of these

Q 21. Pressure in primary dimensions is

(a) M/LT^2 ($ML^{-1}T^{-2}$) (b) M/LT ($ML^{-1}T^{-1}$)

(c) ML/T^2 (MLT^{-2}) (d) None of these

Q 22. A decimeter is

(a) 0.1 m (b) 0.01 m

(c) 0.001 m (d) None of these

Q 23. 100° C in Kelvin scale is nearly

(a) 100° K
(b) 373° K
(c) -173° K
(d) None of these

CHAPTER 2

Q 24. A fluid at rest exerts pressure

(a) Parallel to any surface that it contacts
(b) Perpendicular to any surface that it contacts
(c) In any direction depending upon surface characteristics
(d) None of these

Q 25. Pressure at depth, h below the water surface is given as (ρ is the density of the fluid and g acceleration due to gravity)

(a) ρgh
(b) ρh/g
(c) gh/ρ
(d) None of these

Q 26. Pressure at 1 m below the water surface is (assume ρ as 1000 kg/m³)

(a) 0.00981 N/m^2
(b) 101.937 N/m^2
(c) 9810 N/m^2
(d) None of these

Q 27. Absolute pressure is always

(a) Negative
(b) Can be both positive and negative
(c) Positive
(d) None of these

Q 28. In a hydraulic press, the diameter of larger piston is twice that of smaller piston. If force F1 is exerted on the smaller piston, the force on the larger piston will be

(a) 2F1
(b) 4F1
(c) F1
(d) F1/2

Q 29. A manometer indicates a negative gauge pressure of -50 kN/m^2. The absolute pressure is (assume atmospheric pressure as 101.2 kN/m^2).

(a) 151.2 kN/m^2
(b) 51.2 kN/m^2
(c) 2.02 kN/m^2
(d) None of these

Q 30. Resultant hydrostatic force acts on

(a) Center of mass
(b) Center of gravity
(c) Center of pressure
(d) None of these

Q 31. A rectangular plate submerged in water is near the water surface. In another case, it is much deeper. The center of pressure will be nearer to the center of gravity in

(a) Latter case
(b) Former case
(c) Will be equally away in both cases
(d) None of these

Q 32. The pressure unit torr is equal to

(a) 1 mm of mercury
(b) 760 mm of mercury
(c) 1 m of mercury
(d) None of these

Q 33. In a U-tube inverted differential manometer, normally the manometer fluid used is

(a) Lighter than the working fluids
(b) Heavier than the working fluids
(c) No hard and fast rule, can be lighter or heavier
(d) None of these

CHAPTER 3

Q 34. Center of Buoyancy is the point at which all the forces of buoyancy act in a

(a) Vertically upward direction
(b) Vertically downward direction
(c) Direction depends upon the resultant force
(d) None of these

Q 35. Ratio of the height of the body that submerges in water to the total height of the body is given as (ρ_0 is the average density of the body and ρ the density of water)

(a) ρ/ρ_0
(b) ρ_0/ρ
(c) $\rho_0/(\rho + \rho_0)$
(d) None of these

Q 36. A wooden block having an average density of 500 kg/m^3 is floating in water having density of 1000 kg/m^3. What is the fraction of the total height that submerges in water?

(a) 0.5
(b) 0.75
(c) It gets fully submerged
(d) None of these

Q 37. For stable equilibrium of a floating body (M is metacenter and G center of gravity)

(a) M is above G
(b) M coincides with G
(c) M is below G
(d) Center of buoyancy coincides with G

Q 38. For neutral equilibrium of a floating body (M is metacenter and G center of gravity)

(a) M is above G
(b) M coincides with G
(c) Center of buoyancy coincides with center of gravity
(d) M is below G

Q 39. The equilibrium of a floating body will be unstable, if (M is metacenter and G center of gravity)

(a) M is above G
(b) M coincides with G
(c) Center of buoyancy coincides with G
(d) M is below G

Q 40. For a stable equilibrium to exist in submerged bodies

(a) The center of gravity should be below the center of buoyancy
(b) The center of gravity should be above the center of buoyancy
(c) The center of gravity should coincide with the center of buoyancy
(d) Metacenter should be above the center of gravity

Q 41. In submerged bodies, for neutral equilibrium to exist

(a) The center of gravity should be below the center of buoyancy
(b) The center of gravity should be above the center of buoyancy
(c) The center of gravity should coincide with the center of buoyancy
(d) Metacenter should coincide with the center of gravity

Q 42. In submerged bodies, for unstable equilibrium

(a) The center of gravity should be below the center of buoyancy
(b) The center of gravity should be above the center of buoyancy
(c) The center of gravity should coincide with the center of buoyancy
(d) Metacenter should be above the center of gravity

Q 43. It is possible to change the center of gravity, G by

(a) Adding weights
(b) Reducing weights
(c) Shifting the weights from one place to another within the body
(d) Any or all of these

Q 44. The distance between the center of gravity and the metacenter is

(a) Metacenter radius
(b) Metacenter height
(c) Height of the ship
(d) None of these

Q 45. The distance between the center of buoyancy and the metacenter is

(a) Metacenter radius
(b) Metacenter height
(c) Height of the ship
(d) None of these

Q 46. For a floating body having half of its body submerged in the fluid (ρ_o is average density of the object and ρ is the density of fluid)

(a) $\rho_o < \rho$ (b) $\rho_o > \rho$

(c) $\rho_o = \rho$ (d) None of these

Q 47. A body was given a slight angular displacement. The body remains at rest in the new position. The body is

(a) In stable equilibrium (b) In unstable equilibrium

(c) In neutral equilibrium (d) None of these

CHAPTER 4

Q 48. In the Eulerian description of fluid flow

(a) Individual fluid particles are marked and their motion tracked

(b) A control volume is defined

(c) The kinematic behaviour or properties of fluid particle passing a given point in space is recorded with time

(d) Both (b) and (c)

Q 49. Material derivative is also known as

(a) Total derivative (b) Convective derivative

(c) Substantial derivative (d) All of these

Q 50. Comment upon the statement " A streamline exists in a 3 dimensional flow field but a stream function does not exist"

(a) False (b) True

(c) Can be true or false depending upon certain conditions (d) None of these

Q 51. A stream tube is

(a) A bundle of neighbouring streamlines (b) A bundle of neighbouring streak lines

(c) A bundle of neighbouring path lines (d) None of these

Q 52. A streak line can be drawn from

(a) Time line (b) Path lines

(c) Streamline (d) None of these

Q 53. Two lines that can intersect each other or a single line that can form a loop is known as

(a) Time line (b) Path lines

(c) Stream line (d) None of these

Q 54. An imaginary curve in a flow field tangent to which gives the instantaneous velocity at that point is known as

(a) Time line (b) Path line

(c) Streamline (d) None of these

Q 55. Reynolds transport equation relates the change of a property

(a) Of a system to the change of that property for a control volume

(b) Of a control volume to the change of that property for a system

(c) Both (a) and (b)

(d) None of these

Q 56. If Reynolds number for a pipe flow in a pipe is 3000, the kind of flow is

(a) Laminar (b) Turbulent

(c) Transitional (d) None of these

Q 57. Line of a constant stream function is

(a) Potential line (b) Streamline

(c) Streak line (d) Path line

Q 58 . Flow lines and equipotential lines are always

(a) Parallel (b) Orthognal

(c) Any of A and B (d) None of these

Q 59. In a flow net, strip between two ________ is called a flow channel

(a) Equipotential lines (b) Flow lines

(c) Any of A and B (d) None of these

Q 60. The shape factor (S_f) in a flow net is calculated from the

(a) Nf/Nd (Ratio of the number of flow channels to the number of equipotential drops) (b) Nd/Nf

(c) Nd + Nf (d) Nd - Nf

Q 61. For fast transient problems involving fluid-solid interactions, one prefers to use

(a) Eulerian description (b) Lagrangian description

(c) Coupled Eulerian-Lagrangian description (d) None of these

Q 62. Stretching or shrinking of the fluid particle's sides during its movement in fluid space is

(a) Translation (b) Rotation

(c) Linear strain (d) Shear strain

CHAPTER 5

Q 63. Continuity equation is a statement of

(a) Conservation of mass
(b) Conservation of energy
(c) Conservation of momentum
(d) None of these

Q 64. For a compressible liquid flowing through a pipe, continuity equation at two points is

(a) $A_1V_1 = A_2 V_2$
(b) $\rho_1A_1V_1 = \rho_2A_2 V_2$
(c) $A_1V_1/\rho_1 = A_2 V_2/\rho_2$
(d) None of these

Q 65. Bernoulli's equation is a statement of

(a) Conservation of mass
(b) Conservation of energy
(c) Conservation of momentum
(d) None of these

Q 66. Most appropriate statement for assumption(s) made in deriving Bernoulli's equation

(a) The flow is inviscid or frictionless
(b) The flow is steady, irrotational and continuous
(c) The flow is incompressible (No change in density of the fluid)
(d) All of these

Q 67. In fluid dynamics, a stagnation point is a point in the flow field where the local velocity of the fluid is

(a) Zero
(b) Maximum
(c) Minimum
(d) None of these

Q 68. Hydraulic grade line (HGL) in an open channel flow coincides with

(a) The energy grade line
(b) The level of flowing water
(c) The datum line
(d) None of these

Q 69. The energy grade line in pipe flow is at a distance

(a) $P/\rho g$ above the hydraulic grade line
(b) $Z + V^2/2g$ above the hydraulic grade line
(c) $V^2/2g$ above the hydraulic grade line
(d) None of these

Q 70. A Pitot tube is used to measure flow in

(a) Open channels
(b) Pipe flow
(c) Both (a) and (b)
(d) None of these

Q 71. Dynamic pressure (also called velocity pressure) is given as

(a) Static pressure

(b) Stagnating pressure

(c) Stagnation pressure minus static pressure

(d) Stagnation pressure plus static pressure

Q 72. If a Pitot-static tube is placed in the center of the pipeline, the velocity and hence the calculated discharge will be

(a) Higher than the actual (b) Lower than the actual

(c) Same as the actual (d) None of these

Q 73. Diameter of the venturi meter is least at

(a) At the beginning of the converging section (b) Throat

(c) At the end of the diverging section (d) None of these

Q 74. When an inclined venturi meter with a manometer is used to measure the haed difference, the discharge equation is

(a) Independent of the angle of the venturi meter

(b) Dependent upon the angle of venturi meter

(c) Any of (a) and (b)

(d) None of these

Q 75. Tapered tube of the rotameter is placed

(a) Horizontally (b) Vertically

(c) At an angle to the flow (d) None of these

Q 76. Velocity of flow in a venturimeter is maximum at the

(a) Inlet (b) Outlet

(c) Throat (d) None of these

Q 77. Velocity of flow through a nozzle fitted on the side of a reservoir is

(a) $\sqrt{(2gh)}$ (b) $2\sqrt{(gh)}$

(c) $\sqrt{(gh)}$ (d) 2gh

CHAPTER 6

Q 78. A fully contracted weir is contracted at

(a) On one end (b) On both ends

(c) On both ends and bottom (d) None of these

Q 79. A narrow crested weir is one where height of the water above the weir crest is

(a) Greater than two times the width of the crest of the weir

(b) Less than two times the width of the crest of the weir

(c) Equal to two times the width of the crest of the weir

(d) None of these

Q 80. Which one of the weir is a combination of two shapes

(a) Rectangular weir
(b) Triangular notch/weir
(c) Compound weir
(d) All of these

Q 81. Iteration method is used for

(a) Correction for end contraction
(b) Correction for velocity of approach
(c) Correction for discharge coefficient
(d) None of these

Q 82. Discharge is proportional to $H^{3/2}$ in the case of (H is the head above the weir crest)

(a) Rectangular weir
(b) Triangular weir
(c) Orifice
(d) All of these

Q 83. Discharge is proportional to $H^{1/2}$ in the case of (H is the head above the weir crest)

(a) Rectangular weir
(b) Triangular weir
(c) Orifice
(d) All of these

Q 84. Discharge is proportional to $H^{5/2}$ in the case of (H is the head above the weir crest)

(a) Rectangular weir
(b) Triangular weir
(c) Orifice
(d) All of these

Q 85. Coefficient of discharge in a triangular weir is a function of

(a) Vertex angle of the notch
(b) Energy losses
(c) Contraction of the cross-section of the stream at the bottom and sides
(d) All of these

Q 86. For a V-notch, the discharge equation Q= 1.417 $H^{5/2}$ is valid for (H is head above the notch)

(a) Fully contracted notch
(b) Fully suppressed notch
(c) Valid for both the situations
(d) None of these

Q 87. In a Cipolletti weir, the sides are inclined at

(a) 1H:1V
(b) 4H:1V
(c) 1H:4V
(d) 1H:3V

Q 88. The approximate discharge of a Cipolletti weir can be calculated from the equation of

(a) One side contracted rectangular weir
(b) Two sides contracted rectangular weir
(c) Suppressed rectangular weir
(d) None of these

Q 89. For flow to be maximum in a broad crested weir, the depth of flow after the weir, y_2 should be _____ times the depth of flow before the weir

(a) 2
(b) 3/2
(c) 2/3
(d) 3

Q 90. In a broad crested weir maximum flow is proportional to

(a) (depth of flow before the weir)$^{3/2}$
(b) (depth of flow before the weir)$^{1/2}$
(c) (depth of flow before the weir)^2
(d) depth of flow before the weir

Q 91. Critical depth over the weir is 1 m, what is the critical velocity (g is acceleration due to gravity)

(a) $\sqrt{g}$
(b) g
(c) g^2
(d) None of these

Q 92. Ogee weir is.

(a) Sharp crested weir
(b) Broad crested weir
(c) Neither A nor B
(d) None of these

Q 93. A vena contracts is formed during fluid flow through

(a) A weir
(b) A V notch
(c) An orifice
(d) None of these

Q 94. Velocity of flow in an orifice is maximum

(a) At the inlet
(b) At the vena contracta
(c) At the outlet
(d) None of these

Q 95. Ratio of the area of the jet at vena contracta to the area of the orifice is known as

(a) Coefficient of contraction
(b) Coefficient of velocity
(c) Coefficient of discharge
(d) None of these

Q 96. When an orifice is a 'large orifice'?

(a) If the head of liquid is less than 5 times the depth of orifice
(b) If the head of liquid is less than 2.5 times the depth of orifice
(c) If the head of liquid is less than 4 times the depth of orifice
(d) If the head of liquid is less than 1.5 times the depth of orifice

Q 97. Which is/are Borda's mouthpiece

(a) External mouthpiece
(b) Internal mouthpiece running free
(c) Internal mouthpiece running full
(d) Both (b) and (c)

Q 98. An internal mouthpiece is known as

(a) Borda mouthpiece
(b) Re-entrant mouthpiece
(c) Both (a) and (b)
(d) None of these

Q 99. Coefficient of discharge through a Borda mouthpiece running full is

(a) 0.5
(b) 1.0
(c) 0.707
(d) None of these

Q 100. Separation head of water is 2.5 m absolute. If the atmospheric pressure is 10.3 m, what is the safe head for the Borda mouthpiece running full.

(a) 12.8
(b) 7.8
(c) 4.12
(d) None of these

Q 101. Coefficient of contraction in an external mouthpiece is

(a) 0.855
(b) 1.0
(c) 0.62
(d) None of these

Q 102. For a convergent mouthpiece, the mouthpiece should terminate at

(a) Before the vena contacta
(b) After the vena contacta
(c) At the vena contacta
(d) None of these

Q 103. The diameter of the divergent portion of the convergent-divergent mouthpiece is

(a) Any diameter with no limitation
(b) Less than the diameter where the convergent-divergent portions meet
(c) Limited by the separation head at the meeting point of convergent-divergent portions
(d) None of these

CHAPTER 7

Q 104. Which statement does not characterize an open channel flow

(a) It has a free surface
(b) Free surface is not subject to atmospheric pressure
(c) Driving force is gravity
(d) Hydraulic grade line (HGL) is coincident with the free surface

Q 105. Which statement does not characterize pipe flow

(a) It has no free surface

(b) At all points in the pipe hydraulic pressure exceeds atmospheric pressure

(c) Driving force is mainly the pressure force

(d) Hydraulic grade line (HGL) is within the conduit

Q 106. Vertical depth or distance from the bottom to free surface, y is related to depth of the flow normal to the direction of flow or water surface, d as

(a) $y = d \cos\theta$ (b) $y = d \sin \theta$

(c) $y = d \tan \theta$ (d) None of these

Q 107. For the best hydraulic section of a trapezoidal channel (b is the bottom width, d the depth of flow and θ is the internal angle that side makes with the horizontal)

(a) $b = 2\, d \tan\theta$ (b) $b = 2\, d \tan\theta/2$

(c) $b = d \tan\theta/2$ (d) None of these

Q 108. For the best hydraulic section of a trapezoidal channel (here z is the slope z:1, Horizontal to vertical)

(a) $z = 1/3$ (b) $z = 1/\sqrt{2}$

(c) $z = 1/\sqrt{3}$ (d) None of these

Q 109. For the best hydraulic section of a triangular channel

(a) The sides makes an angle of 60° with the vertical (Half angle between the sides)

(b) The sides makes an angle of 45° with the vertical (Half angle between the sides)

(c) The sides makes an angle of 30° with the vertical (Half angle between the sides)

(d) None of these

Q 110. Flow in an open channel is turbulent if the Reynolds number is

(a) >4000 (b) >1000

(c) <1000 (d) None of these

Q 111. Flow in a channel is super-critical if the Froude number is

(a) >1 (b) <1

(c) Equal to 1 (d) None of these

Q 112. Chézy equation is given as (V is the average velocity of flow, C is a constant, R is the hydraulic radius and S the channel bottom slope)

(a) $V = C\sqrt{(RS)}$ (b) $V = \sqrt{C}RS$

(c) $V = C\sqrt{R}\, S^{2/3}$ (d) None of these

Q 113. Constant of Chézy's equation and Darcy-Weisbach factor are related as (C is the Constant of Chézy's equation, f is Darcy-Weisbach factor and g acceleration due to gravity)

(a) $C = 8g/f$ (b) $C = \sqrt{(8g/f)}$

(c) $C = 1/\sqrt{(8g/f)}$ (d) None of these

Q 114. Constant of Chézy's equation, C and Manning' n are related as (R is the hydraulic radius)

(a) $C = R^{1/2}/n$ (b) $C = R/n$

(c) $C = R^{1/6}/n$ (d) None of these

Q 115. What is the dimension of C in the Chezy's equation?

(a) LT (b) $L^{1/2} T^{-1}$

(c) LT^{-1} (d) $L^{-1} T^{-1}$

Q 116. Velocity limits in a channel are decided by

(a) Prevent sedimentation and vegetative growth

(b) Avoid scouring of the channel bed

(c) Both (a) and (b)

(d) None of these

Q 117. For a rectangular channel, specific energy, E is given as (y is the depth of flow, q is the discharge per unit width of the channel and g acceleration due to gravity)

(a) $E = y + q^2/(2g\,y^2)$ (b) $E = q^2/(2g\,y^2)$

(c) $E = y + q/(2g\,y^2)$ (d) None of these

Q 118. Specific energy at critical flow is

(a) Minimum (b) Maximum

(c) None of A and B (d) None of these

Q 119. Critical flow velocity, v_c at critical depth y_c is given as (g is acceleration due to gravity)

(a) $v_c = \sqrt{(2gy_c)}$ (b) $v_c = g\,y_c$

(c) $v_c = y_c\sqrt{g}$ (d) None of these

Q 120. The critical flow depth, y_c is related with discharge per unit width, q as (g is acceleration due to gravity)

(a) $y_c = (q/g)^{1/3}$ (b) $y_c = (q^2/g)^{1/3}$

(c) $y_c = q\,(g)^{1/3}$ (d) None of these

Q 121. The energy at critical flow, E_c is related to critical flow depth y_c as

(a) $E_c = 2.5\,y_c$ (b) $E_c = 1.5\,y_c$

(c) $E_c = 0.5\,y_c$ (d) None of these

Q 122. At critical depth

(a) Specific force is minimum
(b) Specific force is maximum
(c) None of A and B
(d) None of these

Q 123. Hydraulic jump will occur, if

(a) Froude number is less than 1.0
(b) Froude number is more than 1.0
(c) Froude number is equal to 1.0
(d) None of these

Q 124. Hydraulic jump will occur, if flow changes from

(a) Sub-critical to super-critical
(b) Super-critical to sub-critical
(c) Critical to sub-critical
(d) Critical to super-critical

Q 125. Oscillating jump occurs when Froude number is

(a) 1.0
(b) 4 5 < Fr <9. 0
(c) 2.5 < Fr <4. 5
(d) None of these

Q 126. Strong jump occurs when Froude number is

(a) >9.0
(b) 1.0
(c) 4 5 < Fr <9. 0
(d) 2.5 < Fr <4. 5

Q 127. Ratio of depth of flow before and after the hydraulic jump is a function of

(a) Froude number before the jump
(b) Froude number after the jump
(c) Froude number before and after the jump
(d) None of these

Q 128. Ratio of depth of flow after and before the hydraulic jump is a function of

(a) Froude number before the jump
(b) Froude number after the jump
(c) Froude number before and after the jump
(d) None of these

Q 129. Jump efficiency is a function of

(a) Froude number after the jump
(b) Froude number before the jump
(c) Froude number before and after the jump
(d) None of these

Q 130. Linear momentum is

(a) Mass× Velocity
(b) Mass× Acceleration
(c) Mass× Distance
(d) None of these

CHAPTER 8

Q 131. Velocity profile in a pipe in the fully developed region reveals that

(a) Maximum velocity is at the center of the pipe
(b) At the walls of the pipe
(c) Both (a) and (b)
(d) None of these

Q 132. The dimensionless entrance length in a pipe flow is the ratio of

(a) Length of the entrance region divided by the diameter of the pipe
(b) Diameter of the pipe divided by the length of the entrance region
(c) Diameter of the pipe divided by the half the length of the entrance region
(d) None of these

Q 133. The dimensionless entrance length in a pipe flow is directly proportional to Reynolds number

(a) For laminar flow
(b) Turbulent flow
(c) For transition flow
(d) None of these

Q 134. Flow in pipes turns turbulent at Reynolds number

(a) <1000
(b) >1000 but less than 2000
(c) >4000
(d) None of these

Q 135. The radial (or in the case of, *e.g.*, square ducts, the wall-normal) component of velocity is zero in

(a) Entrance region
(b) In fully developed region
(c) None of A and B
(d) None of these

Q 136. In Poiseuille flow, flow is caused by

(a) Applied pressure difference between stationary plates
(b) Applied pressure difference between moving plates
(c) Due to moving plates
(d) None of these

Q 137. In Couette flow, flow is caused by

(a) Applied pressure difference between stationary plates
(b) Applied pressure difference between moving plates
(c) Due to moving plate
(d) None of these

Q 138. In Poiseuille-Couette flow, flow is caused by

(a) Applied pressure difference between stationary plates

(b) Applied pressure difference between moving plates

(c) Due to moving plates

(d) None of these

Q 139. In the case of Poiseuille flow between two stationary plates, average velocity is

(a) Maximum velocity/2

(b) Maximum velocity/1.5

(c) Equal to maximum velocity

(d) None of these

Q 140. In the case of Poiseuille flow in pipes, average velocity is

(a) Maximum velocity/2

(b) Maximum velocity/1.5

(c) Equal to maximum velocity

(d) None of these

Q 141. In Couette flow, the velocity profile between the stationary plate and the moving plate

(a) Increases linearly

(b) Decreases linearly

(c) Remains constant equal to that of moving plate

(d) Remains constant, equal to the average of the stationary and moving plates

Q 142. In pipe flow, shear stress is zero at

(a) Center line of the pipe

(b) Walls of the pipe

(c) None of A and B

(d) None of these

Q 143. The average velocity of flow in a pipe flow is

(a) At the center of the pipe

(b) At a ring, $r/\sqrt{2}$ distance apart from the walls of the pipe (r is the radius of the pipe)

(c) At a ring, $r/2$ distance apart from the walls of the pipe

(d) None of these

Q 144. In Manning's equation, velocity is proportional to

(a) Directly proportional to Manning's n

(b) Inversely proportional to Manning's n

(c) Directly proportional to square root of Manning's n

(d) None of these

CHAPTER 9

Q 145. Energy of fluid flowing through a pipe is lost because of

(a) Friction between the fluid and the walls of the pipe

(b) Friction between adjacent fluid particles as they move relative to one another

(c) Turbulence caused due to flow being redirected or affected in any way by components such as piping entrances and exits, pumps, valves, flow reducers, and fittings

(d) All of these

Q 146. Minor head losses are caused by

(a) Head loss at the entrance of pipe

(b) Head loss due to obstruction in pipe

(c) Head loss due to pipe fittings

(d) All of these

Q 147. Fanning's co-efficient of friction is ______ of the Darcy-Weisbach coefficient of friction

(a) 1/4

(b) 1/2

(c) 1 or Both are same

(d) None of these

Q 148. Hydraulic diameter of a non-circular duct is given as

(a) 2 × area of flow/perimeter of pipe in contact with fluid

(b) 4 × area of flow/perimeter of pipe in contact with fluid

(c) Area of flow/perimeter of pipe in contact with fluid

(d) None of these

Q 149. Darcy-Weisbach coefficient of friction in laminar flow regime is given as (Re is the Reynolds number)

(a) Re/64

(b) Re/32

(c) 64/Re

(d) 32/Re

Q 150. Hazen-Williams coefficient is

(a) Dimensionless

(b) Has dimension of L

(c) Has dimension of $\sqrt{L}$

(d) None of these

Q 151. Hazen-Williams coefficient is

(a) Higher for a rough pipe

(b) Higher for a smooth pipe

(c) Same for both kinds of pipes

(d) None of these

Q 152. Head loss at the entrance of pipe is more in

(a) Sharp edge pipe
(b) Rounded entrance
(c) Protruding pipe
(d) Equal in all cases A, B and C

Q 153. In the sudden contraction value of resistance coefficient, Ksc approaches_____ when the area of the bigger pipe approaches infinity

(a) 0.5
(b) 0.0
(c) 1.0
(d) None of these

Q 154. Three pipe lengths of L_1, L_2 and L_3 with respective diameters of d_1, d_2 and d_3 are joined in series. The relation of equivalent length or equivalent diameter is given as:

(a) $L_{eq}/d_{eq}^{\ 5} = L_1/d_1^{\ 5} + L_2/d_2^{\ 5} + L_3/d_3^{\ 5}$
(b) $L_{eq}/d_{eq}^{\ 3} = L_1/d_1^{\ 3} + L_2/d_2^{\ 3} + L_3/d_3^{\ 3}$
(c) $(d_{eq}^{\ 5}/L_{eq})^{1/2} = (d_1^{\ 5}/L_1)^{/2} + (d_2^{\ 5}/L_2)^{1/2} + (d_3^{\ 5}/L_3)^{1/2}$
(d) None of these

Q 155. Three pipe lengths of L_1, L_2 and L_3 with respective diameters of d_1, d_2 and d_3 are joined in parallel. The relation of equivalent length or equivalent diameter is given as:

(a) $L_{eq}/d_{eq}^{\ 5} = L_1/d_1^{\ 5} + L_2/d_2^{\ 5} + L_3/d_3^{\ 5}$
(b) $L_{eq}/d_{eq}^{\ 3} = L_1/d_1^{\ 3} + L_2/d_2^{\ 3} + L_3/d_3^{\ 3}$
(c) $(d_{eq}^{\ 5}/L_{eq})^{1/2} = (d_1^{\ 5}/L_1)^{/2} + (d_2^{\ 5}/L_2)^{1/2} + (d_3^{\ 5}/L_3)^{1/2}$
(d) None of these

Q 156. Hydraulic diameter and equivalent diameter are

(a) Same
(b) Not the same
(c) Neither A nor B
(d) None of these

Q 157. Which is not dependent upon Reynolds number

(a) Hazen-Williams coefficient
(b) Darcy's Weisbach coefficient
(c) Both (a) and (b)
(d) None of these

CHAPTER 10

Q 158. Which is not the fundamental or base quantity

(a) Length
(b) Luminous intensity
(c) Mass
(d) Force

Q 159. Which is not the derived quantity

(a) Mass
(b) Acceleration
(c) Shear stress
(d) Force

Q 160. Dimension of pressure in MLT system is given as

(a) ML/T^2 (MLT^{-2})
(b) M/LT^2 ($ML^{-1}T^{-2}$)
(c) M/LT ($ML^{-1}T^{-1}$)
(d) None of these

Q 161. Rayleigh's method of dimensional analysis assumes that functional relationship of variables is

(a) Linear
(b) Exponential
(c) Hyperbolic
(d) None of these

Q 162. In the Bukingham method of dimensional analysis number of pi terms are given as (N is the variables in the equation and m the primary dimensions)

(a) N+m
(b) N - m
(c) N/m
(d) N×m

Q 163. In the Bukingham method of dimensional analysis, number of repeat variables are equal to

(a) Maximum number of primary or reference dimensions of any variable
(b) Minimum number of primary or reference dimensions of any variable
(c) Total number of variables in the functional relation minus 2
(d) None of these

Q 164. In the Bukingham method of dimensional analysis dependent variable should be

(a) Chosen as repeat variable
(b) Should not be chosen as repeat variable
(c) Any of (a) and (b)
(d) None of these

Q 165. Functional relationship for one Pi term in the Bukingham's method is

(a) Equal to Reynolds number
(b) Equal to Froude number
(c) Equal to a constant quantity
(d) None of these

Q 166. Reynolds number expresses the relation between.

(a) Inertia force/viscous force
(b) Viscous force/inertia force
(c) Inertia force/surface tension force
(d) None of these

Q 167. Froude number expresses the relation between.

(a) Inertia force/gravitational force
(b) Viscous force/inertia force
(c) Inertia force/surface tension force
(d) None of these

Q 168. Reynolds number, Re is calculated with the help of equation (ρ is the density, V is the velocity, L is the characteristic length and μ is dynamic viscosity)

(a) $Re = \mu V L/\rho$ (b) $Re = \rho L/\mu V$

(c) $Re = \rho V L/\mu$ (d) None of these

Q 169. Cauchy number is equal to

(a) Mach number (b) Square of the Mach number

(c) Square root of the Mach number (d) None of these

Q 170. A dyamic similarity is possible only if

(a) The prototype and model are geometrically similar

(b) The prototype and model are kinematically similar

(c) There is similarity of forces between model and prototype

(d) All of these

Q 171. All linear dimensions of the model are related to corresponding dimensions of the prototype by a constant scale factor, then the model and prototype are

(a) Geometrically similar (b) Kinematically similar

(c) Dynamically similar (d) All of these

Q 172. For a kinematically similar model and prototype, discharge scale factor, Q_r is related to length scale factor, l_r and time scale factor, T_r as

(a) $Q_r = L_r/T_r$ (b) $Q_r = L_r^3/T_r$

(c) $Q_r = L_r^3 \times T_r$ (d) $Q_r = L_r \times Tr$

Q 173. If the ratios of the corresponding forces acting at corresponding points are equal, then the model and the prototype are said to have

(a) Kinematic similarity (b) Geometric similarity

(c) Dynamic similarity (d) None of these

Q 174. The energy stored by a fluid mass appears in the form of

(a) Potential energy (b) Kinetic energy

(c) Intermolecular energy (d) All of these

Q 175. If the work is done by the fluid on the rotor, then it is

(a) A pump (b) A turbine

(c) A blower (d) None of these

Q 176. If the work is done by the rotor on the liquid then the machines is known as

(a) A pump (b) A turbine

(c) A blower (d) None of these

Q 177. Pelton turbine is a

(a) Impulse turbine
(b) Reaction turbine
(c) None of A and B
(d) None of these

Q 178. In the Pelton wheel turbine number of buckets, N are calculated with the help of an empirical formula N = 15+ 53/A then what is A

(a) Size of buckets
(b) Specific speed of the turbine
(c) Capacity of the turbine
(d) None of these

Q 179. Breaking jet is used to

(a) Increase kinetic energy of the water
(b) Run the runner fast
(c) Stop the runner in shortest possible time
(d) None of these

Q 180. Francis turbine is a

(a) Impulse turbine
(b) Reaction turbine
(c) None of A and B
(d) None of these

Q 181. Type of flow in a Kaplan turbine is

(a) Radial
(b) Mixed
(c) Axial
(d) None of these

Q 182. Francis turbine is

(a) A reaction radial flow turbine
(b) An axial flow turbine
(c) An impulse radial flow turbine
(d) An impulse turbine

Q 183. A turbine is a device which converts

(a) Kinetic energy into mechanical energy
(b) Mechanical energy into hydraulic energy
(c) Hydraulic energy into mechanical energy
(d) None of the above

Q 184. An impulse turbine is used for

(a) Low head of water
(b) High head of water
(c) Medium head high discharge of water
(d) High discharge

Q 185. Maximum power of the water from the reservoir through the pipeline is transmitted when

(a) 1/3 of the head available at the reservoir is lost in transportation through the pipe
(b) 1/2 of the head available at the reservoir is lost in transportation through the pipe
(c) No head lost in transportation
(d) None of these

Q 186. Maximum power transmission efficiency through the pipeline is nearly

(a) 75 per cent (b) 50 per cent
(c) 67 per cent (d) 100 per cent

CHAPTER 12

Q 187. The machines that add energy to the fluid are called

(a) Turbines (b) Pumps
(c) Both (a) and (b) (d) None of these

Q 188. Which are the reciprocating pumps

(a) Piston pumps (b) Rotary pumps
(c) Hydrodynamic pumps (d) Both (a) and (b)

Q 189. Which one is not a rotary pump

(a) Lobe (b) Screw
(c) Plunger (d) Gear

Q 190. Amongst the hydrodynamic pumps, which one is the propeller pump

(a) Axial flow pump (b) Radial flow pump
(c) Mixed flow pump (d) None of these

Q 191. Amongst the hydrodynamic pumps, which in reality is the centrifugal pump

(a) Axial flow pump (b) Radial flow pump
(c) Mixed flow pump (d) None of these

Q 192. Which is most appropriate statement for a positive displacement pump

(a) High head high discharge (b) High head low discharge
(c) Low head high discharge (d) Low head low discharge

Q 193. Which is most appropriate statement for a non-positive displacement pump

(a) High head high discharge (b) High head low discharge
(c) Low head high discharge (d) Low head low discharge

Q 194. Pneumatic pump is a

(a) Positive displacement pump
(b) Rotary positive displacement pump
(c) Non-positive displacement pump
(d) None of these

Q 195. Discharge in a double acting piston pump is nearly

(a) Same as that of a single acting pump

(b) Half that of a single acting pump

(c) Double that of a single acting pump

(d) None of these

Q 196. A diaphragm pump is

(a) Single acting only

(b) Double acting only

(c) Can be single as well as double acting

(d) None of these

Q 197. Crescent is provided in

(a) An internal gear pump

(b) An external gear pump

(c) Both in the internal and external gear pumps

(d) None of these

Q 198. Which of the followings are positive displacement pump(s)

(a) Screw pump

(b) Rotary vane pump

(c) Gear pump

(d) All of the above

Q 199. Theoretical discharge, Qt in m^3/s of a single acting piston pump is given as (A is the cylinder area, m^2, L is the length of stroke, m and N the revolution per minute)

(a) Qt = ALN/60

(b) Qt = ALN

(c) Qt = ALN/3600

(d) None of these

Q 200. The specific weight of water on Earth at 4°C is taken as

(a) 9.81 kN/m^3

(b) 981 kN/m^3

(c) 1000 kN/m^3

(d) None of these

Q 201. Volumetric efficiency is given as (Qt theoretical and Qa actual discharge)

(a) Qt/Qa

(b) Qa/Qt

(c) (Qt –Qa)/Qt

(d) (Qt –Qa)/Qa

Q 202. Slip of a reciprocating pump is given as

(a) (Qt –Qa)/Qa

(b) (Qa –Qt/Qt

(c) (Qt –Qa)/Qt

(d) None of these

Q 203. The volumetric efficiency of a positive displacement pump is 88 per cent. The slip of the pump is

(a) 20 per cent

(b) 12 per cent

(c) 10 per cent

(d) 5 per cent

Q 204. The volumetric and mechanical efficiencies of a positive displacement pump are 90 per cent and 80 per cent respectively. The overall efficiency is

(a) 90 per cent (b) 80 per cent
(c) 72 per cent (d) 88.8 per cent

Q 205. For the net positive suction head (NPSH), following relation should be met (R and A indicate available and required NPSH)

(a) NPSHR >NPSHA (b) NPSHR =NPSHA
(c) NPSHA>NPSHR (d) None of these

Q 206. Air vessel is provided

(a) In the suction pipe only
(b) In the delivery pipe only
(c) Can be provided either in suction or delivery pipes or in both the pipes
(d) None of these

CHAPTER 13

Q 207. In a radial flow pump the fluid enters the pump at the eye axially and exits

(a) Any angle between 0 to 90° to the shaft (b) Right angle to shaft
(c) Along the shaft (d) None of these

Q 208. A propeller pump is used when

(a) High discharge low lifts are required
(b) Low discharge high lifts are required
(c) High discharge high lifts are required
(d) Low discharge low lifts are required

Q 209. A circular chamber is used between the casing and the impeller in

(a) Volute casing (b) Diffuser casing
(c) Vortex casing (d) None of these

Q 210. A high-head pump is capable of lifting the fluid

(a) Up to 20 m (b) Up to 30 m
(c) Up to 40 m (d) More than 40 m

Q 211. Specific speed is directly proportional to

(a) Pump speed (b) Pump discharge
(c) Pump head (d) None of these

Q 212. Specific speed is directly proportional to the square root of

(a) Pump speed
(b) Pump discharge
(c) Pump head
(d) None of these

Q 213. Specific speed of radial flow pump is

(a) Less than mixed flow pump
(b) More than mixed flow pump
(c) More than propeller pump
(d) None of these

Q 214. Which kind of impeller is most suited to handle suspended solids

(a) Open impeller
(b) Closed impeller
(c) Semi-open impeller
(d) All of these

Q 215. Which of the impellers has a back and front wall (plate) around the vanes

(a) Open impeller
(b) Closed impeller
(c) Semi-open impeller
(d) None of these

Q 216. Foot valve is used

(a) On the suction side of a pump
(b) On the delivery side of a pump
(c) Can be used on any of the two sides of a pump
(d) None of these

Q 217. Implosion of vapour bubbles entrapped in the fluid cause

(a) Cavitaion
(b) Pitting of the metals
(c) Both (a) and (b)
(d) None of these

Q 218. Head at zero flow rate is

(a) Shut-off head
(b) Minimum head
(c) Best efficiency point head
(d) None of these

Q 219. Pump efficiency is maximum at

(a) Maximum discharge
(b) Zero discharge
(c) In general between 0.6 to 0.93 of maximum discharge
(d) None of these

Q 220. Pump selection is made on the basis of

(a) Pump characteristics alone
(b) Piping system characteristics alone
(c) Combination of (a) and (b)
(d) None of these

Q 221. Pump discharge is proportional to

(a) Speed of the impeller
(b) Diameter of the impeller
(c) Both (a) and (b)
(d) None of these

Q 222. The parameter(s) are proportional to the square of the speed

(a) Head
(b) Net positive suction head required
(c) Both (a) and (b)
(d) None of these

Q 223. A pump was running at 1000 RPM. Its speed was changed to 1200 RPM. What is the likely discharge at the new speed?

(a) Same as before
(b) 1.2 times the previous discharge
(c) 0.83 times the previous discharge
(d) None of these

Q 224. A pump was running at 1000 RPM. Its speed was changed to 1200 RPM. What is the likely head developed at the new speed?

(a) 1.44 times the previous head
(b) 1.2 times the previous head
(c) Same as the previous head
(d) None of these

Q 225. The diameter of the pump impeller was increased from 12 cm to 18 cm while its speed was kept constant. Power required to run the pump will be

(a) Same as before
(b) 3.375 times the previous pump
(c) 2.25 times the previous pump
(d) None of these

Q 226. Overall efficiency of a pump is

(a) Volumetric × Mechanical efficiencies
(b) Manometric× Volumetric efficiencies
(c) Mechanical× Manometric efficiencies
(d) Volumetric × Manometric × Mechanical efficiencies

Q 227. 1 N-m/s

(a) 1 Watt
(b) 1 HP
(c) 1 kW
(d) None of these

Q 228. 1 Watt is equal to

(a) 0.00134 horsepower
(b) 0.0134 horsepower
(c) 0.134 horsepower
(d) None of these

Q 229. 1 HP

(a) 746 Watts
(b) 0.746 watts
(c) 7.46 Watts
(d) None of these

Q 230. The drive efficiency is 100 per cent, if pump and motor are

(a) Gear driven
(b) Direct driven
(c) Belt driven
(d) None of these

Q 231. Pumps in series are used

(a) For high discharge at the same head
(b) For same discharge at high heads
(c) For same discharge at low heads
(d) None of these

Q 232. Pumps in parallel are used

(a) For high discharge at same head
(b) For same discharge at high heads
(c) For low discharge at low heads
(d) None of these

Q 233. Under similar conditions, the suction lift of a pump at higher altitude is

(a) Less
(b) More
(c) Same
(d) None of these

Q 234. Net positive suction head available decreases with

(a) Increasing altitude
(b) Increasing temperature of the fluid
(c) Increasing discharge
(d) All of these

Q 235. Water hammer in the pump system is caused due to

(a) Sudden reversal of momentum
(b) Sudden closure of pump
(c) Sudden closure of foot valve
(d) All of these

Q 236. Cavitation is a result of

(a) Net positive suction head available being less than net positive suction head required
(b) Net positive suction head available being more than net positive suction head required
(c) Is not related to net positive suction head available and net positive suction head required
(d) None of these

Q 237. Net positive suction head available can be increased by

(a) Increasing the pipe diameter of the suction pipe
(b) Reducing the length of suction pipe and number of pipe fittings on suction side
(c) By reducing the distance between the water level and the pump center line
(d) Any or all of these

Q 238. Priming is required in

(a) Centrifugal pump
(b) Turbine pump
(c) Submersible pump
(d) All of these

Q 239. Which is not the character of a turbine pump

(a) Is normally a multi-stage pump
(b) Motor is placed above ground
(c) Pump is under water
(d) Motor is under water

Q 240. Which is the true character of a submersible pump

(a) Both the motor and pump are under water
(b) Motor is placed above ground
(c) Only pump is under water
(d) None of these

Q 241. Power required to run a pump is proportional to

(a) Cube of the impeller diameter
(b) Cube of the impeller speed
(c) Both (a) and (b)
(d) None of these

Answers

1	(d)	2	(d)	3	(c)	4	(d)	5	(b)
6	(a)	7	(a)	8	(d)	9	(d)	10	(b)
11	(b)	12	(c)	13	(a)	14	(b)	15	(c)
16	(d)	17	(b)	18	(b)	19	(a)	20	(b)
21	(a)	22	(a)	23	(b)	24	(b)	25	(a)
26	(c)	27	(c)	28	(b)	29	(b)	30	(c)
31	(a)	32	(a)	33	(a)	34	(a)	35	(b)
36	(a)	37	(a)	38	(b)	39	(d)	40	(a)
41	(c)	42	(b)	43	(d)	44	(b)	45	(a)
46	(a)	47	(c)	48	(d)	49	(d)	50	(b)
51	(a)	52	(b)	53	(b)	54	(c)	55	(a)
56	(c)	57	(b)	58	(b)	59	(b)	60	(a)
61	(c)	62	(c)	63	(a)	64	(b)	65	(b)
66	(d)	67	(a)	68	(b)	69	(c)	70	(c)
71	(c)	72	(a)	73	(b)	74	(a)	75	(b)
76	(c)	77	(a)	78	(c)	79	(a)	80	(c)
81	(b)	82	(a)	83	(c)	84	(b)	85	(d)
86	(a)	87	(c)	88	(c)	89	(c)	90	(a)
91	(a)	92	(b)	93	(c)	94	(b)	95	(a)
96	(a)	97	(d)	98	(c)	99	(c)	100	(b)
101	(b)	102	(c)	103	(c)	104	(b)	105	(d)
106	(a)	107	(b)	108	(c)	109	(b)	110	(b)
111	(a)	112	(a)	113	(b)	114	(c)	115	(b)
116	(c)	117	(a)	118	(a)	119	(c)	120	(b)

121	(b)	122	(a)	123	(b)	124	(b)	125	(c)
126	(a)	127	(b)	128	(a)	129	(b)	130	(a)
131	(a)	132	(a)	133	(a)	134	(c)	135	(b)
136	(a)	137	(c)	138	(b)	139	(b)	140	(a)
141	(a)	142	(a)	143	(b)	144	(b)	145	(d)
146	(d)	147	(a)	148	(b)	149	(c)	150	(a)
151	(b)	152	(c)	153	(a)	154	(a)	155	(c)
156	(b)	157	(a)	158	(d)	159	(a)	160	(b)
161	(b)	162	(b)	163	(a)	164	(b)	165	(c)
166	(a)	167	(a)	168	(c)	169	(b)	170	(d)
171	(a)	172	(b)	173	(c)	174	(d)	175	(b)
176	(a)	177	(a)	178	(b)	179	(c)	180	(b)
181	(c)	182	(a)	183	(c)	184	(b)	185	(a)
186	(c)	187	(b)	188	(d)	189	(c)	190	(a)
191	(b)	192	(b)	193	(c)	194	(a)	195	(c)
196	(c)	197	(b)	198	(d)	199	(a)	200	(a)
201	(b)	202	(c)	203	(b)	204	(c)	205	(c)
206	(c)	207	(b)	208	(a)	209	(c)	210	(d)
211	(a)	212	(b)	213	(a)	214	(a)	215	(b)
216	(a)	217	(c)	218	(a)	219	(c)	220	(c)
221	(c)	222	(c)	223	(b)	224	(a)	225	(b)
226	(d)	227	(a)	228	(a)	229	(a)	230	(b)
231	(b)	232	(a)	233	(a)	234	(d)	235	(d)
236	(a)	237	(d)	238	(a)	239	(d)	240	(a)
241	(c)								

Annexure 1

Useful Information and Conversion Factors

Meter: The meter is the length of the path travelled by light in vacuum during a time interval of 1/299,792,458 second.

Second: The second is the duration of 9,192,631,770 periods of the radiation corresponding to the transition between the two hyperfine levers (F=4, m_F=0 to F=3, m_F=0) of the ground state of the cesium 133 atom.

Mass: The kilogram (kg) is equal to the mass of the international platinum prototype kept by Bureau International des Poids et Measures (BIPM) at Sevres. 1 kg = 1,000 g (grams)

There is no SI unit of 'weight'.

Weight means the force due to gravity acting on a mass, the mass (kg) must be multiplied by g (9.81 m/s^2) to give the force in Newton's (N).

Density: The mega gram per cubic meter (Mg/m^3) is the density unit adopted for soil mechanics. 1 Mg/m^3 = 1 g/cm^3 = 1,000 kg/m^3

Force: The Newton (N) is that force which applied to a mass of 1 kilogram, gives it an acceleration of 1 m/s^2. 1 N = 1 kg m/s^2, 1 kN = 1,000 N = approximately 0.1 ton f

Pressure and stress: The Pascal (Pa) is the pressure produced by a force of 1 Newton, applied uniformly distributed over an area of 1 m^2. The Pascal is the unit for pressure and stress, and is exactly equal to N/m^2.

Viscosity: 1 CentiPoises (cP) = 1 CentiStokes (cSt)/Specific gravity

Kelvin: The Kelvin, unit of thermodynamic temperature is the fraction 1/273.16 of the thermodynamic temperature of the triple point of water.

Ampere: The ampere is the constant current, which when maintained in two straight parallel conductors of infinite length, of negligible circular cross-section, and placed 1 meter apart in vacuum, would produce between these conductors a force equal to 2×10^{-7} Newton per meter of length

Mole: The mole is the amount of substance of a system which contains as many elementary entities as there are atoms in 0.012 kg of carbon 12 (about 6.022×10^{23} atoms). When the mole is used, specify the elementary entities, may be atoms, molecules, ions, electrons, other particles, or specified groups of such particles.

Candela: The candela is the luminous intensity in a given direction of a source that emits monochromatic radiation of frequency 540×10^{12} Hz and that has a radiant intensity in that direction of 1/683 watt per steradian.

Standard gravity, $g = 9.80665 \sim 9.81 m/s^2$

Conversion Factors (Length, Area, Volume, Velocity and Discharge)

Length				
inch	*ft*	*cm*	*meter*	*km*
1	0.0833	2.54	0.0254	2.54E-05
12	1	30.48	0.3048	3.05E-04
0.3937	0.0328	1	0.01	1.00E-05
39.37	3.28	100	1	0.001
39360	3280	1.000E+05	1000	1
Area				
in²	*ft²*	*cm²*	*m²*	*ha*
1	0.0069	6.452	6.452E-04	6.452E-08
144	1	929	0.0929	9.290E-06
.1550	1.07E-03	1	1.000E-04	1.000E-08
1550	10.764	1.0 E+4	1	1.000E-04
1.550E+07	1.07E+05	1.0E+8	10,000	1
Volume				
in³	*ft³*	*gallon*	*liter*	*m³*
1	0.00058	0.0043	0.0164	1.639E-05
1728	1	7.4805	28.3168	0.0283
231	0.1337	1	3.7854	0.0038
61.0237	0.0353	0.2642	1	0.001
61024	35.314	264.17	1000	1
Velocity				
in/s	*ft/s*	*cm/s*	*m/s*	*km/h*
1	0.0833	2.54	0.0254	9.1E-02
12	1	30.48	0.3048	1.0973
0.3937	0.0328	1	0.01	0.036
39.37	3.28	100	1	3.6
39360	3280	1.000E+05	1000	3.6E+03

Discharge				
in³/s	*ft³/s (cusecs)*	*gallon/s*	*L/s*	*m³/s (cumecs)*
1	0.00058	0.0043	0.0164	1.639E-05
1728	1	7.4805	28.3168	0.0283
231	0.1337	1	3.7854	0.0038
61.0237	0.0353	0.2642	1	0.001
61024	35.314	264.17	1000	1

Conversion Factors (Mass, Force, Power, Energy, and Surface tension)

Mass			
ounce	*pound*	*kg*	*tonne (metric)*
1	0.0626	0.0283	2.835E-05
16	1	0.4536	4.536E-04
35.2741	2.2046	1	0.0010
35,840	2,240	1,016	1.0160
35,274	2,205	1,000	1
Note: 1 ton = 1.016 tonne (Mg)			
Force			
Pound force	*Newton*	*kgf*	*gf*
1 lbf	4.448	0.4536	453.6
Power			
BTU/hr	ft-lb/min	Kcal/hr	kW
1	12.9695	0.2520	2.931E-04
0.0771	1	0.0194	2.260E-05
3.9683	51.4663	1	0.001163
2,544	33,000	641.196	0.7457 (1 HP)
33,470	434,089	8,434	9.8091
3,412	44,254	859.859	1
BTU = British thermal unit, gf = gram fore			
Energy			
BTU	*kCal*	*hp-hr*	*kW-hr*
1	0.25200	3.930E-04	2.931E-04
0.0013	0.0003	5.051E-07	3.766E-07
3.9683	1	0.0016	0.0012
0.9478	0.2388 (1 kJ)	3.725E-04	2.778E-04
2,544	641.19	1	0.7457
3,412	859.85	1.341	1
Surface tension			
Pound force/in	*N/m*	*gf/cm*	*dyne/cm*
1	175.1268	0.9807	0.001
0.0057	1.0	1.0197	1000

Conversion Table: Pressure

Units	*Pascal (Pa)*	*Bar (bar)*	*Atmosphere Metric (at)*	*Atmosphere Standard (atm)*	*Torr (Torr)*	*Pound-force per Square Inch (psi)*
1 Pa	1 N/m^2	10^{-5}	1.0197×10^{-5}	9.8692×10^{-6}	7.5006×10^{-3}	145.04×10^{-6}
1 bar	100,000	10^6 $dyne/cm^2$	1.0197	0.98692	750.06	14.5037744
1 at	98,066.5	0.980665	1 kgf/cm^2	0.96784	735.56	14.223
1 atm	101,325	1.01325	1.0332	1 atm	760	14.696
1 torr	133.322	1.3332×10^{-3}	1.3595×10^{-3}	1.3158×10^{-3}	≡ 1 Torr; ≈ 1 mmHg	19.337×10^{-3}
1 psi	6.894×10^3	68.948×10^{-3}	70.307×10^{-3}	68.046×10^{-3}	51.715	1
1 mm H_2O	9.8067	9.81×10^{-5}	1.0×10^{-4}	9.7×10^{-5}	0.0736	1.421×10^{-3}

Conversion Factors (Kinematic and Dynamic Viscosity)

Kinematic viscosity	
1 cSt	1×10^{-6} m^2/s
1 cm^2/s	1×10^{-4} m^2/s
1 ft^2/h	2.58064×10^{-5} m^2/s
1 ft^2/s	9.2903×10^{-2} m^2/s
1 in^2/h	1.79211×10^{-7} m^2/s
1 in^2/s	6.4516×10^{-4} m^2/s
1 m^2/h	2.77778×10^{-4} m^2/s
1 m^2/s	1 m^2/s
1 St	1×10^{-4} m^2/s
Viscosity (dynamic)	
1 lbf s/ft^2	47.88 Ns/m^2
1 g/(cm s)	0.1 N s/m^2
1 Poise	0.1 N s/m^2
0.1 Poise	1 kg/ms (Ns/m^2)

Conversion Table: Temperature

Degrees Celsius (C) to degrees Kelvin (K)	(C) + 273.15 = (K)
Degrees Celsius (C) to degrees Fahrenheit (F)	[(C) × 1.8] + 32 = (F)
Degrees Celsius (C) to degrees Rankine (R)	[(C) × 1.8] + 491.67 = (R)
Degrees Fahrenheit (F) to degrees Kelvin (K)	[(F) × 0.555556] + 255.37 = (K)
Degrees Fahrenheit (F) to degrees Rankine (R)	(F) + 459.67 = (R)
Degrees Kelvin (K) to degrees Rankine (R)	(K) × 1.8 = (R)
Degrees Rankine (R) to degrees Celsius (C)	[(R) - 491.67] × 0.555556 = (C)
Degrees Rankine (R) to degrees Kelvin (K)	(R) × 0.555556 = (K)

References

Admin. 2009. Lab Manual | Kaplan Turbine. http://engineering.myindialist.com/2009/lab-manual-kaplan-turbine/#.XRmxG-szbIV. Opened on 30.06.2019.

Agrawal, S. K. 2006. Fluid Mechanics and Machinery. Tata McGraw Hill Publishing Company Limited, New Delhi. 665 p.

Anonymous. http://ecoursesonline.iasri.res.in/course/view.php?id=27. Opened on 1.11.2018.

Bansal, R. K. 2009. Fluid Mechanics and Hydraulic Machines, Laxmi Publications (P) Ltd, New Delhi.

Batchelor, G. 2000. An Introduction to Fluid Dynamics. Cambridge University Press, Cambridge. 615 p.

BIS. 2000. Indian Standard: Hydraulic Turbines for Medium and Large Power Houses-Guidelines for Selection. IS 12837:1989. Bureau of Indian Standards, New Delhi. 4 p.

Bos, M. G. (Ed). 1989. Discharge Measurement Structures. Publication 20, Third Revised Edition, International Institute for Land Reclamation and Improvement, Wageningen. 401 p.

Boyle, G. 2004. Renewable Energy: Power for a Sustainable Future, 2nd Ed. Oxford University Press, Oxford, UK.

Bradley, J. N. and Peterka, A. J. 1957. The hydraulic design of stilling basins: hydraulic jumps on a horizontal apron (basin I). Journal of the Hydraulics Division. 83: 1-24.

Buckingham, E. 1914. On physically similar systems; illustrations of the use of dimensional equations. Phys. Rev. 4: 345

Buckingham, E. 1915. The principle of similitude. Nature. 96: 396–397.

Chanson, H. 2004. The Hydraulics of Open Channel Flow. 1st Ed. Elsevier Butterworth Heinemann, Amsterdam.

Chaudhry, M. H. 1993. Open Channel Flow. Prentice-Hall, Inc. Englewood Cliffs.

Che´zy, A. 1897. Rapport sur le canal de lYvette. As quoted in Herschel, C. On the origin of the Che´zy formula. J. Assoc. Eng. Soc.18: 363–365.

Chin, D. A. 2007. Water Resources Engineering, 3nd Ed. Prentice Hall. Yorkshire, England.

Chow, V. T. 1959. Open-Channel Hydraulics. McGraw-Hill Book Company, New York. 680 p.

Chow V. T. 1955. Integrating the equation of gradually varied flow. In: Proceedings Paper No. 838, ASCE, 81: 1–32.

Colebrook, C. F. and White, C. M. 1937. Experiments with fluid friction in roughened pipes. Proc. Roy. Soc. A. Mat. 161: 367-381.

Cowan, W.L. 1956. Estimating hydraulic roughness coefficients. Agricultural Engineering. 37: 473-475.

E-max Group of Institutions. 2017. Fluid Machines Notes. The link of the page opened on 26.10.2019.

Fortier, S. and Scobey, F. C. 1926. Permissible canal velocities. Trans. ASAE. 89: 940-948.

Fox, R. W. and McDonald, A. T. 1994. Introduction to Fluid Mechanics. John Wiley and Sons, 781 p.

Gasoriek, J. M., Douglas, J. F., Swaffield, J. and Jack, L. 2008. Fluid Mechanics. 5th Ed. Pearson Education, Noida – India.

Gülich, J. F. 2007. Centrifugal Pumps. Springer. Heidelberg Dordrech.

Gupta, S. K. and Gupta, I. C. 2020. Drinking Water Quality: Assessment and Management. Scientific Publishers (India), Jodhpur. 273 p.

Harvey, E. H. 1925. The surface tension of crude oils. Ind. Eng. Chem. 17: 85–95.

http://www.engineeringenotes.com/fluids/water-level/measurements-of-water-level-methods-velocity-discharge-and-instruments-fluid-mechanics/47925. Opened on 01.03.2020.

https://neutrium.net/fluid_flow/pressure-loss-from-fittings-equivalent-length-method/

https://nptel.ac.in/courses/112106175/Module per cent 201/Lecture per cent 2011.pdf. Opened on 05.11.2018.

Kindsvater, C. E. and Carter, R. W. C. 1957. Discharge characteristics of rectangular thin plate weirs. J. Hydraulic Div. ASCE. 6: 83.

Kumar, R. and Singh, J. 2005. Textbook of Drainage Engineering. Indian Council of Agricultural Research, New Delhi. 342 p.

Macagno, E. O. 1969. Flow Visualization in Liquids. IIHR Report No. 114. Iowa Institute of Hydraulic Research. The University of Iowa, Iowa City, Iowa U.S. 21 p+ Annexure.

Manning, R. 1891. On the flow of water in open channels and pipes. Trans Inst. Civil Engr., Ireland. 20:161–207.

Matthew J. 2015. Introduction to Fluid Mechanics. matthewjuniper@learnfluidmechanics.org Opened on 12.07.2019.

Mays, L. 1999. Hydraulic Design Handbook.1st Ed. McGraw-Hill, New York.

McDonough, J. M. 2009. Lectures in Elementary Fluid Dynamics: Physics, Mathematics and Applications. Department of Mechanical Engineering and Mathematics, University of Kentucky, Lexington, KY. 158 p.

Michael, A. M. 2008. Irrigation Theory and Practice, 2nd Ed. Vikas Publishing House Pvt. Ltd, Noida, UP India. 772 p.

Michael, A. M., Khepar, S. D. and Sondhi, S. K. 2008. Water Wells and Pumps. McGraw-Hill Education, Columbus, Ohio.

Miller, J. E. 1995. The Reciprocating Pump: Theory, Design and Use. Krieger Publishing Company; 2nd Revised Ed. Florida, USA.

Minz, P. S. ecoursesonline.iasri.res.in/course/view.php?id=450. Opened on 1.11.2018.

Moglen, G. E. 2015. Fundamentals of Open Channel Flow. CRC Press, Taylor and Francis Group, Boca Raton, Florida. 250 p.

Moody, L. F. 1947. An approximate formula for pipe friction factors. Trans ASME, 69: 1005-1006.

Nagpurwala, Q. H. 2019.Hydraulic Pumps. REMP:RMD 2501. M.S. Ramaiah School of Advance studies. Bengaluru. 164.100.133.129 › 15-Hydraulic Pumps [Compatibility Mode] Opened on 28.10.2019.

National Environmental Monitoring Standards. 2013. Water Level Recording: Measurement, Processing and Archiving of Water Level Data. Version: 1.0. The National Environmental Monitoring Standards, New Zealand. 105 p.

Nelson, R. E. and Sundberg, D. M. 1973. Flow Net Construction and Use. Soil Mechanics Note No. 5. U. S. Department of Agriculture Soil Conservation Service, Washington. 45 p.

Poso F. Jr. 2015. Bouyancy- The Archimedes Principle. https://www.grin.com/document/441711. 12 p. Opened on 1.06.2019.

Powell, R. W. 1950. Resistance to flow in rough channels. Trans. Am. Geophys. Union. 31. 575-582.

Pritchard, P. J. and Leylegian, John C. 2011. Fox and Mcdonald's Introduction To Fluid Mechanics. 8th Ed. John Wiley and Sons, Inc. Hoboken, NJ. 875 p.

Sahoo, N. 2009. Dynamics of Fluid flow. IIT, Guwahato. NPTEL course https://nptel.ac.in/courses/105103095/50. Opened on 15.07.2019.

Sahu, G. K. 2017. Pumps: Theory, Design and Applications. New Age International Publishers, Lucknow.

Salih, A. 2017. Dimensionless Numbers. Dept. of Aerospace Engineering IIST, Thiruvananthapuram. (https://www.iist.ac.in/sites/default/files/people/numbers.html). 6 p.

Schwab, G. O., Fangmeier, D. D., Elliot, W. J., and Frevert, R. K. 1993. Soil and Water Conservation Engineering. John Willey and Sons, Inc., New York, USA. 269 p.

Shrotri, P. V. 2017. Fluid Mechanics. Nirali Parkashan, Pune. 484 p.

Smedema, L. K. and Rycroft, D. W. 1983. Land Drainage: Planning and Design of Agricultural Drainage Systems. Cornell University Press, Ithaca, New York. 375p.

Swamee, P. K. and Jain, A. K. 1976. Explicit equations for pipe flow problems. J. Hydraulics Div., ASCE. 102: 657–664.

Tackett, H. H., Cripe, J. A. and Dyson, G. 2008. Positive displacement reciprocating pump fundamentals. Proceedings of the 24th International Pump Users Symposium, Texas A and M University, Texas. 2008. 45-58.

Towler, B. 2012. Weir flow using the Rehbock equation. University of Massachusetts, Amherst. www. Smath.com/wiki/GetFile.aspx?File=Example/Weir Opened on 12.07.2019.

Tracy, H. J. 1957. Discharge Characteristics of Broad-Crested Weirs. United States Department of the Interior. Geological Survey Circular 397, Washington, DC. 21 p.

USBR (United States Bureau of Reclamation). 1955. Research studies on stilling basins, energy dissipators and associated appurtenances. US Bureau of Reclamation, Hydraulic Laboratory Report No. Hyd-399. USBR, Denver.

US Government Printing Press. 1961. Design Charts for Open-Channel Flow. U.S. Department of Transportation, Federal Highway Administration, Hydraulic Design Series No. 3, U.S. Government Printing Office, Washington, D.C.

White, F. M. 2011. Fluid Mechanics. 7th Edition. McGraw-Hill Companies, Inc., New York, NY. 862 p.

www.nptel.iitm.ac.in/courses/105107059/module1/./lecture1.pdf Opened on 13.02.2018).

Index